Green Energy and Technology

Climate change, environmental impact and the limited natural resources urge scientific research and novel technical solutions. The monograph series Green Energy and Technology serves as a publishing platform for scientific and technological approaches to “green”—i.e. environmentally friendly and sustainable—technologies. While a focus lies on energy and power supply, it also covers "green" solutions in industrial engineering and engineering design. Green Energy and Technology addresses researchers, advanced students, technical consultants as well as decision makers in industries and politics. Hence, the level of presentation spans from instructional to highly technical.

Indexed in Scopus.

Indexed in Ei Compendex.

Weijun Gao
Editor

Distributed Energy Resources

Solutions for a Low Carbon Society

Springer

Editor
Weijun Gao
Faculty of Environmental Engineering
The University of Kitakyushu
Kitakyushu, Japan

ISSN 1865-3529 ISSN 1865-3537 (electronic)
Green Energy and Technology
ISBN 978-3-031-21096-9 ISBN 978-3-031-21097-6 (eBook)
https://doi.org/10.1007/978-3-031-21097-6

This Springer imprint is published by the registered company Springer Nature Switzerland AG
The registered company address is: Gewerbestrasse 11, 6330 Cham, Switzerland

Preface

The future of the distributed energy resource (DER) market looks promising with opportunities in the residential and commercial and industrial sectors. The global distributed energy generation market is expected to grow fast in the next ten years. The major drivers for this market are increasing awareness on clean sources of energy, greenhouse gas (GHG) emission reduction targets, and increasing demand for energy.

On the other hand, with the Paris Agreement as an opportunity, various efforts are being made around the world toward the realization of a decarbonized society. There are several trends that are likely to revolutionize the traditional electricity supply and demand structure, such as a sharp drop in solar power costs, development of digital technology, appearance of new regional power, and strengthening the power supply system due to frequent natural disasters (improving resilience). Distributed power generation is one of solutions for low carbon society. In the energy-related field, various possibilities are expected, such as sophistication of supply and demand forecasting using AI and IoT, optimization of power plant operation, aggregation, and optimal control of distributed power generation by demand response, and virtual power plant (VPP).

The editor has been researching the urban environment in Waseda University since 1990, and after the transfer to the University of Kitakyushu since 2001, the distributed energy resource system has been introduced in ecological campus of Kitakyushu, which became the main subject of the editor's research. Up to now, the editor has sent the following 19 doctors in the study of the distributed energy resource among the 60 doctoral students.

This book is the latest scientific technology in this field, compiled by these doctoral students, and summaries of the editor in this field for more than 20 years.

1. RUAN, Yujun: Integration study on distributed energy resource and distribution system, PhD in 2006
2. REN, Hongbo: Integrated plan and evaluation of distributed energy systems taking into consideration renewable resources, PhD in 2009

3. YANG, Yongwen: Study on distributed energy resource and optimization choice of technology, PhD in 2009
4. XUAN, Ji: Estimation study of energy consumption and the introduction effects of co-generation system for the building sector in China, PhD in 2010
5. FAN, Liyang: Integrated plan and evaluation of distributed energy systems by area energy network in low carbon community, PhD in 2013
6. GU, Qunyin: Feasibility assessment of introducing distributed energy systems in Shanghai of China, PhD in 2013
7. SHI, Xingzhi: Study on multi-objective optimal method in the planning of distributed generation and heat source technologies, PhD in 2013
8. KRITSANAWONGHONG, Suapphong: Integrated evaluation of energy use by introducing the distributed energy resources in Thailand's commercial buildings, PhD in 2014
9. WU, Qiong: Integrated assessment of building distributed energy systems in different climate zones: Japan-China comparison, PhD in 2015
10. XU, Lianping: Environmental and economic evaluation of distributed energy resources technology in buildings, PhD in 2015
11. ZHANG,Yao: Integrated assessment of electricity dynamic pricing in buildings, PhD in 2017
12. LI, Yanxue: Modelling and evaluation for power supply system with consideration of supply and demand sides, PhD in 2017
13. JIANG Jinming: Study on maintenance management and reliability in distributed energy resource system, PhD in 2019
14. QIAN, Fanyue: Study on the economy potential and implication of hydrogen energy system with carbon tax introduction, PhD in 2020
15. ZHANG, Liting: Multi-criteria evaluation of a distributed energy system focusing on grid stabilization and carbon emission reduction, PhD in 2021
16. WEN, Daoyuan: The impact of renewable energy policies on solar photovoltaic energy: comparison of China, Germany, Japan, and the United States of America, PhD in 2021
17. LIU, Zhonghui: Study on equipment maintenance and system optimization of distributed energy resource, PhD in 2022
18. XU, Tingting: Study on the limitation of renewable energy penetration and its impact on public grids under different power supply systems, PhD in 2022
19. ZHAO, Xueyuan: Comparison study and economic optimization of different energy system composition in smart house, PhD in 2022

During the past 20 years, the editor as a Principal Investigator has received a lot of funds from the government and private institutions, especially

2002–2004 Field study of an introduction effect of distributed energy system in Kitakyushu Science and Research Park, Grant-in-Aid for Scientific Research (Contact No. 14550591)

2005–2007 Design and management support system for promoting the local distributed energy resources, Grant-in-Aid for Scientific Research (Contact No. 17560535)

2008–2010 Research on energy supply in large-scale non-residential buildings in China and transformation of energy saving technology of Japan, Grant-in-Aid for Scientific Research (Contact No. 20404014)

2009–2011 Initiative and strategy of district distributed energy technology based on field study, Grant-in-Aid for Scientific Research (Contact No. 21560618)

2012–2015 Field study to validate the effect of distributed energy resource and battery system in residential house with hydrogen pipeline, Grant-in-Aid for Scientific Research (Contact No. 24560724)

2017–2019 Field study of lifestyle design with energy conservation in smart community, Grant-in-Aid for Scientific Research (Contact No. 17K06719)

The researches in this book also has been supported by the following funds.

2019–2023 Key technology research and development, integration and demonstration of coastal green city based on the deep integration of smart-environment-recycle, Shandong Province

2019–2023 Improvement of energy efficiency and health performance of buildings based on lifecycle carbon reduction, Key Projects of International Cooperation in Science, Technology and Innovation, National Key R&D Program (Contact No. 2018YFE0106100)

2021–2023 Research on international standards and application of zero-emission smart building industrialization system, International Science and Technology Cooperation Project, Housing and Urban-Rural Development (Contact No. H20200014)

In Chapter 1, current status and bottleneck of international energy development will be introduced, and also the reason of the distributed energy solution for low carbon society has been explained. Chapter 2 introduces one practice of distributed energy system (DES) in Japan with more than 20-years collected data. Chapter 3 investigates the key policies affecting the development of PV technology from the perspective of solar PV Research and Development (R&D), industry, and market development in China, Germany, Japan, and the United States. Chapter 4 seeks the maximum penetration of renewable energy and analyzes its impact on Japan public grids. In Chapter 5, a planning and evaluation tool is developed to support the introduction and optimization of the DER system. Chapter 6 presents a new scheme about area energy network, which emphasizes the collaborative energy use between distributed energy plants in the neighborhood community. Chapter 7 analyzes the basic composition of residential multi-energy system and the possible economic and energy-saving effects with battery storage, PV, and CHP systems. Chapter 8 introduces the reliability calculation method and maintenance strategy of DER system. Chapter 9 introduces a multi-criteria evaluation of a distributed energy system focusing on grid stabilization and carbon emission reduction. Chapter 10 studies application potential and implication of hydrogen energy in distributed energy system. Chapter 11 introduces two systems, virtual power plant, and smart grid as future prospect of distributed energy system.

How to create a sustainable and low carbon society and realize the sustainable development goals (SDGs) of the United Nations (UN) is one of the biggest challenges of this century, even of the next centuries. The covered subject areas of this book aim at finding a way to push SDGs forward by collecting the related knowledge between distributed energy resource and low carbon technology. Specifically, the book focuses on UN SDG 7 to ensure access to affordable, reliable, sustainable, and modern energy, SDG 9 (innovation and new infrastructure), and SDG 13 (climate action).

This is a practical book and a good introduction for researchers and students to understand distributed energy resource for the low carbon society.

Weijun Gao
Kitakyushu, Fukuoka, Japan

Contents

Contributors

Liyang Fan Dr. Liyang Fan is a consultant of Nikken Sekkei Research Institute, Japan. Her research interests include smart city planning, city environment planning, and distributed energy system.

Weijun Gao Dr. Weijun Gao is a tenured professor at the University of Kitakyushu, Japan and academia professor at Qingdao University of Technology, China. His research interests include the science, engineering, management, and dissemination of city environment planning, distributed energy system, building material recycle, health and environmental impacts of energy generation, geographic information system, and climate change, especially in urban area.

Jinming Jiang Dr. Jinming Jiang is a lecturer at Qingdao University of Technology, China and Postdoctoral fellow at the University of Kitakyushu, Japan. Dr. Jiang's research interests include the science, engineering, management, and dissemination of city environment planning, distributed energy resource system, building material recycle, and low-carbon technology for buildings.

Yanxue Li Dr. Yanxue Li is an associate professor at Qingdao University of Technology. Dr. Li's research interests mainly include distributed energy system modeling, flexible energy system optimization, and energy system model predictive control.

Zhonghui Liu Dr. Zhonghui Liu graduated from the University of Kitakyushu. She has published two SCI papers and participated in many international conferences. Dr. Liu's research interests include distributed energy system, maintenance and management of energy system equipment, environmental and economic impacts of energy use, project management, and engineering cost.

Fanyue Qian Dr. Fanyue Qian is a postdoctoral fellow of Tongji University in China. He graduated from the University of Kitakyushu in Japan, and has published academic achievements in many well-known journals in the energy field, such as *Energy* and *Energy Conversion and Management*. Dr. Qian's research interests include hydrogen, the application of machine learning in distributed energy field, urban energy planning, regional energy system design, operation and maintenance, etc.

Hongbo Ren Dr. Hongbo Ren is a professor of Energy and Mechanical Engineering, Shanghai University of Electric Power. He received the PhD degree in Environmental Engineering at The University of Kitakyushu in 2009. His current research interests include plan and design of distributed and integrated energy system, simulation and analysis of urban energy and environmental system, etc.

Yingjun Ruan Dr. Yingjun Ruan is an associate professor at the School of Mechanical and Energy Engineering, Tongji University, China. Dr. Ruan's research interests include distributed energy system and CCHP, optimization and design on district heating and cooling, building new energy utilization, and building energy efficiency, especially in planning, design, and running optimization of district integrated energy.

Xingzhi Shi Dr. Shi has been engaged in the field of climate change and low-carbon development, mainly including distributed energy system simulation, climate change planning, carbon trading market, greenhouse gas emission reduction, etc.

Yafei Wang Dr. Yafei Wang is a PhD candidate at The University of Kitakyushu, Japan. She is committed to the research of economic optimization and carbon reduction effect of virtual power plant and optimization of distributed energy and renewable energy system.

Daoyuan Wen Dr. Daoyuan Wen is a lecturer at Shanghai Dianji University, China. Dr. Wen's research interests include distributed energy systems, solar PV energy, renewable energy policy, energy policy impacts, renewable energy analysis, and forecast.

Qiong Wu Dr. Qiong Wu is an associate professor of Energy and Mechanical Engineering at Shanghai University of Electric Power. She received the PhD degree in Environmental Engineering at The University of Kitakyushu in 2015. Her current research interests include optimal planning, comprehensive evaluation and benefit coordination of integrated energy system, etc.

Tingting Xu Dr. Tingting Xu is a student at The University of Kitakyushu, Japan. The research area focuses on optimizing the utilization of renewable energy and analyzing its development potential in the public grids. Dr. Xu have published SCI papers in the *Journal of Energy* and *Journal of Renewable and Sustainable Energy*.

Yongwen Yang Dr. Yongwen Yang is a lecturer at Shanghai University of Electric Power, China. He is the deputy director of Energy and Electric Power Technology Innovation Center at the University, mainly engaged in planning, design, and project implementation in the field of integrated energy, microgrid, renewable energy, and energy efficiency management.

Liting Zhang Dr. Liting Zhang is a lecturer at Shanghai University of Electric Power. Her research interests include the planning, design, and management of regional distributed energy system, smart microgrid, solar energy utilization, and building energy conservation.

Xiaoyi Zhang Ms. Xiaoyi Zhang is a doctoral student at The University of Kitakyushu, Japan. Ms. Zhang's research interests mainly include renewable energy and zero energy house.

Yao Zhang Dr. Yao Zhang is an associate professor at the School of Mechanical and Electrical Engineering, Shaanxi University of Science and Technology, China. Dr. Zhang's research interests include smart gird system and evaluation, carbon emission calculation and forecast, distributed energy system, and renewable energy utilization.

Xueyuan Zhao Dr. Xueyuan Zhao is a student at The University of Kitakyushu, Japan, and has published academic achievements in well-known journals in the energy field, such as *Energy* and *Energy and Buildings*. Dr. Zhao's research interests include smart house, distributed energy, renewable energy, etc. Dr. Zhao is committed to the research of economic optimization and carbon saving effect of smart house energy system and optimization of distributed energy and renewable energy system.

Chapter 1
Why Do We Need the Distributed Energy Solution for Low-Carbon Society?

Yafei Wang and Weijun Gao

1.1 Background

Energy plays an important role as a cornerstone in the process of social development. Energy sources such as electricity, natural gas, and gasoline support the development of our society and are ubiquitous in our lives. With the development of economic globalization and technological advances, the world energy sector is undergoing radical changes.

1.1.1 Current Status and Bottleneck of International Energy Development

Energy is an important material basis for human survival and civilization development and is a matter of national planning and livelihood and national strategic competitiveness. Currently, economic globalization is facing a new situation: the global energy production and consumption revolution is emerging, in which energy science and technology innovation plays a central leading role. The rational

Y. Wang (✉)
Faculty of Environmental Engineering, The University of Kitakyushu, Kitakyushu, Japan

W. Gao
The Engineering Academy of Japan, Chiyoda City, Tokyo, Japan

The University of Kitakyushu, Kitakyushu, Japan

Qingdao University of Technology, Qingdao, China

International Research Center for Green Buildings and Low-Carbon Cities, Zhejiang University, Hangzhou, China

Xi'an Jiaotong University, Xi'an, China

W. Gao (ed.), *Distributed Energy Resources*, Green Energy and Technology,
https://doi.org/10.1007/978-3-031-21097-6_1

development and scientific use of energy is a necessary guarantee for sustainable development. With the development of society, the energy demand has increased dramatically. However, the energy consumption structure dominated by coal, oil, and other fossil fuels has triggered a series of energy crises while promoting social progress and development.

Firstly, the world will face a huge challenge for the continued and stable supply of energy. In 2011, the global population exceeded 7 billion and is expected to reach 9 billion by 2045 according to the United Nations [27]. World energy demand will continue to increase with socioeconomic development and as world population continues to grow. The share of fossil energy in the world's primary energy structure has remained at over 85% for a long time. According to the IEA's world coal consumption report, total world coal production decreased by 4.8% in 2020, after 3 years of growth (Fig. 1.1). China was the only major producer that increased coal production in 2020, up by 1.1%. The declines that started at the beginning of the century in the USA and the European Union continued, most pronounced in Germany, Poland, and Greece. Production growth in Russia, Indonesia, India, and Turkey recently peaked and is now negative. Globally, the equivalent of more than 11 billion tons of oil is currently consumed annually from fossil fuels. Crude oil reserves are disappearing at a rate of over 4 billion tons every year, and at this rate,

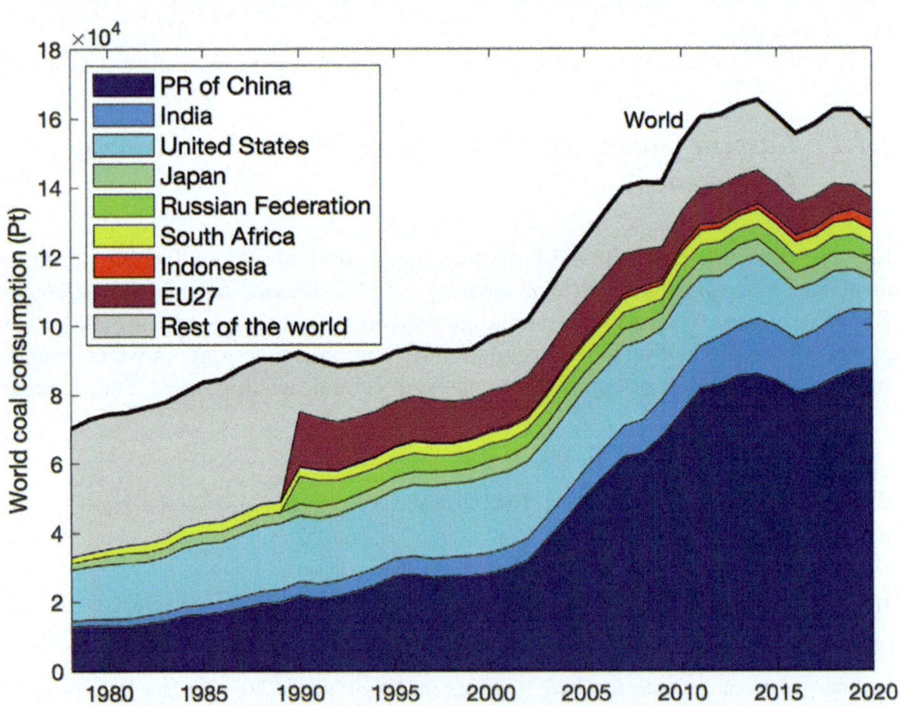

Fig. 1.1 World total coal production, 1971–2020 [14]

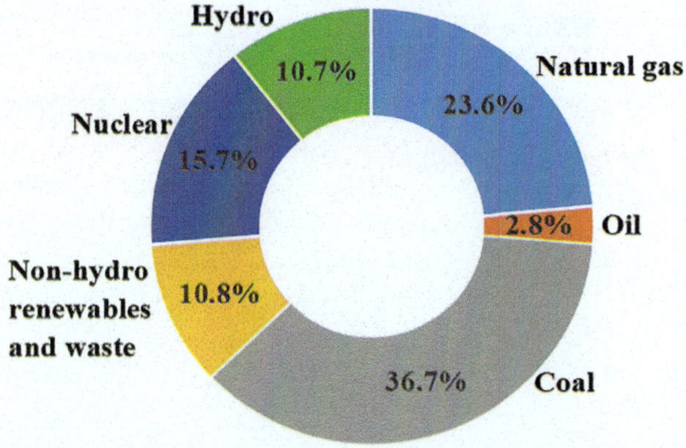

Fig. 1.2 Global share of electricity generation by different energy sources in 2019

oil reserves we have already known could be exhausted in just over 53 years. As fossil energy reserves dwindle, the pressure on the world's sustainable supply of energy increases in the long term.

Secondly, the exploitation of fossil energy sources also poses a series of challenges to the environment. The development and use of energy can cause problems such as water pollution, including wastewater discharge from coal utilization and ocean and groundwater pollution due to oil and gas extraction. The use of fossil energy also emits large amounts of air pollutants. Actually, energy production is the main source of CO_2 [2]. The continued increase in particulate matter emissions from thermal power, transport, and other industries also causes widespread haze, which will threaten human health.

In response to the many challenges facing the world's energy development, changing the way traditional energy is developed and utilized, promoting the application of new energy technologies, and building a new energy system will become the main direction of the world's energy development. Shaping a secure and sustainable energy future for the world continues to be a development theme today. Currently, coal is still the main source of electricity generation, as shown in Fig. 1.2. The British Petroleum (BP) predicted that renewable energy is growing the most rapidly of all energy sources, contributing 40% of primary energy growth. To achieve the net zero emission by 2050, renewable energy will play an increasingly important role in low-carbon power generation. The increasing share of uncertain renewables such as solar and wind means that the electricity system should become more flexible. At the same time, the utilization of conventional power plants will gradually decrease, as decarbonization requires a reduction in use of primary energy. The development of renewable energy and efficient use of energy will be the key direction of future energy system development.

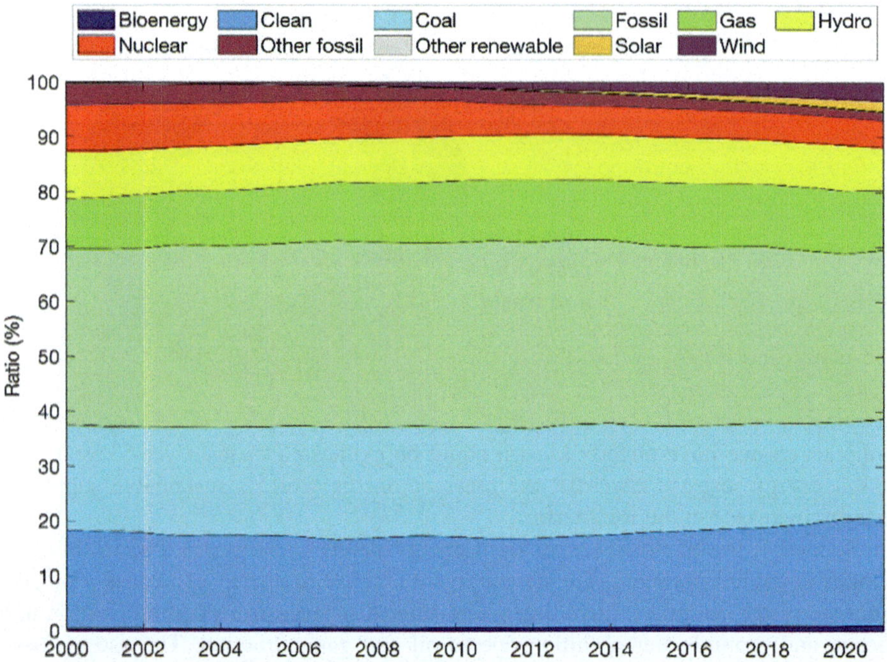

Fig. 1.3 Share of global electricity generation by source [4]

1.1.2 World Energy Demand Trend Forecast

The world's primary energy structure has been in the process of shifting from high-carbon to low-carbon fossil energy sources. Electricity takes center stage in the energy sector. And it also plays a key role in all other sectors, such as the transportation, buildings, and industrial sectors. Electricity generation will need to achieve net zero emissions globally by 2040, and successfully supply almost half of the total energy demand. This requires a significant increase in the flexibility of power systems, such as batteries, hydrogen-based fuels, hydropower, etc., to ensure reliable supply. As shown in Fig. 1.3, the global energy system is transitioning from fossil fuel power generation to renewable power generation, a trend that will accelerate in the coming decades.

Economic development judgments are the basis for energy forecasts. According to the outlook of various institutions, the next 20 years will see a significant slowdown in global population growth and a modest decline in economic growth as a general trend in economic and social development. OPEC is the most optimistic, predicting that the world economy will grow at a rate of 3.5%, while other agencies basically forecast around 3%. By 2040, the world economy will have doubled from its 2015 level to reach US $100–130 trillion, while the population will also reach around 9 billion. However, future energy demand growth and economic growth are not exactly converging. Energy outlooks from different institutions indicate that world energy demand growth between now and 2040 will be between 25% and 35%. The CNPC Economics & Technology Research Institute (ETRI) states that over the next 30 years, primary energy will grow at a much slower rate than economic growth over the same period, with global energy consumption at 36% to support 170% of economic growth. Economic growth is decoupled from energy demand growth, with energy efficiency improvements and declining energy intensity being the main reasons. ExxonMobil sees energy consumption per capita remaining largely unchanged by 2030 and declining by 2040 compared with 2010. The IEA also believes that energy efficiency improvements play a huge role in removing supply-side pressures and that without them the projected increase in final energy consumption would more than double. It is worth noting that China is the largest energy consumer. However, its energy demand is still growing, will no longer be the dominant demand growth country over the next 30 years as the growth rate continues to slow and energy intensity continues to decline with industrial transformation. CNPC ETRI forecasts that China's energy demand will gradually decline after 2035, stabilizing at 23% of the global primary energy share, when energy consumption per unit will be 54% lower than in 2015. The US Energy Information Administration (EIA) predicts that China's energy demand will grow at less than 1% in the future, in contrast to the 8% growth in demand since the beginning of the 21st century.

The world's primary energy consumption structure is tending to be cleaner, lower carbon, and more diversified, and the transition is happening faster than previously expected. The EIA predicts worldwide consumption of fuels other than coal increasing by 2040, and BP's outlook for the last 3 years has raised estimates for installed wind and solar power by 2035 by a significant 150%. With a quarter of the total, clean energy will account for more than 54%. Meanwhile, oil and gas will continue to dominate in the future, with several reports predicting a 55% share in 2040.

In the process of structural transformation, changes in energy consumption in different sectors are equally significant. BP, ETRI, and IEA all note a gradual slowdown in the growth of energy consumption in the industrial sector globally, with the buildings sector being the fastest-growing sector. BP believes that natural gas and electricity will meet the future incremental energy demand in the industrial sector, becoming the main source of energy for the industrial sector in 2040. The IEA stated that electricity is a rising force among the various end uses of energy around the world and will account for 40% of incremental final energy consumption by 2040, which is the share of oil in the growth of energy consumption over the last 25 years.

The most representative country for the transformation of its energy structure is China. ETRI points out that China's energy consumption has entered a period of transition between old and new dynamics. BP and IEA are also concerned about the increasing share of services in China's economic structure and the increasing share of clean energy in its energy structure. The IEA pointed out that, unlike the previous 10 years, the industrial sector will overtake the power sector as the main driver of natural gas demand in the next 10 years. This is mainly due to industrial growth in Asia, where natural gas is increasingly being processed as an energy source as well as a feedstock. In North America and the Middle East, developments in the chemical sector have also contributed to growing demand for natural gas. In the past decade, the power sector contributed half of the increase in natural gas consumption, but in the future, the industrial sector will account for 40% of the increase in natural gas consumption.

1.1.3 Current Status and Bottleneck of Japan Energy Development

Japan is a country that lacks resources such as oil and LNG (liquefied natural gas) and needs to take various measures to ensure a stable supply of energy. Japan relies on imports from abroad for most of its demand for fossil fuels. In 2018, the proportion of fossil fuels dependent on imports was 99.7% for oil, 97.5% for liquefied natural gas (LNG), and 99.3% for coal. The energy self-sufficiency rate of Japan in 2017 was 9.6% [19], which is lower compared with other OECD countries. Low energy self-sufficiency leads to dependence on the resources of other countries. This makes a country vulnerable to international situations, making it difficult to obtain energy in a stable manner.

In the year prior to the Fukushima Crisis, Japan's dependence on fossil fuels accounted for 81.2% of its total primary energy supply. And after the Fukushima Crisis, Japan's energy self-sufficiency rate has fallen sharply, as low as 6.4% in 2014, and then slowly increased, shown in Fig. 1.4. And dependence on imported energy rose to 87.4% in 2017 as thermal power generation was increasingly used to compensate for power shortages caused by nuclear plant closures. Oil still accounts for about 40% of Japan's primary energy supply, and more than 80% of imported oil comes from the politically unstable Middle East. Moreover, prospects for importing electricity from neighboring countries are very poor because Japan is an island nation. In addition, there is an urgent need for global warming countermeasures

Fig. 1.4 Japan energy self-sufficiency rate (FY 2017 actual figures from the "Comprehensive Energy Statistics of Japan") [19]

such as reduction of carbon dioxide emissions from the use of energy. To ensure Japan's stable electricity supply, it is crucial to establish an optimal combination of power sources that can concurrently deliver energy security, economic efficiency, and environmental conservation while placing top priority on safety. In 2018, the dependence rate of fossil fuels on imports was 99.7% for oil, 97.5% for liquefied natural gas (LNG), and 99.3% for coal [13]. It has also brought electricity prices to unsustainable levels.

Another question is where Japan imports resources from. About 88% of crude oil is imported from the politically unstable Middle East. As the Middle East is one of the world's most important energy suppliers, ensuring the safety of navigation in the region is vital for Japan and the international energy market. Regarding coal, there is a high level of dependence on Australia. On the other hand, LNG (liquefied natural gas) is being sourced from diverse regions such as Australia, Asia, Russia, and the Middle East.

Since the Fukushima Crisis, electricity prices have risen several times. This is due to the increased use of thermal power generation to mitigate the effects caused by the closure of nuclear power plants. This was also due to an increase in fuel prices until 2014. Compared with the period before the Great East Japan Earthquake, household electricity bills increased by approximately 16%, and industrial electricity bills increased by approximately 21% in 2017.

Faced with these challenges, the government of Japan has revised its energy policy in recent years to focus on further diversifying its energy mix (less use of fossil fuels, more reliance on renewable energy, restarting nuclear plants when declared safe) and curbing carbon emissions. Building on these plans, Japan has outlined ambitious goals to cut greenhouse gas emissions by 26% between 2013 and 2030. This emission reduction commitment requires a balancing act between energy security, economic efficiency, environmental protection, and safety. The 2016 in-depth review of Japan's policies highlights three areas that are critical to its success: energy efficiency, increasing renewable energy supply, and restarting nuclear power generation.

Japan has commitment to reduce greenhouse gas emissions by at least 46% by 2030, and achieving net zero emissions by 2050 is one of the most laudable climate targets in the world. Unlike many other countries, Japan is not rich in renewable energy resources, and its high population density, mountainous terrain, and steep coastline are serious barriers to expanding the resources it has, particularly because many of its few plains are already heavily covered with solar panels. It has been a world leader in energy efficiency for decades, and much of the potential offered by this fast and efficient way of decarbonizing the economy has already been realized. Japan has already made great strides in reducing energy consumption through behavioral and lifestyle changes, such as reducing the use of air conditioning in the summer and supporting public transport. Its geology is not conducive to carbon storage, and this technology will play an important role in some other parts of the world. In the long term, Japan will need a broader range of new technologies to continue to achieve net zero emissions. Fortunately, this plays to the country's strengths, as it has long been a global leader in energy innovation through technologies such as hybrid electric vehicles, solar photovoltaics, smart grids, high-speed trains, and robotics. Offshore wind looks particularly promising, although Japan needs to help push the frontiers that could reap the major benefits it offers. The world's largest offshore wind project today, currently being developed off the coast of the UK, will generate just 2% of Japan's electricity needs [15].

1.2 General Structure of a Distributed Energy System

A distributed energy system (DES) is a term that encompasses a diverse array of generation, storage, and energy monitoring and control solutions. DESs can be tailored to very specific requirements and users' applications including cost reductions, energy efficiency, security of supply, and carbon reduction. Currently, practice shows that a DES offers great advantages. Fonseca et al. defined distributed generation, storage, electric vehicles, and demand responses as distributed energy resources (DERs) [7]. A DES is a system that determines the unit configuration and capacity scale by optimizing the resources, environment, and economic benefits. It pursues the maximization of terminal energy utilization efficiency and adopts demand-responsive design and modular combination configuration, which can meet various energy needs of users and optimize and integrate the supply and demand of resources. A DES is a complex system, which mainly can be reflected in the following points. (1) Various energy resources as input and multiple-energy output are a reason for its complexity. For example, the input resources can include fossil energy (oil, coal, natural gas, etc.), hydrogen (H_2), biomass, solar energy, wind energy, and so on; the multiple-energy output may include electricity, heating (for space heating, hot water, etc.), and cooling. A DES is more complex than a conventional power plant, which only uses one resource for power generation, or a thermal plant, which only uses one resource for thermal generation. (2) A DES may consist of multiple devices and components. For example, power generation can adopt a variety of devices, like a gas engine, gas turbine, fuel cell, reciprocating engine, and so on; if the system should meet the heating and cooling demand, it must have heat recovery devices, absorption chiller, adsorption chiller, electrical chiller, solar thermal and geothermal gas engine, and so on; in order to overcome the fluctuation of energy supply, the power system must have certain energy storage capacity, which can either be an electrical storage device or thermal storage device.

In addition to power generation, thermal generation, thermal convection, and energy storage devices, some auxiliary devices and components also constitute the complexity of the system, like a DC-DC converter, DC-AC converter, pump, fan, pipe, wire, and so on. DERs always can be divided into distributed generation technologies and energy storage technologies [6, 7], and some main technologies are shown in Fig. 1.5. Figure 1.6 is a schematic diagram of a distributed energy system. And in the following sections, some main technologies will be introduced.

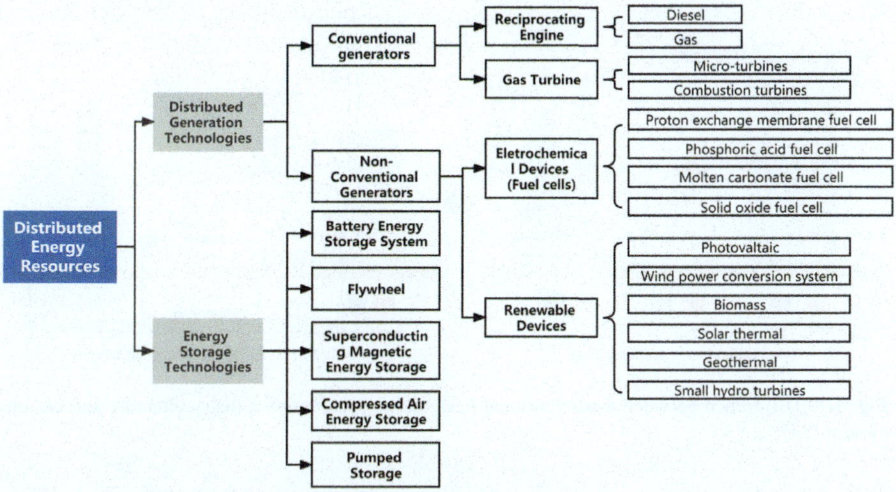

Fig. 1.5 Technologies of a distributed energy resource system

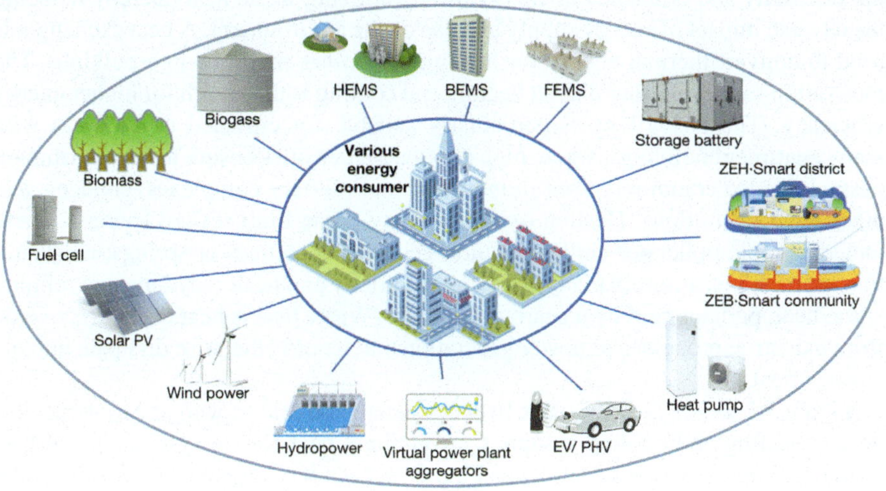

Fig. 1.6 Schematic diagram of a distributed energy system [21]

1.2.1 District Cooling and Heating

A district heating and cooling (DHC) system is a kind of typical distributed thermal source. It is an energy efficiency link between thermal wastes from some sources. These may include industrial processes, power generation, waste incineration, or renewable or sustainable energy. The waste heat generated from these processes is used to heat or cool water, which will be transported into the building through a

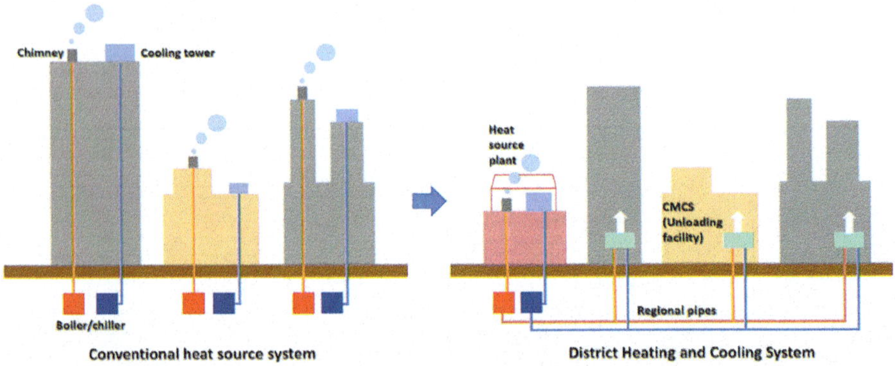

Fig. 1.7 Difference between a conventional heat source system and a district heating and cooling system [24]

piping system. It is also piped to buildings for space heating or cooling, as well as heating domestic water. Therefore, a DHC system is an efficient system to increase the flexibility and efficiency of energy use by utilizing waste heat, thereby reducing the adverse impact of energy supply and use on the environment. A heat exchanger is used to convert thermal energy for heating or cooling purposes in a building. The motivation for developing district heating and cooling is the search for higher energy efficiency. The technology is particularly suitable for crowded urban areas with many apartment buildings, where high population densities ensure lower distribution costs. It is also commonly used in industrial and military complexes, colleges, and other large institutions. Heating or cooling may come from several sources. These include simple boilers or chillers; natural geothermal sources or springs; cogeneration sources, i.e., a central plant providing electricity and heat; recycling and refining waste heat, perhaps sold by a nearby plant; etc. (Waste heat is heat energy recovered from industrial processes or power generation that would otherwise dissipate into the environment.)

Figure 1.7 shows the difference between a conventional heat source system and a district heating and cooling system. Conventional heat source systems install heat source equipment such as a chiller and boiler in each building to produce cold water and steam (hot water) for cooling and heating. However, district heating and cooling systems are a heat source plant that installs chillers and boilers for a group of neighboring buildings centrally for heating and cooling in district units. The cold water and steam (hot water) produced by the heat source plant are supplied to each building through regional pipes built inside the district to use for cooling and heating.

1.2.2 *Combined Heat and Power*

Combined heat and power (CHP) is also called cogeneration. Cogeneration is the thermodynamically efficient use of fuels. In stand-alone electricity production, some of the energy must be discarded as waste heat, but in cogeneration some of this thermal energy is used. All heat emitted during the generation of electricity from a CHP plant can be released into the natural environment via cooling towers, flue gases, or by other means. Conversely, CHP captures some or all of the by-products used for heating, either very close to the plant or, especially in Scandinavia and Eastern Europe, as hot water for domestic area heating in the temperature range of about 80–130 °C. Small-scale cogeneration plants are an example of distributed power generation.

CHP is an efficient energy technology that generates electricity while capturing otherwise waste heat to provide useful thermal energy – such as steam or hot water – that can be used for space heating, cooling, domestic water heating, and industrial processes. A CHP plant can be located in a single facility or building, or it can be a district energy or utility resource. A CHP plant is often located in facilities where there is a demand for both electricity and heat. Nearly two-thirds of the energy used in conventional power generation is wasted in the form of heat energy emitted into the atmosphere. Even more energy is wasted in the distribution of electricity to end users. By capturing and utilizing heat that would otherwise be wasted and by avoiding distribution losses, CHP can achieve efficiencies of over 80%, compared with 50% for typical technologies (i.e., conventional power generation and on-site boilers).

The most common CHP system configurations include a steam boiler with a steam turbine (Fig. 1.8) and a combustion turbine, or reciprocating engine, with a heat recovery unit (Fig. 1.9). Combustion turbine or reciprocating engine CHP systems burn fuel (natural gas, oil, or biogas) to turn generators to produce electricity and use heat recovery devices to capture the heat from the turbine or engine. This heat is converted into useful thermal energy, usually in the form of steam or hot water. With steam turbines, the process begins by producing steam in a boiler. The steam is then used to turn a turbine to run a generator to produce electricity. The steam leaving the turbine can be used to produce useful thermal energy. These systems can use a variety of fuels, such as natural gas, oil, biomass, and coal.

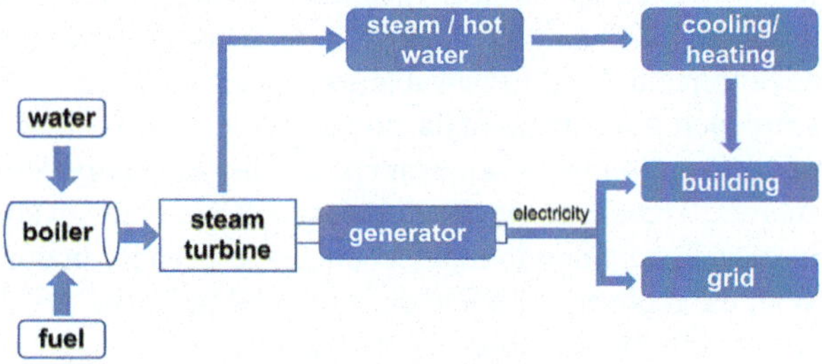

Fig. 1.8 Steam boiler with a steam turbine [5]

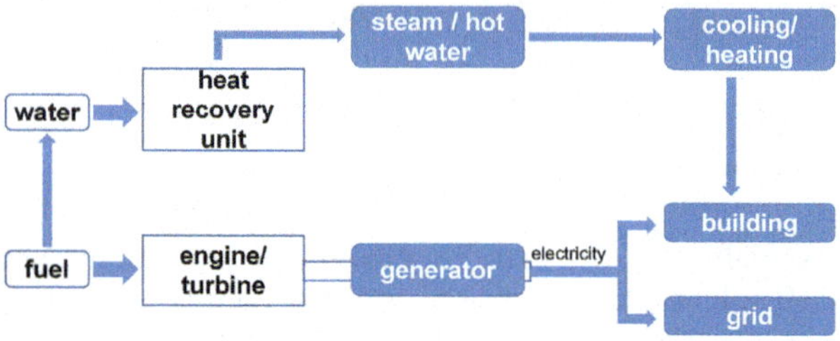

Fig. 1.9 Combustion turbine, or reciprocating engine, with a heat recovery unit [5]

CHP requires less fuel to produce a certain energy output and avoids the transmission and distribution losses that occur when electricity is transmitted over power lines. In the USA, the average efficiency of fossil fuel power plants is 33%. This means that two-thirds of the energy used to generate electricity in most US power plants is wasted in the form of emissions into the atmosphere. By recovering this waste heat, CHP systems typically achieve total system efficiencies of between 60% and 80% in producing electricity and useful thermal energy. Some systems achieve efficiencies close to 90% [8].

1.2.3 Gas Turbine (GT)

A gas turbine, otherwise known as a combustion turbine, is a rotary engine that extracts energy from a flow of combustion gas. It has a combustion chamber in between the upstream compressor coupled to a downstream turbine. Gas turbines are generally divided into three main categories, namely, heavy frame, aeroderivative, and micro-turbine. Energy is added to the gas stream in the combustor, where air is mixed with fuel and ignited. Combustion increases the temperature, velocity, and volume of the gas flow. This is directed through a nozzle over the turbine's blades, spinning the turbine and powering the compressor.

1.2.4 Renewable and Unutilized Energy Resources

Renewable energy is clean, naturally renewable, regionally distributed, low in energy density, and intermittent. And it remains an underutilized resource within urban environments [23]. Renewable energy technologies like solar and wind are essential for reducing emissions of the power sector, which currently accounts for a large amount of GHG emissions. There is no possibility of energy depletion from renewable energy sources. Therefore, the development and use of renewable energy sources is receiving increasing attention in many countries, especially in countries with energy shortages. With the recovery of nuclear energy and the rapid development of renewable energy worldwide, the development of clean energy is on a year-on-year upward trend, and its growth rate is second only to that of natural gas. According to statistics, global renewable energy consumption increased by 16% in 2017 compared with 2016 and maintained a double-digit growth rate. Of this, solar energy grew at 29.6% and wind energy at 15.6%. Taking into account nuclear energy, hydropower, and natural gas, the global share of clean energy consumption reached 38% in 2017, surpassing the 28% of coal consumption and 34% of oil consumption. At the same time, electricity generation structures also changed with the renewable energy development. Among the renewable energy sources, solar and wind power generation is considered to be an important component of a hybrid distributed energy system. Renewable energy technologies play an important role in the energy systems of the future, not only in achieving low-carbon society but also in providing socioeconomic benefits [20].

1.2.4.1 Solar Power Generation

Conversion of solar energy directly to electricity has been technologically possible since the late 1930s, using photovoltaic (PV) systems. These systems are commonly known as solar panels. PV solar panels consist of discrete multiple cells, connected either in series or parallel, that convert light radiation into electricity. PV technology could be stand-alone or connected to the grid. Solar photovoltaic power generation is a power generation method that uses the photovoltaic effect of solids (semiconductors) to directly transform light energy to electrical energy. The solar photovoltaic power generation system consists of three parts: solar panels, batteries, and controllers. With the continuous reduction of manufacturing cost, solar photovoltaic power generation will present a good development prospect. Figure 1.10 shows a solar power station in Japan.

Fig. 1.10 Solar power station in Ikishima, Japan (photo by Yafei Wang)

1.2.4.2 Wind Power Generation

Power generation is the main form of wind energy utilization. Wind turbines can be powered either individually or in combination with other forms of power generation, such as diesel generators or micro-gas turbines, to supply power to a unit or an area or to integrate power into conventional grid operations. Windmills or wind turbines convert the kinetic energy of the streaming air to electric power. Investigation has revealed that power is produced in the wind speed range of 4–25 m/s. The size of the wind turbines has increased rapidly during the last two decades with the largest units now being about 4 MW compared with unit sizes in the 1970s that were below 20 kW. For wind turbines above the 1.0 MW size to overcome mechanical stresses, they are equipped with a variable speed system incorporating power electronics. Single units can normally be integrated to the distribution grid of 10–20 kV, though the present trend is that wind power is being located offshore in larger parks that are connected to high voltage levels, even to the transmission system. The power quality depends on the system design. Direct connection of synchronous generators may result in increased flicker levels and relatively large active power variation. At present, wind energy has been found to be the most competitive among all renewable energy technologies (Fig. 1.11).

Fig. 1.11 Wind power station in Ikishima, Japan (photo by Yafei Wang)

Table 1.1 Value of battery energy storage systems [3]

Utilities	• Increase renewable energy integration. • Reduce dependence on fossil fuel peaker plants. • Reduce operating expenses.
Grid operators	• Balance electricity supply and demand. • Improve power quality and reliability. • Avoid costly system upgrades.
Commercial consumers	• Keep critical equipment online during power disruptions. • Reduce utility bills and generate revenue.
Residential consumers	• Reliable backup power during severe weather and other blackouts. • Reduce utility bills and generate revenue.

1.2.5 Energy Storage System (ESS)

The rapid growth in the use and development of renewable energy sources in today's power grids requires the development of energy storage technologies to eliminate intermittent power disparities [25]. And the clean energy group has provided a summary of the important role of battery storage systems in all areas and applications, as shown in Table 1.1. Energy storage technology meets the demand for electricity or heating/cooling energy over a period by storing electricity, with functions such as peak shaving, frequency and voltage regulation, smooth transition, and reduction of grid fluctuations. Energy storage technology can solve the problem of intermittent renewable energy limited by environmental factors and ensure the balance of supply and demand of energy systems. As mentioned earlier, global energy consumption is steadily increasing, some of which is due to the increased consumption of inefficient peak plants to accommodate industrial facilities. One promising way to reduce the power demand of facilities is to discharge energy storage system equipment during peak hours and recharge it during off-peak hours; this is known as "peaking" and "valley filling." Energy can be stored in different forms, including electrical, electrochemical, magnetic, thermal, and mechanical. A classification by form of energy storage is shown in Table 1.2.

These different functions of ESS will only expand over time, making battery storage technology so important for clean energy and climate change. Although we are in the earliest stages of this technology's development, storage could be a key transformative energy technology of this century. With the advancement of technology, energy storage systems have become less and less expensive in recent years. It is believed that as prices fall, ESSs will become more popular.

1.2.6 Fuel Cell

A fuel cell is a kind of power generation device that can convert the chemical energy of hydrogen and other fuels into electrical energy directly through electrochemical reaction without combustion. Because a fuel cell does not involve combustion and is not limited by the Carnot cycle, the energy conversion rate is high. In addition, a fuel

Table 1.2 Overview of energy storage technologies

Energy storage	Electrical and electrochemical	Battery	Li-ion
			NaS
			Lead-acid
			Flow
		Capacitor	Supercapacitor
	Thermal	Sensible	
		Latent	
		Thermochemical	
	Mechanical	Flywheel	High speed
			Low speed
		Pumped hydro	
		Compressed air	Conventional
			Adiabatic
			Isothermal
			Variable pressure ratio
	Magnetic		

cell does not use mechanical transmission parts and has no noise pollution; the reaction products are mainly electricity, heat, and water, and the emission of harmful gases is very little. Therefore, a fuel cell is an efficient, environment-friendly, high-reliability, quiet energy conversion mode, which is one of the research hotspots in the field of energy.

1.2.7 Hydrogen Energy System

As hydrogen must be produced from hydrogen-containing substances such as water and fossil fuels, it is a secondary energy source. The two secondary energy sources, electricity and hydrogen, have much in common: they are both technology driven; they are both produced from any available primary energy source; once produced, they are environmentally and climatically clean throughout their respective conversion chains, from production to utilization; they are electrochemically interchangeable through electrolysis and fuel cells [28]. The production of hydrogen with high efficiency and low cost is currently a focus of attention worldwide. The production of hydrogen from renewable energy sources reduces production costs and serves the purpose of protecting the environment. It is the most effective way to produce hydrogen. At the same time, this approach can effectively alleviate the current consumption problems caused by the continuous development of renewable energy sources. Hydrogen and electricity can be said to complement each other in the transformation of energy systems. The use of electrolysis devices to produce hydrogen from renewable electricity facilitates the integration of highly volatile renewable electricity into the energy system. Meanwhile, the large-scale production of hydrogen from renewable energy sources such as solar and wind power, and the development of large-scale, low-cost volatile renewable energy dedicated to hydrogen production in marginal areas rich in solar or wind energy resources, will enable the reuse of wind and light and energy conversion, improve the utilization rate of renewable energy, and reduce the waste of clean energy. When used in conjunction with fuel cells, it is a key solution for efficient energy production and effective decarbonization of the energy sector. Researchers [6] have already explored the role that hydrogen can play in distributed energy systems. There are also an increasing number of studies showing the advantages of hydrogen fuel cell vehicle-to-grid systems [16, 22].

1.3 Possibility and Challenge of Distributed Energy Resources

Distributed energy systems (DESs) are small-scale generation technologies that use renewable energy, energy storage, etc., close to the load demand user, while multiple technologies combine to complement each other. With the rapid development of photovoltaic and wind power generation, the phenomenon of wind and light abandonment is becoming more and more serious. Developed countries such as Germany and the USA experienced this phenomenon earlier and took early measures to deal with it, mainly including changing the operation mode of the electricity market, constructing power transmission channels (including mutual aid with neighboring countries), improving the electricity price mechanism (such as negative electricity prices), and adding flexible units such as hydropower and gas power. Some results have been achieved, which promoted the consumption of photovoltaic and wind power, but at the same time caused the slowdown of the development of renewable energy.

While batteries and demand-side measures can provide short-term flexibility, hydrogen is the only large-scale technology that can be used for long-term energy storage. It can make use of existing gas networks, salt caverns, and barren gas fields to store energy for the long term at a low cost. With hydrogen produced from renewable energy sources, large amounts of renewable energy can be channeled from the power sector to the end use sector. Renewable electricity can be used to produce hydrogen, which in turn can provide energy for sectors that are difficult to decarbonize through electrification, enabling sustainable energy development. The development of flexible distributed energy systems with abundant renewable energy technologies for efficient consumption is an important strategic choice for all countries. However, renewable energy sources are characterized by significant peaks and valleys, intermittency, and randomness, increasing the difficulty of coordinating the dispatch of distributed units [11].

1.3.1 Problems of Introducing Distributed Energy Resources

The government has made great efforts to develop distributed energy systems because of their advantages, such as high efficiency, energy saving, and environmental protection. However, the practice of distributed energy systems has shown that in many cases, their actual operation is not satisfactory. There are some major barriers.

1.3.1.1 Economic Problem

Even though lower fuel and operating costs may make a DES cost-competitive over the life cycle, higher initial capital costs may mean that a DES provides less installed capacity per dollar invested than conventional energy systems. As a result, investments in a DES typically require more capital for the same amount of capacity. Depending on the circumstances, capital markets may require a premium on loan rates for DES projects, as more capital must be risked up front compared with conventional energy projects.

1.3.1.2 Technology Problem

The unjustified capacity of the DES is the most important issue. DESs are available in a variety of optional system forms, with main and auxiliary equipment and their capacity. There is no universally applicable technical solution. Its configuration is closely related to the climatic characteristics, load demand, and energy prices and availability of the customer's region, which place high requirements for the system configuration determination. For the optimal configuration of a regional distributed energy supply system, the main task is to determine the system structure and form reasonably; optimize the type, capacity, and number of main equipment; and obtain the comprehensive performance of economic, environmental, and other aspects of the whole year, so as to provide a decision-making reference for owners, provide selection basis for design, and provide guidance for operation strategy formulation. Improper configuration of a distributed energy supply system will lead to waste of equipment investment, failure to give full play to economic benefits, low system operation efficiency, and other problems and even to system failure in extreme events.

1.3.2 Design Concept of Distributed Energy Resources

Distributed energy systems can utilize a wide range of energy sources, including natural gas; biomass; wind, solar, and geothermal energy; etc. These can also be combined with waste heat, residual pressure, residual gas, and other forms of energy. DESs come in various forms and structures because of different energy forms. As a systematic and complex energy-saving and emission reduction scheme, the issue of optimal system planning and design has been troubling since the concept of distributed energy was introduced, attracting sufficient attention from researchers.

In terms of optimization of DESs, there are many researches. L. Blackhall et al. [1] have analyzed the value of optimizing distributed energy from various technical, economic, and social perspectives. ISF Gomes et al. [9] explored the synergy of solar PV, batteries, and electric vehicles in a distributed energy system and explored the impact of different retail tariff designs on private investment incentives and cost shifting in the system in the California scenario. G. Wu et al. [30] have proposed an energy-reserve co-optimization model for electricity and natural gas systems with multi-type reserve resources. L. Li et al. [17] developed a collaborative hierarchical framework to coordinate electrical and thermal interactions and explored the impact of carbon tax, electricity, and heat demand response on the outcome of multi-stakeholder interaction issues. F. Tooryan et al. [26] presented an optimization solution to reduce the operational cost for a hybrid residential microgrid consisting of a diesel generator, wind turbine and photovoltaic array, and battery energy storage system. D. Wu et al. [29] proposed a two-stage stochastic method for jointly sizing microgrid assets considering both economic benefits and resilience performance. X. Luo et al. [18] proposed a mixed-integer linear programming model for optimizing the structure and operation of a distributed energy system with a district energy network on a virtual island in the South China Sea and allocated system costs among the various stakeholders based on a cost allocation analysis of cooperative game theory. S. Guo et al. [10] presented a new method of optimizing the distributed energy of AC/DC hybrid microgrids with a power electronic transformer. Another important issue is the operation strategy of the DES. The economic operation is essential for the DES to promote its configuration. M. García-Plaza et al. [8] proposed a new peak shaving algorithm in combination with a continuous battery peak power estimation algorithm for a battery energy storage system, which is aimed at avoiding energy exchanges when the output power is considered to be too high. JD Fonseca et al. [7] proposed a modeling and multi-criterion optimization strategy for the design and operation of decentralized power plants that include different energy vectors. The modeling approach considers the time-varying operation of energy conversion units in response to electricity and hydrogen demand, as well as the seasonal behavior of energy storage systems. Y. Huang et al. break through the idealized demand-side distribution and build an optimization model of regional distributed energy systems under discrete conditions, pointing out that the main influence of energy station location is the cost difference caused by the layout of the pipe network [12].

Previous studies have evaluated various DESs from different perspectives. Most papers use a single economic performance metric to evaluate DESs. Some researchers have also evaluated the environmental performance to improve the advantages of DESs. Grid stability performance is also essential when comparing various feasible DES technology combinations and their corresponding scales.

Based on the aforementioned studies, we can conclude some design concept of a DES. It is generally accepted that the inherent instability and unpredictability of renewable energy generation can have an impact on the operation of the grid, particularly when connected to large-scale renewable energy sources. In many countries, the lack of clarity about the technical requirements and responsibility for grid investment and construction has delayed the progress of projects when it comes to grid access for renewable energy generation. However, there is some successful international experience in planning the operation of wind farms on the grid, and solutions have been proposed that have increased the utilization of renewable energy by being implemented in many countries, for example, in many European countries and in states in the USA.

PV System

The design of a PV power station should take into account sunlight conditions, land and building conditions, as well as installation and transport conditions. It needs to meet the requirements of safety, reliability, economy, environmental protection, aesthetics, ease of installation, and maintenance. PV systems installed on buildings must not lower the sunlight standards of adjacent buildings. The choice of site for a PV power station should be in line with the national medium and long-term development plan for renewable energy, taking into account regional natural conditions, solar energy resources, transportation, access to the power grid, regional economic development planning, and other factors.

Wind Power Station

Carrying out a sound wind resource assessment is essential for siting wind farm developments and identifying project-specific needs. Understanding the characteristics of wind energy is also important for building an efficient wind farm. Wind speed is the most important indicator, and this cannot be overstated. For example, a wind turbine installed on a wind farm with an annual average wind speed of 9.0 m/s will produce twice as much power as one installed at an annual average wind speed of 6.5 m/s. If insufficient wind measurement is done, this can lead to unreliable estimates of power generation and pose a risk to investment. Use of technically mature and certified turbines is very important for offshore wind projects, as solving any small problem with offshore wind power can be very costly. International experience shows that the use of technically immature wind turbine equipment has cost some offshore wind projects dearly.

Figure 1.12 shows the design flow of distributed energy systems. There are three steps in the design of distributed energy systems. Designers should understand the energy usage status and equipment operation status in the building and region, automatically perform optimal operation control in consideration of the load based on the demand forecast, monitor and control the energy supply equipment and demand equipment, and perform optimal driving while making predictions. Some other design ideas for distributed energy systems will be mentioned in subsequent sections.

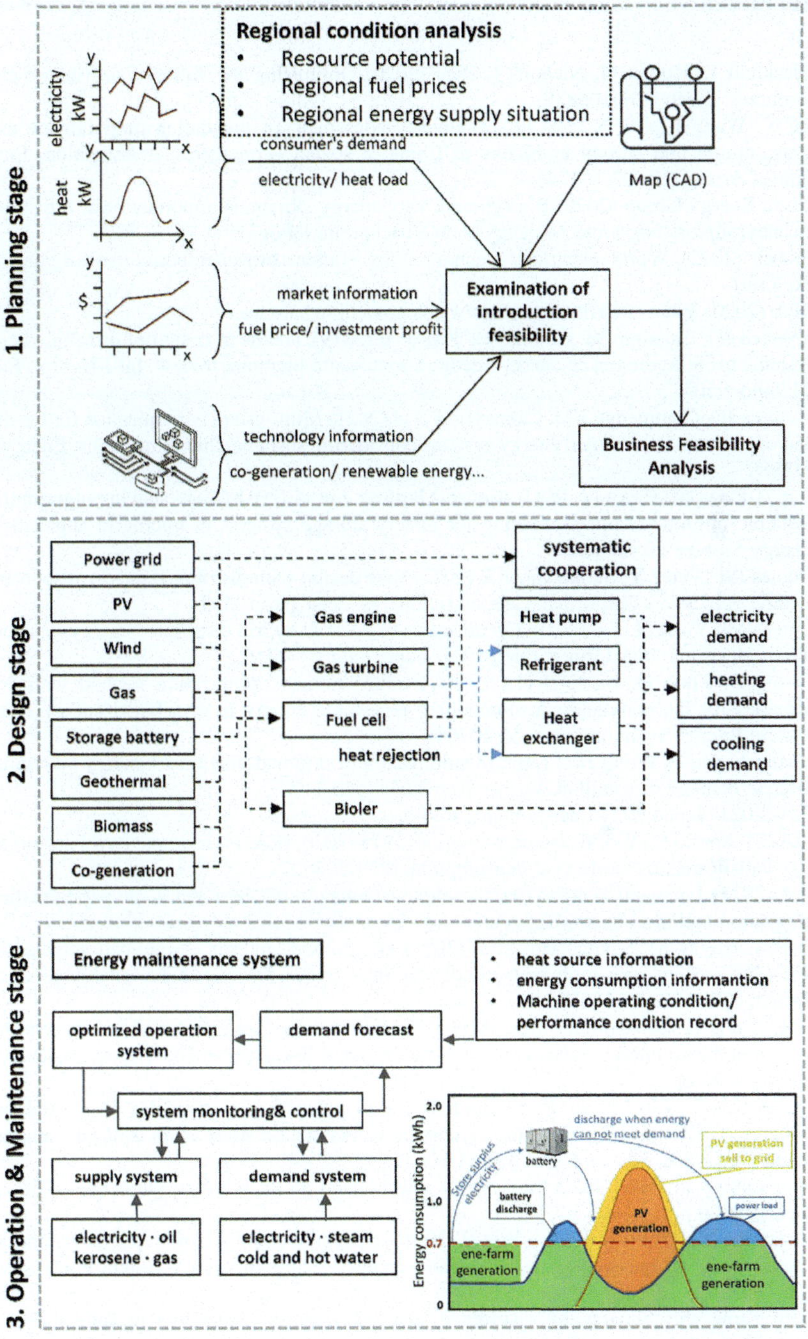

Fig. 1.12 Design flow of distributed energy systems

References

1. Blackhall L, Kuiper G, Nicholls L et al (2020) Optimising the value of distributed energy resources. Electr J 33:106838
2. Bu Y, Wang E, Qiu Y et al (2022) Impact assessment of population migration on energy consumption and carbon emissions in China: a spatial econometric investigation. Environ Impact Assess Rev 93:106744
3. Clean Energy Group (2020) Background: why energy storage is important now. https://www.cleanegroup.org/ceg-projects/energystorage/background/#toggle-id-1
4. Ember (2022) World electricity generation by source. https://ember-climate.org/data/data-explorer/
5. EPA (2020) What is CHP? https://www.epa.gov/chp/what-chp
6. Fonseca JD, Camargo M, Commenge J-M et al (2019) Trends in design of distributed energy systems using hydrogen as energy vector: a systematic literature review. Int J Hydrog Energy 44:9486–9504
7. Fonseca JD, Commenge J-M, Camargo M et al (2021) Multi-criteria optimization for the design and operation of distributed energy systems considering sustainability dimensions. Energy 214: 118989
8. García-Plaza M, Carrasco JE-G, Alonso-Martínez J et al (2018) Peak shaving algorithm with dynamic minimum voltage tracking for battery storage systems in microgrid applications. J Energy Storage 20:41–48
9. Gomes ISF, Perez Y, Suomalainen E (2021) Rate design with distributed energy resources and electric vehicles: a Californian case study. Energy Econ 102:105501
10. Guo S, Mu Y, Jia H et al (2019) Optimization of AC/DC hybrid distributed energy system with power electronic transformer. Energy Procedia 158:6687–6692
11. Huang D, Zhou D, Jia X et al (2022) A mixed integer optimization method with double penalties for the complete consumption of renewable energy in distributed energy systems. Sustain Energy Technol Assess 52:102061
12. Huang Y, Xu P, Wu Q et al (2022) Optimization of regional distributed energy system under discrete demand distribution. Energy Rep 8:634–640
13. IEA (2021) Japan 2021 Energy Policy Review
14. IEA (21 Apr 2022) World coal consumption, 1978–2020, IEA, Paris. https://www.iea.org/data-and-statistics/charts/world-coal-consumption-1978-2020
15. IEA (2021) Japan will have to tread a unique pathway to net zero, but it can get there through innovation and investment. IEA, Paris
16. İnci M, Büyük M, Savrun MM et al (2021) Design and analysis of fuel cell vehicle-to-grid (FCV2G) system with high voltage conversion interface for sustainable energy production. Sustain Cities Soc 67:102753
17. Li L, Yu S (2020) Optimal management of multi-stakeholder distributed energy systems in low-carbon communities considering demand response resources and carbon tax. Sustain Cities Soc 61:102230
18. Luo X, Liu Y, Liu J et al (2019) Optimal design and cost allocation of a distributed energy resource (DER) system with district energy networks: a case study of an isolated island in the South China Sea. Sustain Cities Soc 51:101726
19. Meti Energy in Japan 2020. https://www.enecho.meti.go.jp/en/category/brochures/pdf/japan_energy_2020.pdf
20. Nagatomo Y, Ozawa A, Kudoh Y et al (2021) Impacts of employment in power generation on renewable-based energy systems in Japan—analysis using an energy system model. Energy 226:120350
21. Oati Tame the impacts of distributed energy resources. https://www.oati.com/Solution/SmartEnergy/distributed-energy-resource-management

22. Qian F, Gao W, Yang Y et al (2020) Economic optimization and potential analysis of fuel cell vehicle-to-grid (FCV2G) system with large-scale buildings. Energy Convers Manag 205: 112463
23. Sait MA, Chigbu UE, Hamiduddin I et al (2018) Renewable energy as an underutilised resource in cities: Germany's 'Energiewende'and lessons for post-brexit cities in the United Kingdom. Resources 8:7
24. Shinryo District heating and cooling systems to realize the optimization of energy supply. https://www.shinryo.com/en/tech_service/dhc.html
25. Tan KM, Babu TS, Ramachandaramurthy VK et al (2021) Empowering smart grid: a comprehensive review of energy storage technology and application with renewable energy integration. J Energy Storage 39:102591
26. Tooryan F, Hassanzadehfard H, Collins ER et al (2020) Optimization and energy management of distributed energy resources for a hybrid residential microgrid. J Energy Storage 30:101556
27. UNDESA. UN DESA annual highlights report 2020–2021. https://www.un.org/en/desa/highlights-report-2020-2021
28. Winter C-J (2009) Hydrogen energy—abundant, efficient, clean: a debate over the energy-system-of-change. Int J Hydrog Energy 34:S1–S52
29. Wu D, Ma X, Huang S et al (2020) Stochastic optimal sizing of distributed energy resources for a cost-effective and resilient Microgrid. Energy 198:117284
30. Wu G, Xiang Y, Liu J et al (2020) Distributed energy-reserve co-optimization of electricity and natural gas systems with multi-type reserve resources. Energy 207:118229

Chapter 2
Integration and Application of Distributed Energy Resources and the Distributed Energy System in Japan

Zhonghui Liu and Yingjun Ruan

2.1 Present Condition of Distributed Energy Systems in Japan

2.1.1 Concept of a Distributed Energy System

In recent years, fossil fuels have been rapidly depleted, and environmental pollution has been severe. Therefore, there is an urgent need to find alternatives to fossil fuels and to utilize state-of-the-art technologies to improve energy efficiency. A distributed energy system (DES) has now attracted widespread attention. Unlike conventional energy supply systems where production is usually far from the user, a DES is an energy system where energy is produced close to end use, typically relying on a number of modular and small-scale technologies [1]. The poly-generation systems can be combined heat and power (CHP) systems; combined cooling, heating, and power (CCHP) systems, and so on [2]. CCHP systems use waste heat from on-site electricity generation to meet the thermal demand of the facility [3]. Energy cascading can be realized by the poly-generation process in DESs [4].

A DES is a faster, less expensive alternative to the construction of large, centralized power plants and high-voltage transmission lines. It offers consumers the potential for lower cost, higher service reliability, high power quality, increased energy efficiency, and energy independence. The use of renewable distributed

Z. Liu (✉)
School of Municipal and Environmental Engineering, Jilin Jianzhu University, Changchun, China

Y. Ruan
School of Mechanical and Energy Engineering, Tongji University, Shanghai, China
e-mail: Ruanyj@tongji.edu.cn

W. Gao (ed.), *Distributed Energy Resources*, Green Energy and Technology,
https://doi.org/10.1007/978-3-031-21097-6_2

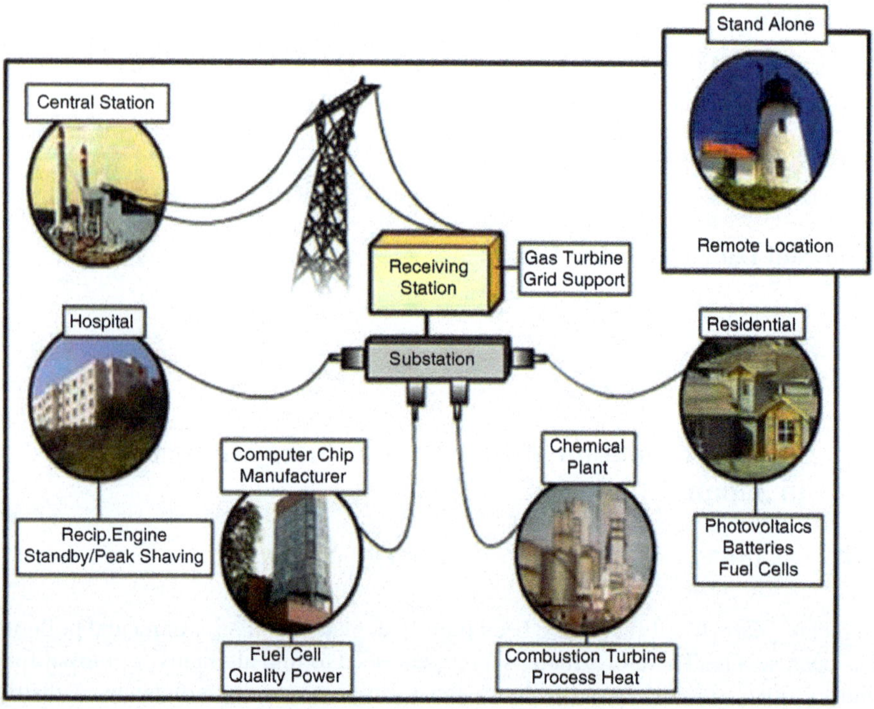

Fig. 2.1 Schematic diagram of a DES [5]

energy generation technologies and "green power" such as wind, photovoltaic, geothermal, biomass, or hydroelectric power can also provide a significant environmental benefit. Figure 2.1 shows the schematic diagram of a DES.

2.1.2 Combined Heat and Power

Combined heat and power (CHP) has grown more important and is widely expected to spread for the efficient use of energy and for the prevention of global environmental problems. The concurrent production of electricity or mechanical power and useful thermal energy (heating and/or cooling) from a single source of energy is a type of distributed generation, which, unlike central station generation, is located at or near the point of consumption. A suite of technologies can use a variety of fuels to generate electricity or power at the point of use, allowing the heat that would normally be lost in the power generation process to be recovered to provide needed heating and/or cooling. CHP technology can be deployed quickly, cost-effectively, and with few geographic limitations. CHP can use a variety of fuels, both fossil based and renewable. It has been employed for many years, mostly in industrial, large commercial, and institutional applications. Figure 2.2 shows the schematic diagram of a CHP.

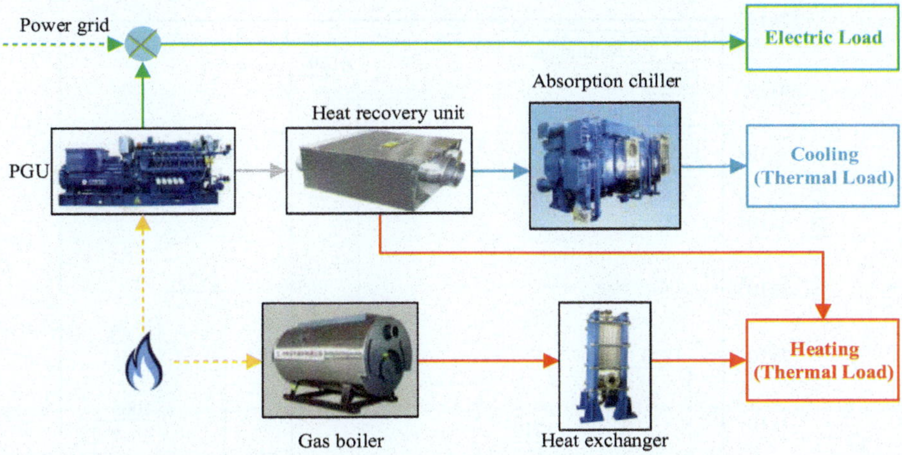

Fig. 2.2 Schematic diagram of a typical CHP system

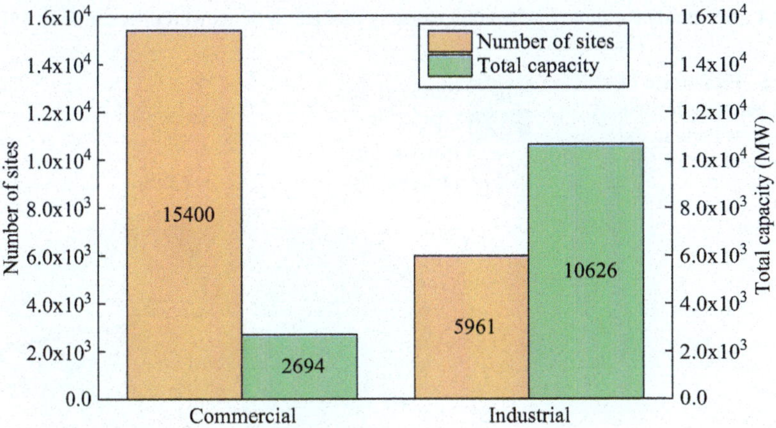

Fig. 2.3 Generation capacity and number of installation sites of CHP systems in Japan [6]

In Japan, CHP systems have been developed rapidly. As shown in Fig. 2.3, the total generation capacity of CHP systems has reached 13,320 MW as of March 2021, including 2694 MW in the commercial sector and 10,626 MW in the industrial sector. These CHP systems have been installed on 21,361 sites, with 15,400 commercial facilities and 5961 industrial facilities. Figure 2.4 shows the accumulative power generation capacity of CHP systems with different fuels. Figures 2.5 and 2.6 show the proportion of power generation capacity of different fuels in the commercial and industrial sectors. It is clear that power generation using natural gas is highest. Natural gas accounts for 67.7% of power generation in the commercial sector and accounts for 58.1% of power generation in the industrial sector. This shows that CHP systems mainly depend on natural gas in Japan. Table 2.1 shows the prime mover of CHP systems in each fiscal year.

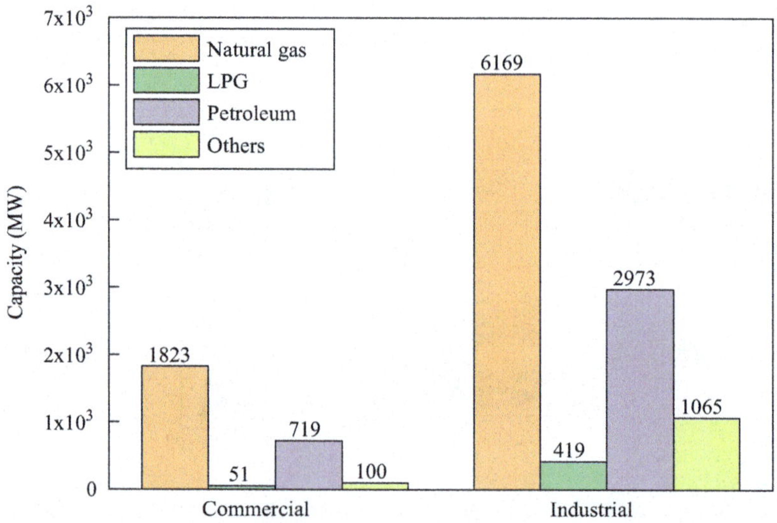

Fig. 2.4 Accumulative power generation capacity of CHP systems with different fuels [6]

Fig. 2.5 Proportion of power generation capacity of different fuels in the commercial sector [6]

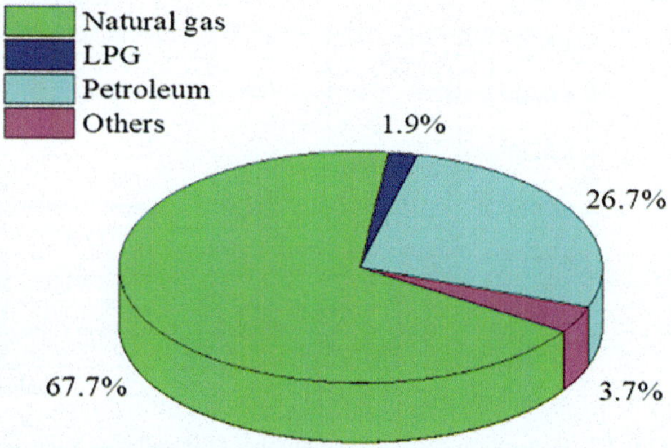

2.2 Distributed Energy System in KSRP

A DES is used to meet electricity demand, heating demand, cooling demand, and hot water demand. DESs are medium- and small-scale energy conversion and utilization systems that are directly oriented to customers, produce and supply energy locally according to their needs, and have multiple functions to meet multiple objectives. At the Kitakyushu Science and Research Park (KSRP), a DES had been installed to supply the electricity demand, heating demand, cooling demand, and hot water demand.

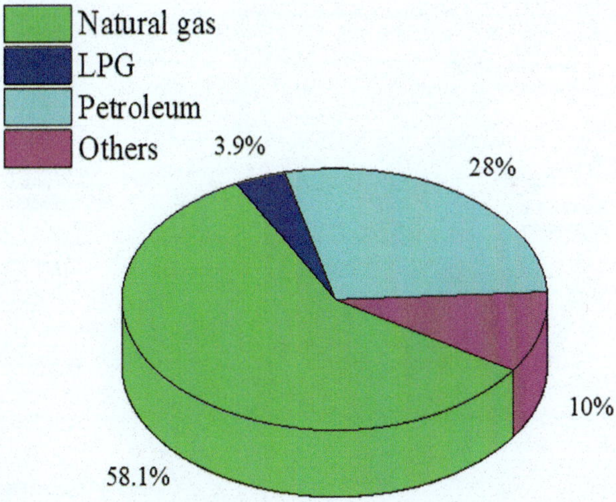

Fig. 2.6 Proportion of power generation capacity of different fuels in the industrial sector [6]

Table 2.1 Prime mover of CHP systems in each fiscal year [6]

Prime mover		Number of installations (unit)	Generation capacity (MW)	Capacity per installation unit (kW/Unit)
Gas turbine	Commercial	594	535	901
	Industrial	1024	4974	4858
	Total	1618	5509	3405
Gas engine	Commercial	12,482	1422	114
	Industrial	2433	2697	1109
	Total	14,915	4119	276
Diesel engine	Commercial	2117	714	337
	Industrial	2405	2547	1059
	Total	4522	3261	721

2.2.1 Introduction of KSRP

The Kitakyushu Science and Research Park (KSRP) is located in Kitakyushu City, Fukuoka, Japan. Kitakyushu is an industrial city in Japan. KSRP is located in the western part of Wakamatsu ward and the northwestern part of Yahatanishi ward, with a total development area of approximately 335 ha. Figure 2.7 shows the joint-use facilities at KSRP. KSRP has a number of industrial-academic collaboration facilities, with four universities: Kitakyushu City University School of Environmental Engineering, Kyushu Institute of Technology, Waseda University Graduate School of Information, Production and Systems, and Fukuoka University. The eco-campus is a campus of environmental symbiosis, water recycling, power generation, and heating. The Park has a number of research institutes and companies. Besides these, it has some buildings such as a collaboration center, semiconductor center, library, technology development and communication center, and so on.

Fig. 2.7 Joint-use facilities at KSRP [7]

In KSRP, several technologies and measures were used to reduce energy consumptions and improve energy use efficiency and water use efficiency because KSRP is composed of universities, research institutes, and companies focused on science and engineering, the future of technology, and sustainable development. Especially, in the Faculty of Environmental Engineering of the University of Kitakyushu, some technologies were used, such as natural wind, natural light, green roofs and walls, underground heat storage systems for air conditioning and heating, generation of electricity and heat, water recycling systems, and so on.

2.2.2 Distributed Energy System at KSRP

The Kitakyushu Science and Research Park (KSRP) comprises a lot of universities, colleges, and research institutions. Here, to supply necessary energy and water for the educational research activities efficiently, many technologies considering low environmental load have been introduced. Specially, at KSRP, a distributed energy system had been installed to supply the energy demand of the end use since 2001 and discontinued in 2016, as well as all the energy supply from the energy center at KSRP.

The energy center is the main part of the environmentally proactive campus that supports the educational research activities in KSRP. The function of the energy center is to supply energy (electricity, heating, cooling, and hot water) and water and to dispose of sewage water. Therefore, complex equipment and many of the facilities are located in the energy center, for instance, power generation system, cooling and

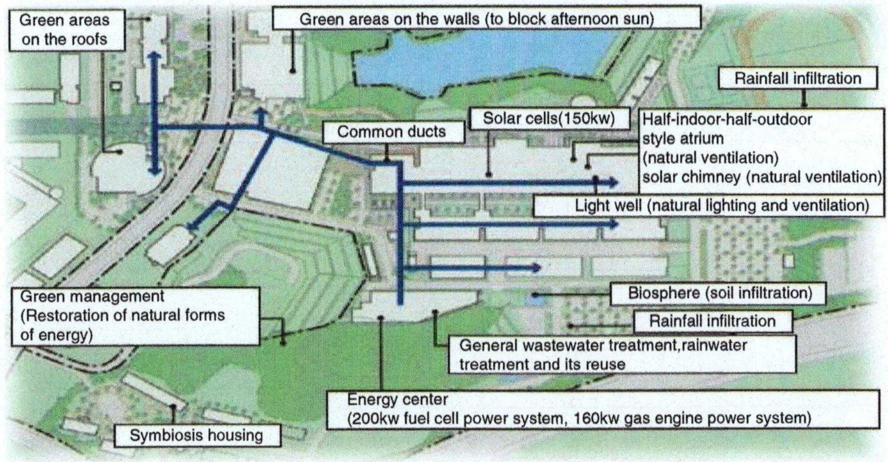

Fig. 2.8 The eco-campus planning of KSRP [7]

heating system, equipment monitoring system, middle water treatment system, water supply system, power exchange system, standby power equipment, maintenance centers, component stores, and so on.

The energy center provides electricity, heating, cooling, and hot water to a number of buildings and facilities within KSRP. It provides energy to only a few buildings and facilities, not to all buildings and facilities in KSRP. As shown in Fig. 2.8, the blue line is the energy supply line for the energy center.

A DES has the highest energy use efficiency and low environmental load. In the KSRP DES, electricity is provided by a gas engine (160 kW capacity; the gas engine was discontinued in 2016) and fuel cell (200 kW capacity; the fuel cell was discontinued in 2011), photovoltaics, and the Kyushu grid utility. The waste heat from the gas boiler, gas engine, and fuel cell meets the heat load. The waste heat from the gas engine and fuel cell meets the cooling load.

Figure 2.9 shows the schematic diagram of the DES installed in KSRP. It is known that the energy load of a building or campus includes electrical load, space cooling load, space heating load, and hot water load. The system at the KSRP includes photovoltaics, fuel cell, gas engine, gas absorption chiller, heat exchanger, and gas boiler. In addition to the main equipment, the system includes a large number of auxiliary equipment, such as various pumps (for cooling water, heating, cooling supply, circulation, etc.), cooling towers, ejectors, valves, piping, etc.

In this system, the city gas is used to supply the gas engine and the fuel cell to produce the electricity. In order to reduce the fossil energy consumption and carbon dioxide (CO_2) emissions and improve energy efficiency, the gas engine and fuel cell are used to produce the electricity as a small-scale power generation system. The gas engine has 160 kW capacity as shown in Fig. 2.10, and the fuel cell has 200 kW

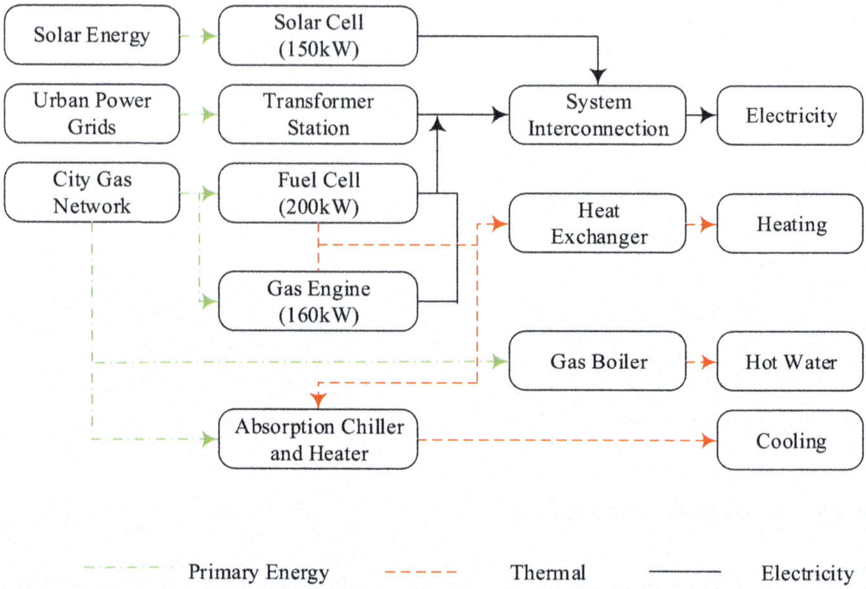

Fig. 2.9 Schematic illustration of the distributed energy system in KSRP

Fig. 2.10 The gas engine system

capacity as shown in Fig. 2.11. The capacity of the 150 kW solar PV is used to meet some electricity demand in the system. When the electricity load is low or the electricity production is not enough, the electricity from grid utilities is used to meet the load. A part of waste heat is sent to the absorption chiller to meet the cooling load, and another part is sent to the heat exchanger unit to meet the space heating load. And the gas boiler is used to meet the hot water load in this system.

Fig. 2.11 The fuel cell system

Fig. 2.12 Multi-crystal silicon solar cell

A solar cell consists of either single-crystal silicon sliced from a single crystal of high-purity silicon or polycrystalline silicon obtained by poly-crystallizing metallic silicon with a mold in order to make the manufacturing cost lower. There are other types of amorphous silicon solar cells, as well as a hybrid type combined with a crystal type. The Kitakyushu Science and Research Park has installed 156 single-crystal molds (250 cm × 75 cm) and 864 polycrystalline molds (132 cm × 89.5 cm) on the roof sloping table in the eaves of the north building as shown in Figs. 2.12 and 2.13.

Table 2.2 provides details of heat recovery efficiency and generator set efficiency of the gas engine and fuel cell in the distributed energy system when the equipment system is running at full capacity.

Fig. 2.13 Single-crystal silicon solar cell

Table 2.2 Details of DES

Equipment system	Gas cogeneration system		Solar energy cell	
Type	Fuel cell	Gas engine	Polysilicon	Monocrystalline silicon
Capacity	200 kW	160 kW	129.6 kW	23.4 kW
Power generation efficiency (with 100% load)	40%	28.70%	13.30%	7.20%
Heat recovery efficiency	(90 °C hot water) 20%(50 °C hot water) 20%	47.7% (90° C hot water)	None	
Gas cost	43.3 Nm^3/h	44.1 Nm^3/h	None	
Operation mode	24 h/day	Run for 8:00 ~ 22:00	All the year	
Waste heat utilization equipment	Heat exchanger, absorption chiller and heater, hot water tank		None	

The capacity of the fuel cell is 200 kW; the two circuits have a generation efficiency of 40% and a heat recovery efficiency of 20%, with a high temperature of 90 °C for one circuit and 50 °C for the other. The high-temperature hot water circuit preheats the hot water supply scheme. The system operates continuously for 24 h throughout the year. The gas-fired unit has a capacity of 160 kW, a generation efficiency of 28.7%, and a heat recovery efficiency of 47.7%.

Figure 2.14 is a simplified process flow diagram for the gas engine. The gas engine uses city gas to generate electricity, and the waste heat is sent to the heat exchanger to supply high-temperature water to the heat recovery unit (heat exchanger/absorption chiller for heating or cooling or hot water heat exchanger). Then the returned water is sent to the condenser for cooling, and the water is circulated to the gas engine.

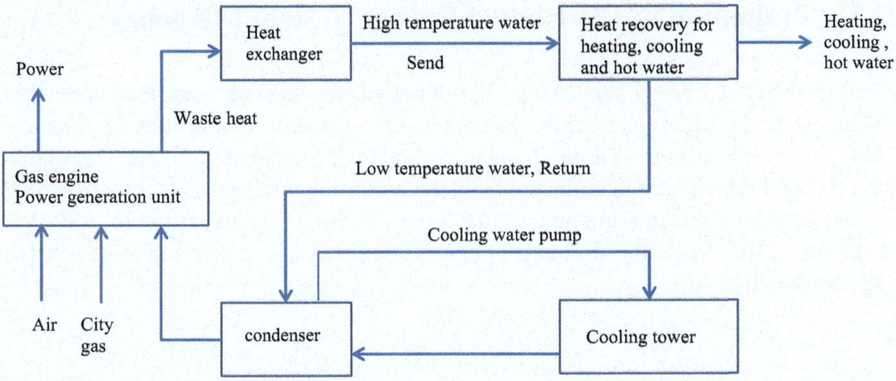

Fig. 2.14 A simplified process flow diagram for the gas engine

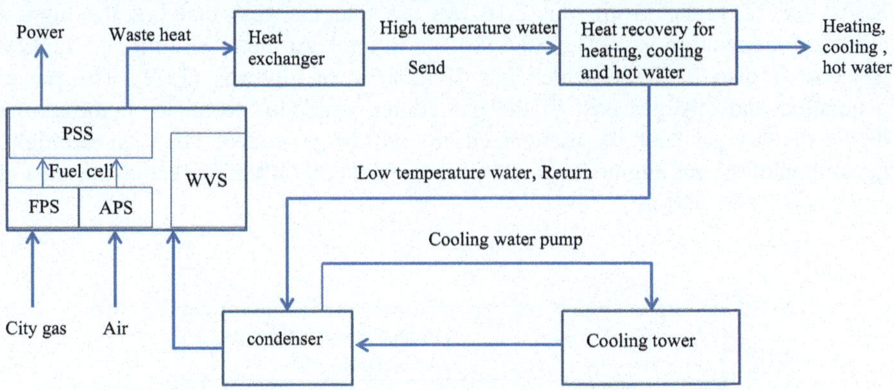

PSS: power section system; FPS: fuel processing system; APS: air processing system; WVS: water vapor separator.

Fig. 2.15 A simplified process flow diagram for the fuel cell

Figure 2.15 is a simplified process flow diagram for the fuel cell. The fuel cell produces power by use of city gas. First, the city gas is changed to H2 through the fuel processing system. And the air will be cleared and changed by the air processing system. Then the H2 and the changed air are sent to the power section system to produce electricity. And the waste heat will be sent to the heat exchanger to supply high-temperature water to the heat recovery unit. And the returned low-temperature water is recycled to the condenser. Then it will be sent back to the fuel cell.

2.3 Evaluation of Distributed Energy System Efficiency

The distributed energy system in the Kitakyushu Science and Research Park includes the fuel cell, gas engine, and solar cell. Because of the lack of solar cell data, in this research, the gas engine and fuel cell will be the main focus. Gas engine and fuel cell data has been collected from the environmental energy center of KSRP. The data shows that the gas engine has operated for 15 years, from July 2001 to February 2016, while the fuel cell operated for about 10 years from June 2001 to November 2011.

2.3.1 *Electricity Generation and Efficiency of the Gas Engine*

Figure 2.16 shows the electricity generation and city gas cost of the gas engine in KSRP for 15 years. From Fig. 2.16, we are told the year that has the highest electricity generation is 2004, 780,644 kWh in total. At the same time, the highest gas cost is also in 2004, which has 235,286 $m^3$of city gas (13A). The power generation and city gas cost of the gas engine generally correlate. It means the higher the city gas cost, the more electricity will be generated. The total electricity generation of the gas engine for the 15 years is 8,824,397 kWh, while the total cost of city gas is 2,691,890 m^3.

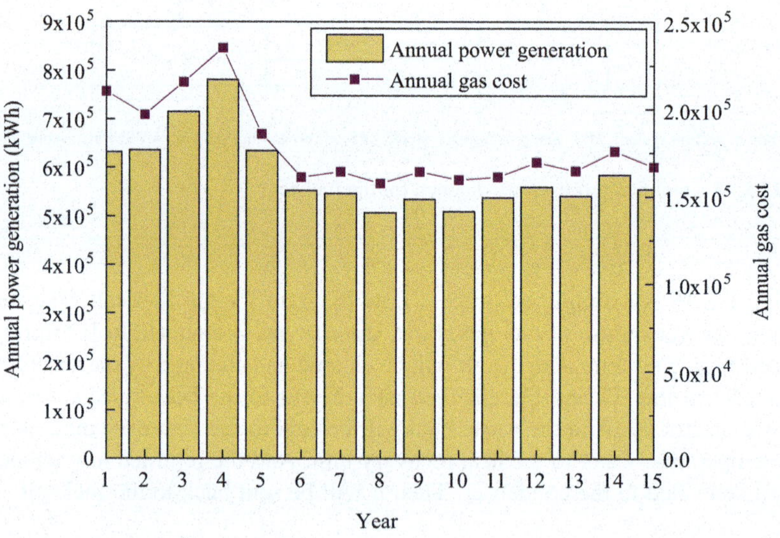

Fig. 2.16 Electricity generation and city gas cost of the gas engine

The power generation efficiency of the gas engine was calculated using the city gas consumption and the power generation data. The data was collected from the environmental energy center in the Kitakyushu Science and Research Park. And the power generation data was recorded every hour from July 2001 to February 2016.

The power generation efficiency calculation of the gas engine is shown as follows:

$$\eta_g = \frac{E_g \times 0.86}{V_g \times 11 \times 0.1} \times 100\% \tag{2.1}$$

where η_g is the power generation efficiency of the gas engine. E_g is the power generation of the gas engine (GJ). V_g is the city gas consumption of the gas engine (GJ).

In accordance with Eq. 2.1, the efficiency of the electricity generation of the gas engine is shown in Fig. 2.17, for every year during the 15 years. Looking at the gas engine electricity generation efficiency during the 15 years, it is obvious that the efficiency of the first year was very unstable, because the gas engine was still at the debugging phase. In the next 3 years, its electricity generation efficiency tended to be stable, in the range 25–25.9%. There was a shock between 2008 and 2010, while the year 2006 had the highest efficiency during the 15 years. It was caused that the gas engine was begun the interim period stoppage. The efficiency line was interrupted in this figure.

The heat recovery of the gas engine was calculated using these data: temperature of the forwarded hot water, temperature of the returned hot water, and instantaneous flow rate of the hot water. These data were collected from the environmental energy

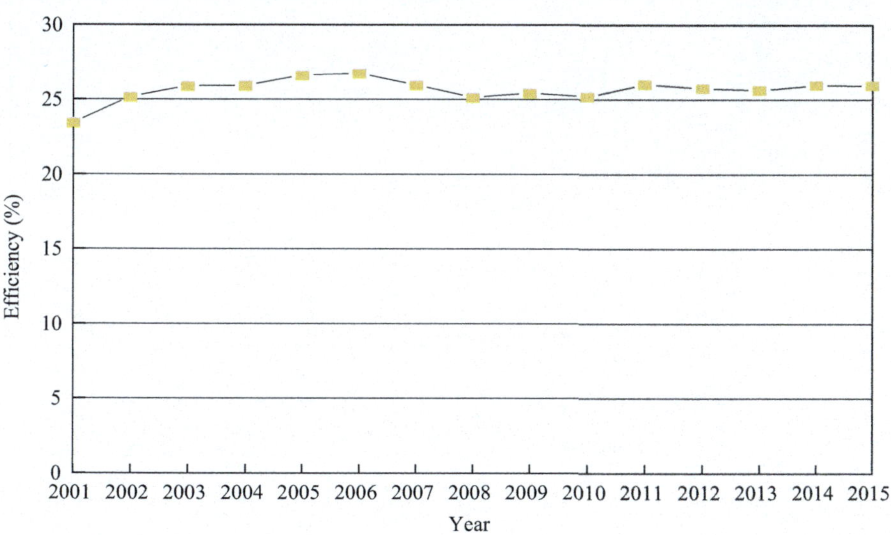

Fig. 2.17 Efficiency of the gas engine electricity generation

center in the Kitakyushu Science and Research Park (KSRP). The temperature of the forwarded hot water, temperature of the returned hot water, and instantaneous flow rate of the hot water were recorded every hour from July 2001 to February 2016.

The heat recovery calculation of the gas engine is shown as follows:

$$Q_r = (T_f - T_r) \times V_f \tag{2.2}$$

where Q_r is the heat recovery of the gas engine (heat/Mcal). T_f is the temperature of the forwarded hot water (temperature/°C). T_r is the temperature of the returned hot water (temperature/°C). V_f is the instantaneous flow rate of the hot water (flow/m^3/h).

Furthermore, the heat recovery efficiency of the gas engine was calculated using the city gas cost and the heat recovery data; the data was collected from the environmental energy center in the Kitakyushu Science and Research Park.

The heat recovery efficiency calculation of the gas engine is shown as follows:

$$\eta_h = \frac{Q_r}{N_g \times 11} \times 100\% \tag{2.3}$$

where η_h is the heat recovery efficiency of the gas engine. Q_r is the heat recovery of the gas engine (heat/Mcal). N_g is the city gas consumption of the gas engine (unit/m^3).

Figure 2.18 is the heat recovery of the gas engine during the 15 years. Because the heat recovery and power generation were recorded at the same time, the shapes of the heat recovery and power generation graphs are roughly the same. The highest heat

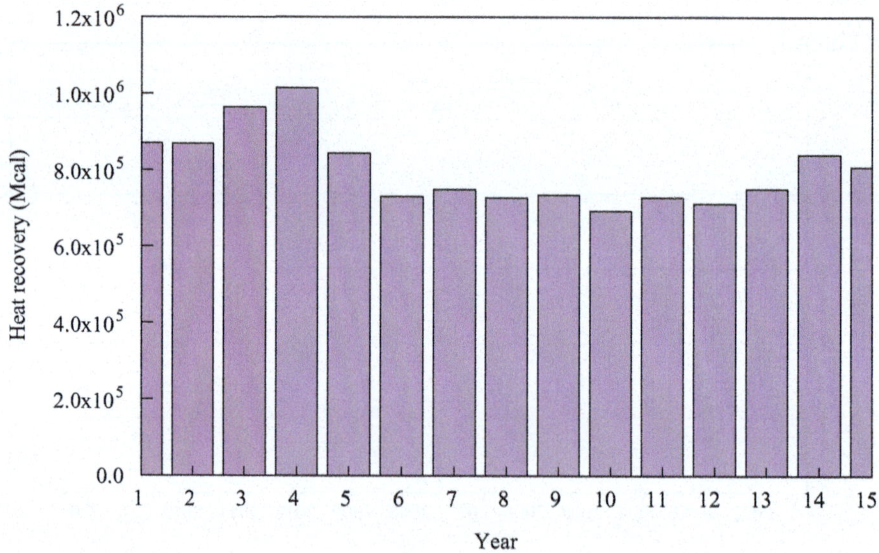

Fig. 2.18 Heat recovery of the gas engine during the 15 years

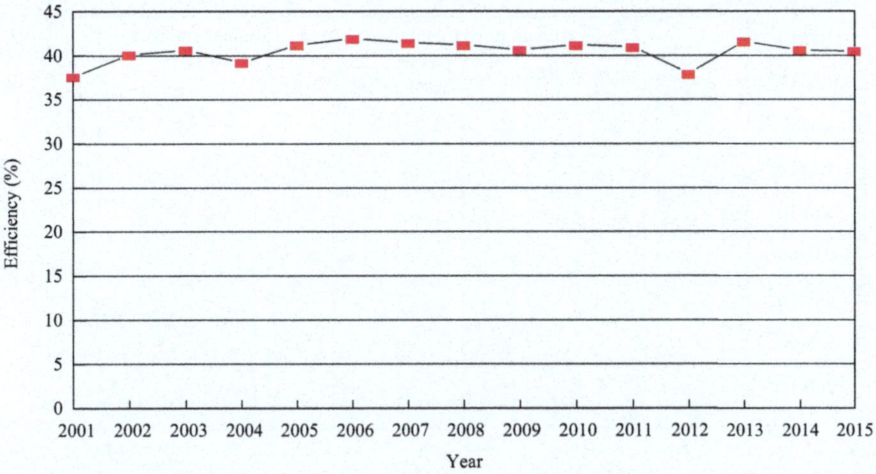

Fig. 2.19 Heat recovery efficiency of the gas engine during the 15 years

recovery is in 2004, 1,014,852.89 Mcal in total. At the same time, the year with the highest gas cost is also 2004; it has 235,286 m^3 of city gas (13A). The heat recovery and city gas consumption of the gas engine are generally correlated. It means the higher the city gas cost, the more electricity will be generated, and the more heat could be reused. At the same time, the waste heat also increased. The total heat recovery of the gas engine during the 15 years is 12,034,602 Mcal, while the total cost of city gas is 2,691,890 m^3.

Figure 2.19 shows the heat recovery efficiency of the gas engine. It is noticed that in 2006, the efficiency is highest. It was caused that the gas engine was begun the interim period stoppage. The operation status and power generation of the gas engine have been shown in this part.

2.3.2 Electricity Generation and Efficiency of the Fuel Cell

Figure 2.20 shows the power generation and city gas cost of the fuel cell in the Kitakyushu Science and Research Park for 11 years. From Fig. 2.20, we are told the year that has the highest power generation is 2006, 1,697,149 kWh in total. At the same time, the highest gas cost is also in 2006; it has 419,003 m^3 of city gas (13A). The power generation and city gas cost of the fuel cell are generally correlated. It means the higher the city gas cost, the more power will be generated. The total power generation of the fuel cell for the 11 years is 15,185,001 kWh, while the total cost of city gas is 3,694,043 m^3.

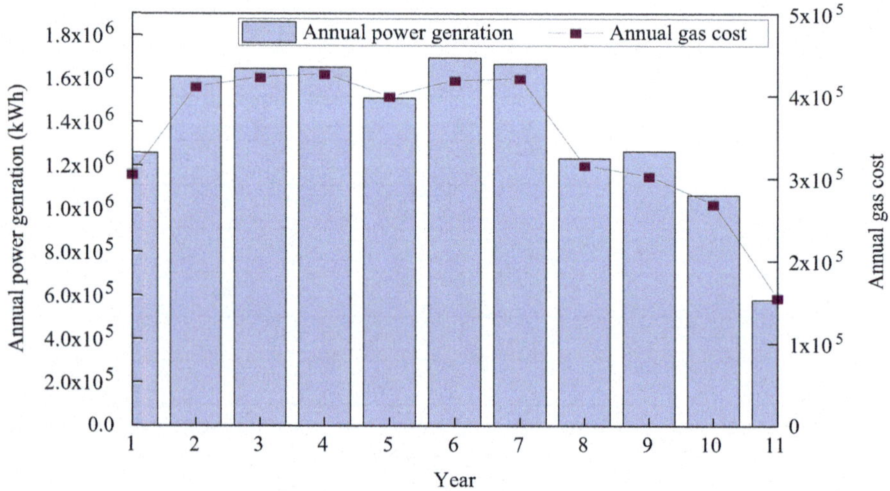

Fig. 2.20 Electricity generation and city gas cost of the fuel cell

The power generation efficiency calculation of the fuel cell is shown as follows:

$$\eta = \frac{E_f \times 0.86}{V_f \times 0.11} \times \frac{273 + 20}{273 + T} \times 100\% \quad (2.4)$$

where η is the power generation efficiency of the fuel cell (%). E_f is the power generation of the fuel cell (electricity/kWh). V_f is the city gas consumption of the fuel cell (unit/m^3). T is the temperature of the energy center (temperature/°C).

In accordance with Eq. 2.4, the efficiency of the electricity generation of the fuel cell is shown in Fig. 2.21, for every year during the 11 years. Looking at the electricity generation efficiency of the fuel cell for 11 years, it is obvious that the efficiency of the first year was very unstable, because the fuel cell was still at the debugging phase.

The heat recovery of the fuel cell was calculated using these data: temperature of the forwarded hot water, temperature of the returned hot water, and instantaneous flow rate of the hot water. These data were collected from the environmental energy center in the Kitakyushu Science and Research Park.

The temperature of the forwarded hot water, temperature of the returned hot water, and instantaneous flow rate of the hot water were recorded every hour from June 2001 to January 2010.

The heat recovery efficiency of the fuel cell was calculated using the city gas cost and the heat recovery data; the data was collected from the environmental energy center in the Kitakyushu Science and Research Park.

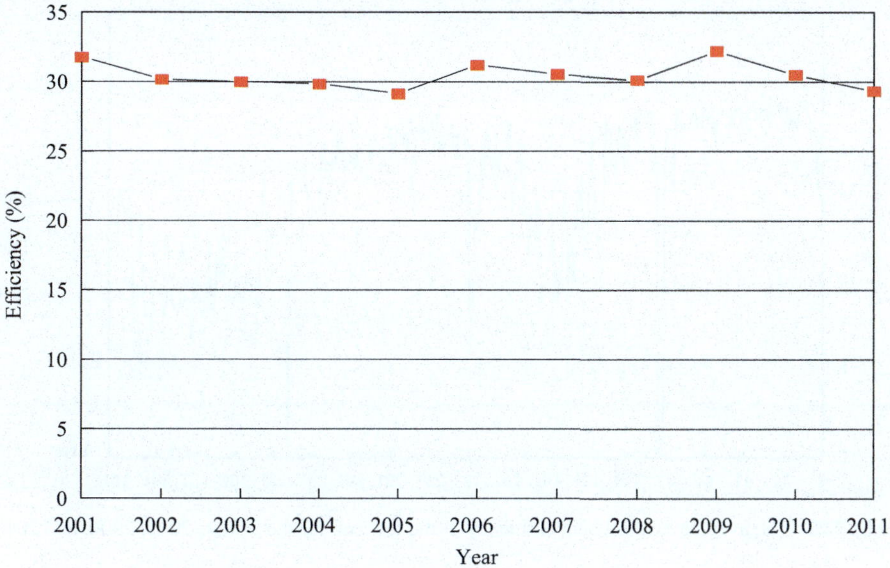

Fig. 2.21 The average power generation efficiency of the fuel cell

The heat recovery efficiency calculation of the fuel cell is shown as follows:

$$\eta_H = \frac{V_i \times (t_f - t_r) \times 0.01}{V_f \times 0.11 + C \times 0.025} \times 100 \qquad (2.5)$$

where η_H is the heat recovery efficiency of the fuel cell. t_f is the temperature of the forwarded hot water (temperature /°C). t_r is the temperature of the returned hot water (temperature/°C). V_i is the instantaneous flow rate of the hot water (flow/m^3/h). C is the cost of electricity (electricity/kWh). V_f is the city gas consumption of the fuel cell (unit/m^3).

The exhaust heat data of the years 2002, 2006, and 2010 are used in this paper. Figure 2.22 is the comparison of the 3 years' average exhaust heat efficiency. The impact of 0 exhaust heat efficiency has been removed. According to Fig. 2.22, the exhaust heat efficiency has an obvious decreasing trend in the 3 years. The efficiency of 2002 is about 15.0%, 2006 is 12.3%, and 2010 is 6.78%. It means that the exhaust heat capacity gradually weakens as time goes by. It is noticeable that after September 2002, the exhaust heat efficiency is lower than that in 2006. It is because the fuel cell was still in the debugging stage at that time. Its operation, power production, and exhaust heat were not stable enough. Compared with the exhaust heat efficiency of 2002, the efficiencies of 2006 and 2010 are more stable. In other words, the fuel cell has been kept in an inefficient operating condition in the second half of its life.

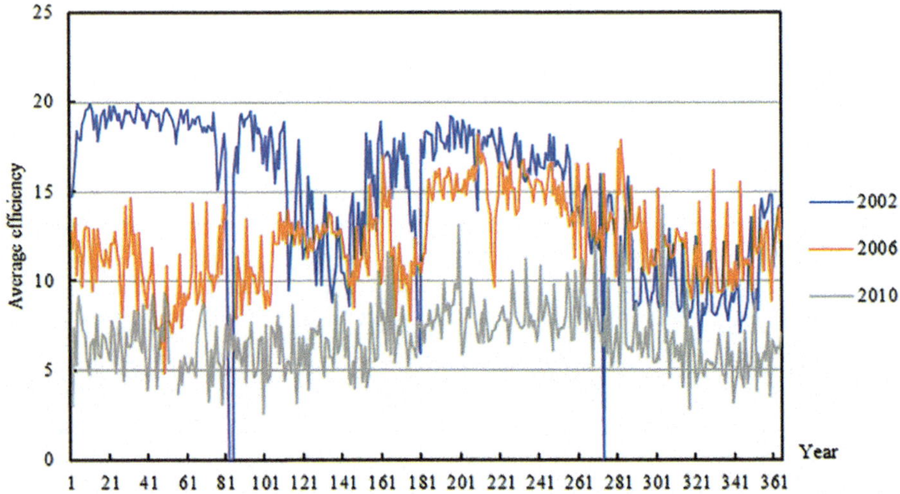

Fig. 2.22 The average exhaust heat efficiency of the fuel cell for the 3 years 2002, 2006, and 2010

References

1. Zhou Z, Liu P, Li Z, Ni W (2013) An engineering approach to the optimal design of distributed energy systems in China. Appl Therm Eng 53(2):387–396. https://doi.org/10.1016/j.applthermaleng.2012.01.067
2. Han J, Ouyang L, Xu Y, Zeng R, Kang S, Zhang G (2016) Current status of distributed energy system in China. Renew Sust Energ Rev 55:288–297. https://doi.org/10.1016/j.rser.2015.10.147
3. Abdollahi G, Meratizaman M (2011) Multi-objective approach in thermoenvironomic optimization of a small-scale distributed CCHP system with risk analysis. Energ Buildings 43(11): 3144–3153. https://doi.org/10.1016/j.enbuild.2011.08.010
4. Ren H, Zhou W, Nakagami K, Gao W, Wu Q (2010) Multi-objective optimization for the operation of distributed energy systems considering economic and environmental aspects. Appl Energy 87(12):3642–3651. https://doi.org/10.1016/j.apenergy.2010.06.013
5. Distributed Energy Resources (DER) | WBDG – whole building design guide
6. https://www.ace.or.jp/web/works/works_0010.html
7. https://www.ksrp.or.jp/e/access/map.html

Chapter 3
Impact of Renewable Energy Policies on Solar Photovoltaic Energy: Comparison of China, Germany, Japan, and the United States of America

Daoyuan Wen and Weijun Gao

3.1 Support Policies and PV Technology Development in China, Germany, Japan, and the USA

3.1.1 Introduction

Since the 1970s, due to the limited supply of fossil energy and increasing pressure regarding environmental protection, numerous countries worldwide have begun to exploit and utilize renewable energy. Among all renewable energy sources, solar photovoltaic (PV) technology has a huge potential in alleviating pollution, reducing CO_2 emissions, and addressing energy demand pressures [1]. Therefore, promoting solar PV technology has become a vital part of sustainable development strategies worldwide. In the last few decades, driven by advanced technology and improved regulations, solar PV technology has experienced growth rapidly [2].

The first PV device was invented by Bell Labs in the USA in 1954 and mainly applied to space satellites [3]. After the 1970s, the USA took the lead in introducing policies to support the solar PV ground deployment. In the following decades, the

D. Wen (✉)
School of Electrical Engineering, Shanghai Dianji University, Shanghai, China

W. Gao
The Engineering Academy of Japan, Chiyoda City, Tokyo, Japan

The University of Kitakyushu, Kitakyushu, Japan

Qingdao University of Technology, Qingdao, China

International Research Center for Green Buildings and Low-Carbon Cities, Zhejiang University, Hangzhou, China

Xi'an Jiaotong University, Xi'an, China

W. Gao (ed.), *Distributed Energy Resources*, Green Energy and Technology,
https://doi.org/10.1007/978-3-031-21097-6_3

USA gradually developed a niche market for PV technology. Since the 1990s, Japan and Germany had started launching several programs for installing PV system products on the rooftops. However, high costs had limited the scale of PV deployment, and PV had remained a niche market worldwide. The situation improved in the year 2000, when market incentive policy gradually became a dominant support scheme along with the first feed-in tariff (FiT) law introduced in Germany. Successful Germany PV market expansion promoted the diffusion of the FiT to other countries in the following years. Specifically, after 2010, the Chinese market became the fastest-growing solar PV market in the world. The global solar PV market has seen remarkable growth, with global cumulative capacity increasing from 1.2 GW in 2000 to 760 GW in 2020. The top four countries by solar PV cumulative capacity were China (253.4 GW), the USA (93.2 GW), Japan (71.4 GW), and Germany (53.9 GW), who shared more than 62% of the total solar PV installed capacity worldwide [4].

Many scholars have explored the key factors behind the successful development of PV technology by investigating these typical countries. In this chapter, a detailed analysis of the rise of solar PV technology in China, Germany, Japan, and the USA is presented, along with how PV development is influenced by policies in different periods in these four countries as study cases. The effects of different incentive policies implemented over the past decades on PV development in these four leading countries are demonstrated. At different development periods, some special external factors may have guided the introduced policy, and the type of policy implemented may vary across different countries. The role of policy instruments and international factors is investigated. Thereafter, different policies are identified, and how they have driven PV development in China, Germany, Japan, and the USA is examined.

3.1.2 Investigation of PV Incentive Policies in China, Germany, Japan, and the USA

Table 3.1 shows the PV incentive policies in China, Germany, Japan, and the USA from 2000. Consistent public funding for PV R&D has helped the USA become the technology leader in the solar PV industry. Until 2006, the DOE was appropriated USD 5.8 billion for solar research [5]. The "US Photovoltaics Industry Roadmap," which was refined in December 2000 and updated in 2004, unifies the long-term (2000–2020) strategies and goals for the PV industry in the country [6, 7]. The production targets of the US PV industry roadmap reveal that 70% of the production capacities are aimed for export. This series of efforts by the policy instruments facilitated expansion of the PV industry in the USA [8, 9]. In 2005, the "Energy Policy Act 2005 (ITC)" was introduced to promote PV market development, which provided a 30% investment tax credit to those who invested in PV systems. The ITC has proven to be one of the most important federal policy mechanisms to incentivize PV development in the USA. This Act was complemented by accelerated depreciation, which added approximately 26% to the tax benefit, thus reducing the system cost by approximately 56% over a six-year period for many investors ([10]; Stegman

Table 3.1 The PV incentive policies in China, Germany, Japan, and the USA

Year	The USA	Germany	Japan	China
2000		Feed-in tariff law (EEG)		
2001			The new five-year plan for PV power generation technology R&D	10th Five-Year Plan
2003			New monitoring program for residential PV systems; Renewable portfolio standard	
2004		EEG amended		
2005	Energy Policy Act 2005 (ITC)			
2006	Solar America	Funding for Solar Power Development Center	PV roadmap toward 2030	Catalog of Chinese high-technology products for export; Renewable energy law
2008	Solar America Initiative; Extension ITC	EEG amended		
2009			R&D for high-performance PV generation systems; Subsidy for residential PV systems; Feed-in tariff law	The "Golden Sun" demonstration project
2010		The Innovation Alliance PV; EEG amended		The BIPV subsidy program
2011				973 program; 863 program; Solar PV feed-in tariff
2012			New feed-in tariff law	The new "Golden Sun" demonstration project
2013				Feed-in tariff support for solar PV; PV electricity grant
2014		EEG amended	NEDO PV challenges	
2015	Extension ITC			The top runner program

(continued)

Table 3.1 (continued)

Year	The USA	Germany	Japan	China
2016	Solar Energy Technologies Office/ SunShot	Subsidy for solar PV with storage installations		Notice on solar PV deployment management and introduction of competitive bidding
2017		The Landlord-to-Tenant Electricity Act 2017	Solar PV auctions (FiT amended)	
2018 Extension ITC				Action Plan for the Development of Smart Photovoltaic Industry

and Davis 2016). The residential and commercial ITC has helped the solar PV market to grow significantly since it was implemented, with an average annual growth of 50% over the last decade alone [11]. The ITC Act 2005 was implemented until the end of 2007. Thereafter, the ITC Act was extended in 2008 and 2015 to ensure continued growth of the PV market. In 2007, the Solar America Initiative (SAI) funded up to USD 13.7 million for 11 university-led projects that focused on the development of advanced solar PV technology manufacturing processes and products [8]. During 2009–2011, public funds for PV R&D exceeded USD 400 million in the USA. In 2011, the "SunShot Initiative" was introduced by the Solar Energy Technologies Office (SETO) of the DOE, which aimed to reduce the total cost of PV solar energy systems by 75% by 2020 [12]. As solar PV technology made rapid progress closer to the 2020 targets, the SETO committed to reaching new cost targets for the upcoming decade, supporting greater energy affordability by reducing the cost of solar electricity by an additional 50% between 2020 and 2030. The SunShot 2030 targets were 0.05 USD/kWh for residential PV, 0.04 USD/kWh for commercial PV systems, and 0.03 USD/kWh for utility-scale PV systems [13].

In Germany, the "100,000 Roofs Program" and the EEG (FiT) scheme became an opportunity for rapid growth in the PV market since 2000 [14]. The FiT scheme has driven the rapid growth of the market, which has grown consistently toward the government targets; the growing PV market has become an opportunity for new companies to enter the PV industry [15]. Therefore, the government of Germany reformulated the R&D program emphasizing not only cost reduction but also the consequent utilization of the R&D results in PV production. Since autumn of 2002, the Federal Ministry for the Environment, Nature Conservation and Nature Safety (BMU) has been responsible within the federal government for promoting renewable energy development [16]. In 2006, in addition to BMU grants, the Federal Ministry of Education and Research (BMBF) also provided funding for the development of PV technologies [17]. In 2010, the BMU and BMBF initiated an Innovation Alliance for PV technology. Under this scheme, the R&D projects were funded to support a significant reduction in PV production costs for enhancing the competitiveness of the German PV industry. The BMU and BMBF allocated EUR 100 million to support this initiative. The German PV industry agreed to raise an additional EUR 500 million to accompany the Innovation Alliance [18]. To streamline the German energy policies, the responsibility for all energy-related activities was concentrated within BMWi since the end of 2013 [19]. The EEG has accelerated the growth of the PV market, which has been consistent and has surpassed the government targets. Therefore, the government further fundamentally revised the EEG in 2014 [20].

In Japan, a new R&D program called "the new five-year plan for PV power generation technology R&D" was initiated in 2001. This program focused on four areas: advanced solar cell technologies, comprehensive introduction of common basic PV technologies, innovative next-generation PV power technologies, and advanced manufacturing technology of PV systems. In 2006, the new five-year plan was completed, and then a four-year plan was launched based on the "PV roadmap toward 2030 (PV2030)" plan [21]. The "R&D for high-performance PV generation systems for the future" and "R&D on innovative solar cells" were initiated in 2009; these plans aimed to make a breakthrough in next-generation solar cells were governed by the Ministry of Education, Culture, Sports, Science, and Technology (MEXT) and were promoted by the Japan Science and Technology Agency (JST). A new guidance for technology development based on the "NEDO PV challenges," formulated in 2014, set a target to realize a power generation cost of 14 JPY/kWh by 2020 and 7 JPY/kWh by 2030 [22]. Under the new framework of technological research, NEDO shifted its direction from "strategies to promote dissemination of PV power generation" to "strategies to support the society after penetration of PV power" [23]. On the demand-pull policy side, parallel to a new monitoring program for residential PV systems, the government introduced another renewable energy policy known as the "Renewable Portfolio Standard (RPS)" in 2003 [24]. In addition, the FiT scheme for residential PV was adopted in November 2009 [25]. It was estimated that more than 90% of the PV installations were carried out in residential buildings [26]. With the start of the new FiT Act in 2012, the Japanese PV market entered a new growth phase [27]. For residential PV installations, tariffs with 42 JPY/kWh were paid for 10 years. The non-residential sector had a 40 JPY/kWh paid for 20 years [6]. The FiT policy has thus driven the rapid growth of the PV market in Japan.

In China, the Ministry of Science and Technology (MOST) supports PV R&D in universities and research institutions and provides assistance to enterprises for realizing each of the central government's "Five-Year Plans" [28]. In the Plan for New Energy and Renewable Energy Industry Development in the 10th Five-Year (2001–2005) Plan, renewable energy was viewed as a significant choice to optimize the Chinese energy structure. The public PV R&D funding increased to USD 6 million per year for the 11th Five-Year Plan (2006–2010). In the 12th Five-Year Plan (2011–2015), the support for PV fields covered the entire manufacturing chain. The average annual investment in R&D from the MOST was approximately USD 75 million during this period. In 2006, China began to enact the "Renewable Energy Law." The law was a national framework for promoting renewable energy development. This proved to be a huge driving force for the Chinese PV industry. From 2004, China's PV production increased remarkably [29]. Benefiting from the assistance of the "catalog of Chinese high-technology products for export" in the form of tax rebates, free land for factories, and low-interest government loans, Chinese solar PV product suppliers expanded their production lines rapidly, especially for PV cells and modules [30]. Since 2009, the government has attached importance to the domestic PV market and adopted a range of policies to support its development, such as special funds for renewable energy, feed-in tariff subsidies, preferential income tax for high and new technology enterprises, financial aid for PV applications, and demonstration projects. The "Rooftop Subsidy Program" and "Golden Sun Demonstration Program" were initiated by the MOST and the National Energy Administration (NEA) [31]. In July 2011, the National Development and Reform Commission (NDRC) announced a nationwide FiT policy for the development of solar PV energy [32]. In August 2013, the NDRC issued a "notice on the role of price lever in promoting the healthy development of the PV industry." PV power generation was categorized into either distributed or centralized systems [33]. Concerning centralized power generation, the whole country was further divided into three regions based on the solar resource distribution. In particular, the FiT is to be guaranteed for 20 years. Thus, the FiT policy has driven the rapid growth of the PV market in China. In 2015, a "Top Runner Program" was introduced to encourage Chinese PV companies to invest in PV R&D [34]. With the expansion of the domestic PV market, the PV product capacity in China continues to grow. Until now, the Chinese PV product output and market scale still ranks first worldwide.

3.1.3 Impacts of Incentive Policies on PV Development in China, Germany, Japan, and the USA

Based on the investigation of PV incentive policies mentioned above, their impacts are presented. Figure 3.1 depicts the dramatic change in PV module prices from 2000 to 2018 in the four countries. The PV module prices were influenced by PV

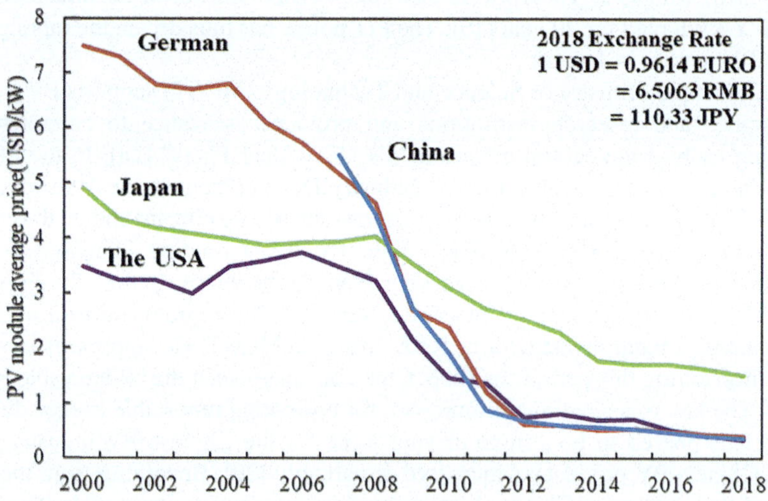

Fig. 3.1 Photovoltaic (PV) module average prices in China, Germany, Japan, and the USA from 2000 to 2018; 2018 prices and exchange rates. (Data source: IEA. PVPS. National Survey Report of PV Power Applications)

R&D activities and the PV industry's production status. The USA and Japan retained a price advantage regarding the PV module compared with other countries until 2008. Furthermore, before 2010, PV module average prices in the USA were lower than in other countries. This can be explained by the large-scale investment in PV R&D activities over a long period. Beginning in 2008, module prices in the four countries declined rapidly. This is partly due to advances in PV technology and partly due to the expansion of the global PV production capacity. In Japan, the decline in module prices has been slow because of their high domestic production costs [35]. In Germany, the expansion of the PV industry in eastern Germany (after 2006) has contributed to a decline in the module prices [36]. PV industrial research collaborations managed to get support from nationally funded R&D collaboration programs. In China, before 2010, even though the production in the PV industry was large-scale, average module prices were still higher than those in the USA and Germany. This can be explained by the lack of systematic investment in PV R&D in China, considering that the other three countries invested much more public funding in PV R&D than China. From 2012, Chinese PV products were enforced by anti-dumping duties and anti-subsidy countervailing duties in both the USA and Europe. Most Chinese manufacturers have increased R&D investment to improve product competitiveness to reduce costs [37]. Chinese PV module costs have decreased rapidly as well. In 2010, the PV module price reductions in Germany and China caught up with those in the USA and Japan [38]. Until now, German and Chinese PV modules have maintained their price advantage among the four countries; China has the lowest module price compared with the other countries. As a developing country, China's PV industry development trajectory is completely different from that of other developed countries. It is important to note that China's PV development has not experienced a long basic technology R&D period, and improvements to technology were only achieved via learning-by-doing strategies. For a long time, most of the technology was imported, mainly from Western countries. To summarize, we can state that compared with the USA, Germany, and Japan, China lacked a long-term PV R&D program and invested less public R&D funds. Additionally, in China, the PV R&D activities and policies were production-oriented to reduce costs, while in Germany, Japan, and the USA, the focus was more on technology improvement.

Figure 3.2 shows the trends in average PV module prices with an increase in the cumulative public R&D funding. The USA, Germany, and Japan maintained long-term PV R&D programs and invested considerable public funds. The USA had the highest public investment for PV R&D compared with the other two countries. As a result, it gained an early advantage in terms of PV module cost reduction, with Japan following closely behind. Compared with these two countries, Germany's PV R&D investments were less and had fewer links between institutions, academia, and the PV industry. After 2006, the industrial research collaborations were supported by nationally funded R&D collaboration programs, which contributed to cost reduction. Even though the German cumulative PV R&D investment was lower than that of the other two countries, the PV module cost reductions have been effective. The three countries' success could be attributed to their long-term stable coordinated public

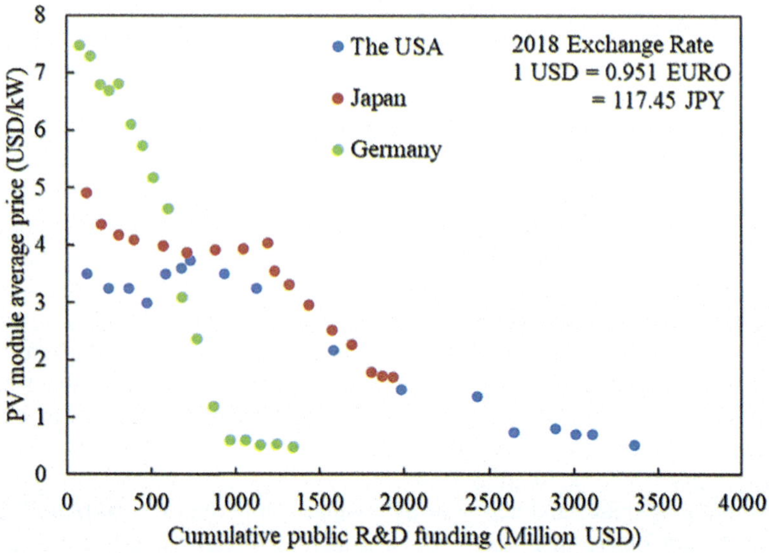

Fig. 3.2 Trends in average photovoltaic (PV) module prices with increase of cumulative public R&D funding. (Data source: IEA. PVPS. National Survey Report of PV Power Applications)

investment in PV technology innovation. And another crucial point is that the policies implemented by these countries provide a high level of collaboration between the PV industry, academia, and research institutions.

Figure 3.3 shows the PV production share of the four countries and the rest of the world from 2000 to 2018. Due to the massive R&D investment before 2000, the US and Japanese PV modules achieved technology and price advantages and occupied most of the PV production market [39]. The PV industry in Japan experienced a period of robust growth, leading to Japan being the leader in the PV industry worldwide. Since 1999, Japan had ranked first in PV production worldwide. Japan dominated the PV cell and module markets and contributed to more than 40% of the world's PV production capacity until 2006 [40]. Due to the establishment of the PV industry in East Germany, the PV production share in Germany increased rapidly since 2005 [41]. From 2005 to 2007, Germany and Japan occupied more than 50% of the market for PV products. Then, the rise of China's PV industry shocked the world. China's current PV production is higher than that of any other country. The highly profitable PV market in Europe has attracted many Chinese companies to enter the PV manufacturing sector. German PV companies have played an important role in the rise of China's PV industry. To find a large PV equipment market, German companies helped China install PV production lines, thereby increasing China's competitiveness. High-tech capabilities and knowledge were embedded in the production line, and the Chinese PV industry obtained technology for large-scale production [42]. On the other hand, Chinese PV manufacturers benefited directly from the investment support measures offered by the Chinese central government

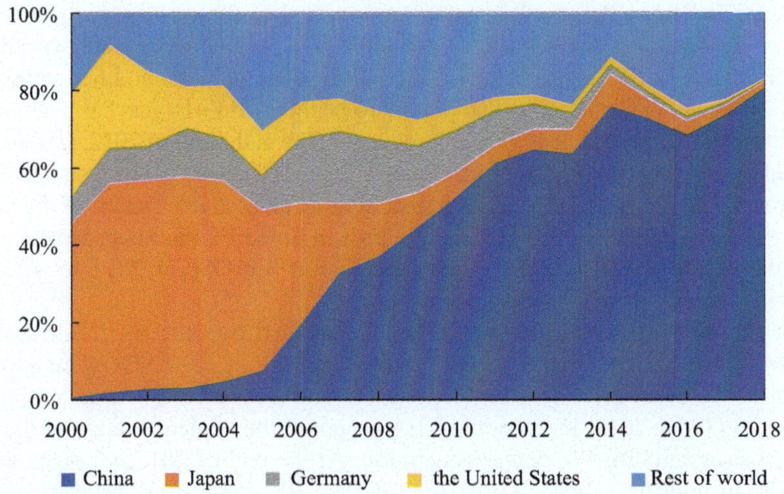

Fig. 3.3 The photovoltaic (PV) production share by countries from 2000 to 2018. (Data source: IEA data service and Fraunhofer ISE)

[37]. As the PV industry is one of the strategic emerging industries, the Chinese government has substantially subsidized the PV manufacturing sector along with related research grants, tax rebates, loans, and lands. Among the four countries, China is the only country that implements supply-push policies for promoted PV production [30], and thus, the Chinese PV industry quickly gained competitive advantage over other countries. China's involvement has greatly affected the structure of the global PV industry. Over time, China has started to dominate the worldwide PV production, and the production of Japan, Germany, and the USA decreased immediately. In 2011, China's PV products accounted for more than 66% of the global production. Subsequently, the USA and the European Union launched anti-dumping and countervailing duties on Chinese PV products, forcing Chinese PV companies to struggle [43]. The restriction on exports caused a decline in Chinese PV cell production in 2012. Therefore, the Chinese government drafted market incentives to improve domestic PV market development. Thereafter, the Chinese domestic market has expanded significantly, and the Chinese PV industry continued to grow. Currently, China's PV production share accounts for more than 70% of the world.

From 2001 to 2009, the USA once again became a major player in the global PV development process, with an average PV market growth rate of approximately 60% per annum, the fastest growth being approximately 100% in 2003 [44]. In 2000, the PV total installed capacity was 138 MW, but the number increased to 1642 MW by the end of 2009. The ITC has contributed to the tremendous growth of the PV market since its implementation. In 2010, compared with 2009, the PV market in the USA grew by 92%. The PV installed capacity exceeded 40 GW from 2010 to 2016, with an average annual growth rate of over 70% [45]. In Germany, from 2000, the subsidy

"100,000 Rooftops Program" became an opportunity for rapid growth in the PV market. The residential PV market continuously increased under stable conditions, and the "Renewable Energy Sources Act (EEG) program" was modified in the form of a FiT. Since 2008, Germany had proven to be the world's largest PV market, with its cumulative installed capacity increased to 34 GW at the end of 2012. In Japan, the annual installed capacity was approximately 290 MW in 2005. The installed capacity grew by more than 200% in 2008, reaching a cumulative capacity of 4.9 GW in 2011. The FiT policy has driven the rapid growth of the PV market in Japan, and the cumulative PV installed capacity increased from 4.9 GW in 2011 to 42.7 GW in 2016.

By the end of 2009, the cumulative PV installed capacity in China was only 300 MW. By 2012, 455 projects with a total capacity of 2872 MW were approved under the Golden Sun demonstration program. The cumulative PV installed capacity reached 3 GW in 2011. In 2015, the NEA proposed the implementation of the "Top Runner Program" for PV power generation. At the end of 2017, 43 projects with 26 GW in total have been approved [46]. The PV cumulative installed capacity increased from 3.5 GW in 2011 to 77 GW in 2016. In 2017, China added 52.83 GW of new PV installed capacity, accounting for over half of all PVs installed worldwide that year.

Figure 3.4 shows the changes in the PV market worldwide from 2000 to 2018. In 2000, the PV market in Germany and Japan shared nearly 60% of the world's PV

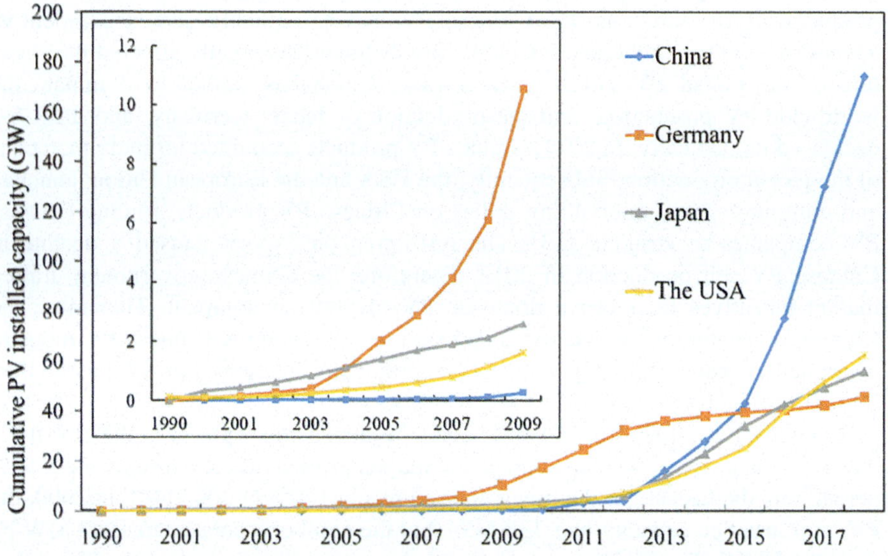

Fig. 3.4 The photovoltaic (PV) market development in China, Germany, Japan, and the USA from 1990 to 2018. (Data source: IEA. PVPS. National Survey Report of PV Power Applications)

market. In Japan, the subsidy program for PV deployment ended in 2005. Thus, the expansion of the PV market in Japan was caught during stagnation. Japan lost its position as the world leader of the PV market share in 2005, and Germany began to rule the world PV market. The German PV market accounted for more than 60% of the world PV installed capacity. In 2012, a new EEG was implemented, and the growth of the German PV market slowed down. In the USA, Congress passed the "Energy Policy Act (ITC)" in 2005, and the PV market has grown rapidly across the country. In China, the Chinese government introduced the first significant measure in 2009, which is "the Golden Sun demonstration program," to promote the development of the domestic PV market. The market grew by over 300% in 2010 and 500% in 2011 [38]. In 2011, China began implementing the FiT scheme, followed by Japan in the following year; corresponding to this scheme, the PV markets in China and Japan expanded significantly. The Chinese PV market ranked first worldwide and accounted for more than 50% of the world PV market in 2018.

3.2 Techno-economic Analysis of Solar PV Energy in China, Germany, Japan, and the USA

3.2.1 Introduction

To achieve energy security goals and reduce greenhouse gas emissions, countries around the world have introduced different policy tools to promote renewable energy installations, such as feed-in tariffs (FiTs), capital investment subsidies (CISs), and investment tax incentives (ITCs). In particular, in China, Germany, Japan, and the USA, four governments have been promoting renewable energy sources, including photovoltaic (PV) energy. These policies have significantly increased the amount of renewable energy installed in the four countries. As a result, governments have become more ambitious about renewable energy development.

In 2019, the world PV energy installation capacity has reached 586 GW. China's PV installation capacity is 205.5 GW, ranking first in the world. Germany PV installed capacity is 49.2 GW, ranking fourth in the world. Japan's installed solar PV capacity reached 63 GW, ranking third in the world. The USA has PV installation capacity of 60.6 GW and ranks second in the world (IRENA 2020). These four countries, from their central government to local governments, all target residential PV systems and large-scale PV power plants and support a policy of diversification.

In this context, investors in residential PV systems could receive a positive return on their investment. This ensures the rapid growth of the PV market in these four countries. However, the explosive growth of PV energy has led to a series of problems, such as substantial net demand changes and the high renewable energy tax burden. The increase in PV penetration has affected the stability of the grid. The daily or seasonal balancing of supply and demand has become a huge challenge. As a result, governments in these four countries have been reducing subsidies for PV systems each year and considering eliminating the subsidy policy.

With the continued reduction or even elimination of incentives, the development of PV energy tends to slow down. The growth rate of PV energy introduction reduced year after year in recent years. Furthermore, continued reliance on the FiT to facilitate reinvestment is not a sustainable approach to expanding the introduction of PV energy. For mitigation of this reduction, governments are trying to achieve renewable energy goals by implementing innovative policy solutions. These policies could improve the utilization and flexibility of PV power generation by introducing battery systems into the residentials and increasing the stability of the grid.

Based on the real measured load data of typical residential users, this section establishes a virtual model of residential PVS to analyze the technical and economic performance in the four cities in each of the selected countries. At the same time, according to the current different policy conditions, the impact of policies on the economic feasibility of PV systems in various regions is analyzed.

3.2.1.1 PV System Model

Figure 3.5 shows the schematic layout of the grid-connected residential PV system in this paper. The main components of the grid-connected system are as follows:

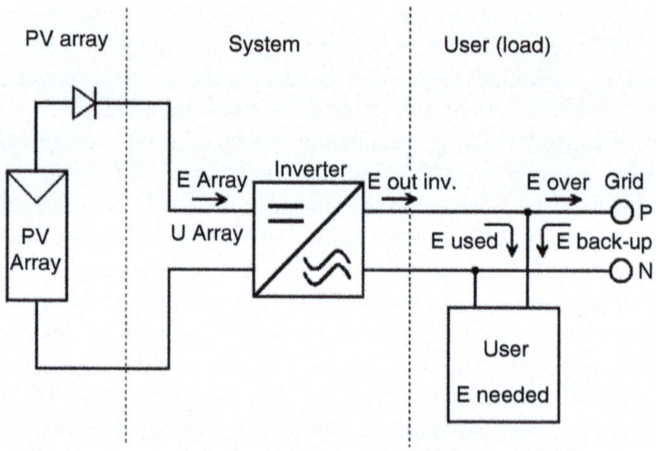

Fig. 3.5 The schematic layout of the grid-connected residential PV system

- Solar panels: The present manufacturing industries of the solar modules produce various types of PV panels depending on the materials utilized. But crystalline solar panels are commonly used in residential PV plant installations.
- Inverters They are used to transform direct current (DC) into alternating current (AC). Based on the PV plant rating, the inverter size is selected.
- Mounting structures: Structures are needed for the positioning of photovoltaic plates, inverters, and several other accessories. The installation of photovoltaic panels is a crucial one to discuss here, ensuring that they are mounted at optimal angles according to the site's specifications.
- Grid connection: Includes substations and their elements such as transformers, net meters, protection devices, etc.
- Cables: DC cables are used to connect the PV array with the inverter, and AC cables are used for connection between the inverter and grid.

In this PV system, if excess power is produced, it can be supplied to the grid. In the USA, it could by net metering be injected into the grid and then used when there is less system power generation. In Japan, Germany, and China, the excess PV energy is injected into the grid, gaining profit through feed-in tariff.

3.2.1.2 The Levelized Cost of Electricity (LCOE)

The PV system economic assessment includes the levelized cost of electricity (LCOE), the net present value (NPV), and the internal rate of return (IRR) as criteria for the evaluation of the profitability of a PV investment.

The LCOE represents the total project lifecycle cost, measured in USD per kilowatt-hour (USD/kWh). Through calculations, it is possible to compare the impact of different technologies on financial feasibility, project size, production capacity, and capital cost. Grid parity is defined as the situation where the LCOE for alternative energy production is the same as the cost of purchasing power from the grid.

$$\text{LCOE} = \frac{\sum_{t=0}^{T} C_t/(1+d)^t}{\sum_{t=0}^{T} E_t/(1+d)^t} \tag{3.1}$$

where C_t is the annual project cash flow including installation, operation and maintenance, financial costs, and fees and E_t is the electricity generated by the system in year t.

In this paper, the LCOE has been compared with the current electric bill, which ignores the future inflation in the prices. The LCOE also depends on investment and operating costs and is greatly affected by investment subsidy policies.

3.2.1.3 Net Present Value (NPV) and Internal Rate of Return (IRR)

A discounted cash flow analysis has been used in this study [47]; the NPV was calculated for different economic scenarios involving a range of electricity prices, solar PV degradation rates, and inverter and battery replacement costs to reproduce the annual cash flow for the lifetime of the solar PV system. The NPV was calculated using this equation:

$$\mathrm{NPV} = \sum\nolimits_{t=0}^{T} \frac{C_t}{(1+d)^t} \tag{3.2}$$

where C_t is the cash flow, t is the number of years, d is the nominal discount rate, and T is the project lifetime. In this study, cash flow analyses were conducted with a discount rate of 4%. The discount rate is the primary factor affecting the NPV calculation. For residential solar projects, the discount rate should be the same as or higher than the target for the return on investment.

The IRR is one of the most useful tools for measuring profitability and is the most used method to calculate the rate of return. It is calculated using this equation:

$$\mathrm{NPV} = \sum\nolimits_{t}^{N} \frac{C_n}{(1+\mathrm{IRR})^t} = 0 \tag{3.3}$$

3.2.1.4 Simulation Parameters

Nowadays, the most common installation capacity of a residential PV system worldwide is mainly between 4.0 and 5.5 kW, due to the limited roof area and limited weight bearing capacity of the building. In this study, the applied PV production profile is a simulation profile of a 5 kW PV system on a residential house in four selected countries. The proposed solar module for the 5 kW residential PV system and PV BESS is the LONGi Solar LR4-60HPH-350 M Si-mono PV with a rating of 350 W. The inverter used for this plant is SB5.0-1SP-US-40 made by SMA. The selected SMA inverter has a rated power of 5.05 kW. Other PV system parameters include the inverter efficiency and module degradation, and operation and maintenance costs and other economic parameters are shown in Table 3.2. The parameters have been chosen according to typical parameters that can be found in various government statistical reports and official websites.

3.2.1.5 Research Location

Four typical cities in China, Germany, Japan, and the USA were selected as case studies based on the distribution of national grids or power companies and PV and

Table 3.2 List of simulation parameters

Category	Assumption
PV system parameters	
Energy yield	1000 kWh/kW
Total AC capacity	5.050 kW
Total inverter DC capacity	5.207
Number of modules	14
Number of strings	2
Inverter efficiency	96.9%
Module degradation	0.05%
Lifetime	25 years
Economic parameters	
Operation and maintenance costs	2%
Discount rate	4%
Electricity inflation rate	2%
Project lifetime	25 years

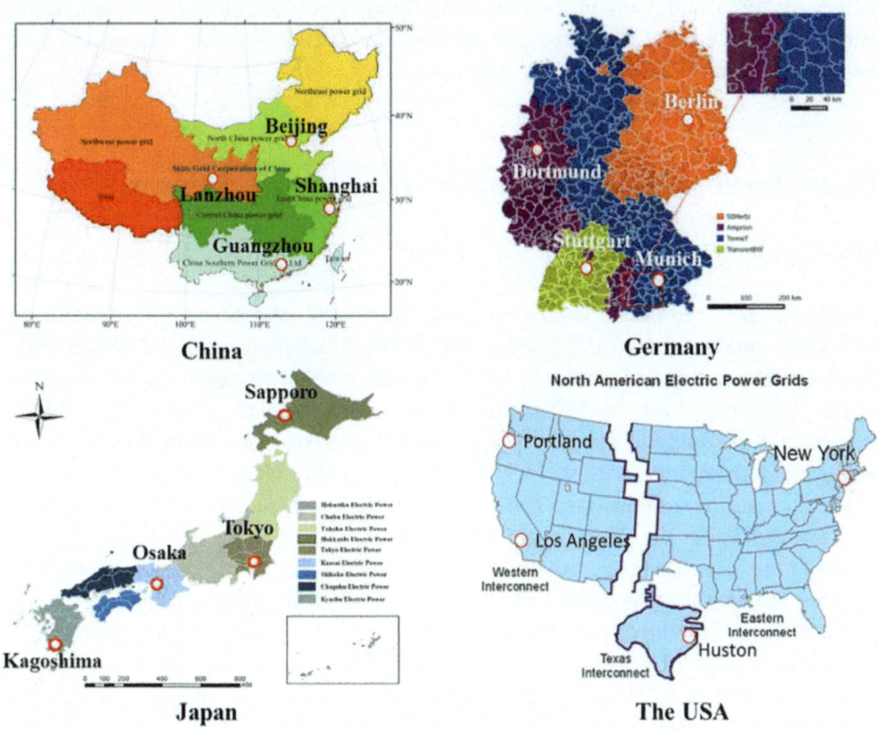

Fig. 3.6 Geographical location of study cities

Table 3.3 Performance parameters in study cities

	Beijing	Shanghai	Guangzhou	Lanzhou
Solar radiation (kWh/m^2/day)	4.86	4.1	3.26	5.07
Energy production (kWh/year 1)	7166	5979	4478	7810
Energy yield (kWh/kW/year 1)	1462	1220	914	1593
Performance ratio (PR)	0.82	0.81	0.79	0.86
Capacity factor (CF)	16.70%	13.90%	10.40%	18.20%

Table 3.4 Incentive policy and electricity tariff

City	Incentives (cents/kWh)	Duration (year)	Tariff (cents/kWh)	Feed-in tariff (cents/kWh)
Shanghai	4.34	5	8.8407	1.1594
Beijing	4.34	5	11.3045	1.1594
Guangzhou	2.17	6	13.6234	1.1594
Lanzhou	–	–	11.7393	1.1594

battery systems and with investment incentives. Each technology combination is analyzed in the case of no increase in the average electricity price. Figure 3.6 shows the geographical location of study cities.

3.2.2 *Techno-economic Analysis Results of PV Systems*

3.2.2.1 China

Among the selected cities, Guangzhou has the lowest annual energy production of 4478 kWh and Lanzhou the highest at 7810 kWh. Solar power plants perform best in Lanzhou because it is located in the solar-rich northwest region with an annual power plant production of 1593 kWh/kW. The average solar radiation of 5.07 kWh/m^2/day is most suitable for photovoltaic power plants. Guangzhou has the lowest power plant production with 914 kWh/kW due to its hot weather and less supportive power generation conditions. The main simulation results for all locations are compared in Table 3.3.

The most important parameters for comparing the performance of different systems are the performance ratio and CF. The simulation results for the four selected locations show that Lanzhou City has the highest energy yield with a performance ratio of 86% and a CF of 18.2%. Guangzhou has the lowest energy yield with a PR of 0.79 and a CF of only 10.4%. In the case of China, the feed-in tariff (FiT) is 1.1594 cents/kWh set by the government in 2020, which is used to calculate the price of electricity to be fed into the grid. Table 3.4 shows the residential electricity prices and local government generation subsidies for the four

Table 3.5 Economic indicators

	Beijing	Guangzhou	Lanzhou	Shanghai
LCOE (nominal)	3.73	7.69	4.9	5.43
LCOE (real)	3.04	6.28	3.99	4.43
Energy bill without system (USD/year 1)	367	443	381	381
Energy bill with system (USD/year 1)	126	215	111	144
Net savings (year 1)	241	227	270	237
NPV (USD)	369	−417	−175	−733
Payback period (PBP)	10.7	15	14.4	16.4

Table 3.6 Performance parameters in study cities

Performance parameters				
	Berlin	Munich	Dortmund	Stuttgart
Solar radiation (kWh/m^2/day)	3.06	3.46	2.88	3.39
Energy production (kWh/year 1)	4542	5150	4269	5021
Energy yield (kWh/kW/year 1)	927	1051	871	1024
Performance ratio	0.82	0.83	0.82	0.83
Capacity factor	10.60%	12.00%	9.90%	11.70%

cities. Shanghai, Beijing, and Guangzhou have generation subsidies for a period of 5–6 years. The results show that Beijing has the highest return on investment due to its higher annual energy production from PVs and local generation subsidies. The study shows that Guangzhou has the lowest NPV because it has the lowest annual energy production and lower local generation subsidy prices.

Table 3.5 shows that Beijing has the lowest LCOE and the shortest payback period due to energy production capacity and policy. In terms of electricity bill savings, Lanzhou has the highest electricity bill savings due to its highest annual generation capacity. Shanghai has the lowest NPV and the highest payback period because of the lower electricity tariff.

3.2.2.2 Germany

Among the selected cities, Dortmund has the lowest energy production of 4269 kWh and Munich the highest at 5150 kWh. Solar power plants perform best in Munich because it is located in the solar-rich northwest region with an annual power plant production of 1051 kWh/kW. The average solar radiation of 3.46 kWh/m^2/day is most suitable for PV systems in this city. Dortmund has the lowest power plant production with 871 kWh/kW, because of its less supportive energy generation environment with 2.88 kWh/m^2/day. The main simulation results for all cities are compared in Table 3.6.

The most important parameters for comparing the performance of different systems are the PR and CF. The simulation results for the four selected cities show

Table 3.7 Incentive policy and electricity tariff

City	Incentives (USD)	Tariff (cents/kWh)	Feed-in tariff (cents/kWh)
Berlin	–	32.1	11.844
Munich	200/kW	31.104	11.844
Dortmund	300/set	32.28	11.844
Stuttgart	350–450/kW	34.452	11.844

Table 3.8 Economic indicators

	Berlin	Munich	Dortmund	Stuttgart
LCOE (nominal) (cents/kWh)	26.12	22.1	26.08	21.78
LCOE (real) (cents/kWh)	21.31	18.03	21.29	17.77
Energy bill without system (USD/year 1)	1439	1394	1447	1544
Energy bill with system (USD/year 1)	578	464	613	568
Net savings (USD/year 1)	860	930	834	976
NPV (USD)	−1673	−71	−1179	1176
PBP (years)	14.8	13.4	14.5	12.3

that Munich City has the highest energy yield with a PR of 83% and a CF of 12%. Dortmund has the lowest energy yield with a PR of 82% and a CF of only 9.9%.

In the case of Germany, the average feed-in tariff (FiT) is 11.844 cents/kWh in 2020, which is used to calculate the price of PV energy to be injected into the grid. Table 3.7 shows the residential electricity tariff and local government PV generation subsidies for the four cities. Munich, Dortmund, and Stuttgart have investment subsidies for residential PV systems, the cash flows for the proposed projects in the selected cities. The results show that Stuttgart has the highest NPV due to its higher residential tariff and local subsidies. The study shows that Berlin has the lowest NPV because there is no local investment subsidy in Berlin.

Table 3.8 shows that Stuttgart has the lowest LCOE and the shortest payback period due to energy production capacity and policy, while Berlin has the lowest NPV and the highest payback period because of the lower electricity tariff and lower energy yield. In terms of electricity bill savings, Stuttgart has the highest electricity bill savings due to its highest annual generation capacity.

3.2.2.3 Japan

The lowest energy production is observed in Sapporo (5273 kWh), with Osaka the highest (6347 kWh). The solar plant gives the best performance in Osaka because it is located in the highest solar resource region with 4.34 kWh/m^2/day normalized production. The plant energy production is lowest in Sapporo due to normalized

Table 3.9 Performance parameters in study cities

Performance parameters				
	Tokyo	Osaka	Sapporo	Kagoshima
Solar radiation ($kWh/m^2/day$)	4.04	4.34	3.53	4.12
Energy production (kWh/year 1)	5959	6347	5273	6009
Energy yield (kWh/kW/year 1)	1216	1295	1076	1226
Performance ratio	0.83	0.82	0.84	0.81
Capacity factor	13.90%	14.80%	12.30%	14.00%

Table 3.10 Incentive policy and electricity tariff

City	Incentives (USD)	Tariff (cents/kWh)	Feed-in tariff (cents/kWh)	
Tokyo	922/set	0.25	0.2	10 years
Osaka	–	0.23	0.2	10 years
Sapporo	337 /kW	0.28	0.2	10 years
Kagoshima	190/kW	0.23	0.2	10 years

production of 3.53 $kWh/m^2/day$ and less supportive weather conditions. The main simulation results are compared for all locations in Table 3.9.

The most important parameters to compare the performance of different systems are PR and CF. Simulation results for the four selected locations have shown the highest normalized production in Osaka City with 82% PR and 14.80% CF, while the lowest normalized production is in Sapporo with 84% PR and 12.30% CF.

In the case of Japan, the feed-in tariff (FiT) is selected at 0.2 cent/kWh to calculate annual savings from the solar plant. Table 3.10 shows the residential electricity tariff and local government PV generation subsidies for the four cities, Tokyo, Osaka, Sapporo, and Kagoshima. It shows that the residential electricity tariff and local government PV generation subsidies are for 10 years. The results show that Kagoshima has the highest NPV due to the higher annual energy production and incentives. The study reveals the lowest NPV at the Osaka site because of the lowest incentives obtained.

Table 3.11 shows that Kagoshima has the lowest LCOE and the shortest payback period due to energy production capacity and policy. In terms of electricity bill savings, Sapporo has the highest electricity bill savings due to its highest electricity tariff. Osaka has the lowest NPV and the highest payback period because of its lowest electricity tariff.

Table 3.11 Economic indicators

	Tokyo	Osaka	Sapporo	Kagoshima
LCOE (nominal) (cents/kWh)	18.36	18.44	19.56	18.17
LCOE (real) (cents/kWh)	14.98	15.04	15.96	14.83
Energy bill without system (USD/year 1)	1409	1297	1578	1297
Energy bill with system (USD/year 1)	143	−18	861	49
Net savings (USD/year 1)	1267	1315	1718	1247
NPV (USD)	613	−23	−1878	1713
PBP (years)	12.8	13.3	12.6	12.3

Table 3.12 Performance parameters in study cities

Performance parameters	New York	Houston	Los Angeles	Portland
Solar radiation (kWh/m²/day)	4.72	5.36	6.25	4.16
Energy production (kWh/year 1)	6961	7298	8878	5932
Energy yield (kWh/kW/year 1)	1420	1489	1811	1210
Performance ratio	0.82	0.78	0.8	0.8
Capacity factor	16.20%	17.00%	20.70%	13.80%

3.2.2.4 USA

The lowest energy production is observed in Portland (5932 kWh) and the highest in Los Angeles (8878 kWh), among the selected cities. The solar plant gives the best performance in Los Angeles because it is located in the highest solar resource region with 6.25 kWh/m^2/day normalized production. The plant energy production is lowest in Portland due to normalized production of 4.16 kWh/m^2/day and less supportive weather conditions. The main simulation results are compared for all locations in Table 3.12.

The most important parameters for comparing the performance of different systems are the performance ratio (PR) and capacity factor (CF). The simulation results for the four selected cities show that Los Angeles City has the highest energy yield with a PR of 80% and a CF of 20.7%. Portland has the lowest energy yield with a PR of 80% and a CF of only 13.8%.

Table 3.13 Incentive policy and electricity tariff

City	Incentives		Tariff (cent/kWh)	Net metering
New York	ITC 26%	State ITC 25%	0.232	Y
Houston	ITC 26%	–	0.135	Y
Los Angeles	ITC 26%	–	0.171	Y
Portland	ITC 26%	State 1500	0.107	Y

Table 3.14 Economic indicators

	New York	Houston	Los Angeles	Portland
LCOE (nominal) (cents/kWh)	11.63	14.25	8.94	15.39
LCOE (real) (cents/kWh)	9.45	11.58	7.29	12.55
Energy bill without system (USD/year 1)	1806	1505	2320	1353
Energy bill with system (USD/year 1)	692	491	573	685
Net savings (USD/year 1)	1114	1014	1757	668
NPV (USD)	6796	2338	16,086	−1325
PBP	7.4	8.2	5.9	15

The economic aspects should be considered to assess the investment benefits of PV systems. For proper economic analysis, parameters such as NPV, LCOE, and payback period ensure the profitability of the PV system investment.

In the case of the USA, the ITC rate was set by the government in 2020, which is used to offset the taxes paid for PV generation. Table 3.13 shows the residential electricity tariff and local government PV generation subsidies for the four cities, the cash flows for the proposed projects in the selected cities. The results show that Los Angeles has the highest NPV due to its highest annual energy production from PVs and local generation subsidies. The study shows that Portland has the lowest NPV because it has the lowest annual energy production.

Table 3.14 shows that Los Angeles has the lowest LCOE and the shortest payback period due to energy production capacity. In terms of electricity bill savings, Los Angeles also has the highest electricity bill savings due to its highest annual energy generation capacity. Portland has the lowest NPV and the highest payback period because of the lower electricity tariff and energy production.

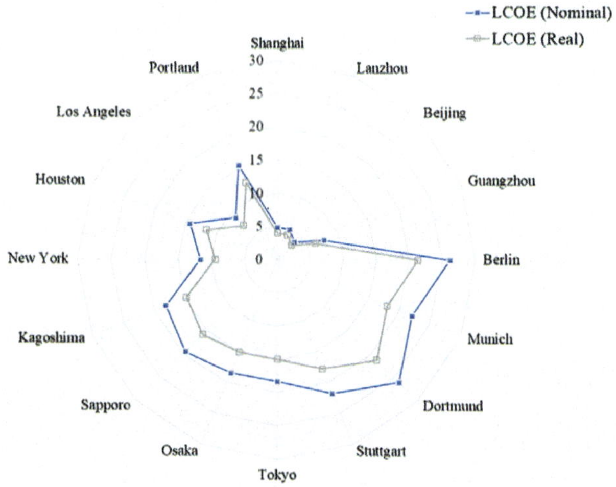

Fig. 3.7 The LCOE of PV energy in four countries' cases

Table 3.15 Comparison of LCOE and average electricity costs

Country	City	Nominal LCOE (cents/ kWh)	Real LCOE (cents/ kWh)	Average tariff (cents/ kWh)
China	Beijing	3.73	3.04	11.3
	Shanghai	4.92	4.01	8.8
	Guangzhou	7.69	6.28	13.6
	Lanzhou	4.9	3.99	11.7
Germany	Berlin	26.12	21.31	32.1
	Munich	22.1	18.03	31.1
	Dortmund	26.08	21.29	32.3
	Stuttgart	21.78	17.77	34.5
Japan	Tokyo	18.36	14.98	25.4
	Osaka	18.44	15.04	23.2
	Sapporo	19.56	15.96	28.1
	Kagoshima	18.17	14.83	23.3
USA	New York	11.63	9.45	23.2
	Houston	14.25	11.58	13.5
	Los Angeles	9.61	7.85	17.1
	Portland	15.39	12.55	10.7

3.2.3 Comparative Analysis of PV System Economic Indicators in Four Countries

3.2.3.1 The LCOE of PV Systems in Four Countries

Figure 3.7 and Table 3.15 show the LCOE of residential PV systems for the four countries in all cases. The lowest LCOE among the four countries is in China, which can be attributed to the fact that China has the lowest residential PV investment cost. At the same time, with the national FiT and local-level PV subsidies, especially in Beijing, the real LCOE for residential PV reaches 3.04 cents/kWh, the lowest of all cities and less than one-third of the residential electricity price. Even Guangzhou, which has the lowest annual PV generation among the four cities, has an LCOE of 6.28 cents/kWh, which is less than half of the residential electricity rate. The country with the highest residential PV LCOE is Germany, due to higher system installation costs and low solar radiation. The highest LCOE is in Berlin, with a real LCOE of 21.31 cents/kWh, which is attributed to the rapid decline in German subsidy prices in recent years, as well as the reduction and elimination of subsidies at the local level. In the last 2 years, Berlin has eliminated its residential PV. This has led to the highest LCOE in Berlin. However, the LCOE of PV in Germany is still at a low level compared with the high residential electricity costs. Japan has the highest PV system cost among the four countries; however, because of local-level investment subsidies, the LCOE of residential PV in Japan is lower than in Germany, at around 14 cents/kWh in all four cities. The US residential PV LCOE is only higher than China's among all countries, and relative to Germany's higher system costs among the four countries, the LCOE is only half of Germany's, especially in Los Angeles and New York, where the LCOE has dropped below 10 cents/kWh, especially in Los Angeles, where it is only 7.85 cents/kWh, compared with the residential electricity rate of 17.1 cents/kWh. The LCOE is less than half of the residential rate. A special case is Portland, where the LCOE is higher than the residential rate among all countries, and the residential rate in Portland is less than half of the residential rate in New York. Overall, the LCOE of residential PV plants is lower than residential electricity rates, with the lowest being in China, followed by the USA, the third highest being in Japan, and the highest being in Germany.

3.2.3.2 Discussion

Except for Germany, governments in the other three countries lacked new long-term goals for PV deployment. Long-term goals can greatly impact the future of PV development. Long-term targets, updated planning, and stable measures are needed to meet the challenges and maintain healthy PV development. Governments in the four countries should rapidly upgrade their long-term policies, including R&D, and supply-push and demand-pull policies, in line with the current state of PV development. Currently, China, Germany, and Japan are scaling back or eliminating subsidies for PV power generation, which increases uncertainty in terms of policy form and market risk. According to the results of the techno-economic analysis in the previous sections, the LCOE of residential PV has been significantly reduced and is lower than residential electricity prices in all four countries. However, the results of the financial analysis show that none of the residential PV systems can achieve the expected returns, especially in Germany and China. This is mainly due to the adjustment of the FiT in recent years. Currently, residential PV systems in all cities and regions studied in the four countries must rely on national and local subsidies if they are to generate revenue. Current PV policies in China and Germany do not provide much support for investment in residential PV systems. Promoting policy reform is particularly important if we want to further promote residential PV in the future. With the LCOE of PV electricity lower than residential electricity prices and as FiT prices continue to decrease, net metering policy becomes more economic. In the USA, for example, residential PV has the highest return among the four countries.

China's FiT prices for residential PV have fallen rapidly in the last 2 years. Moreover, unscheduled subsidy price reductions are not conducive to the development of residential PV. Japan's FiT fixed price is still higher than those of other countries, and it is necessary to reduce the fixed price more frequently and set an annual upper limit for the capacity of PV plants of different sizes. In addition to the tender system, China and Japan could design a predetermined declining rate for fixed prices, taking a cue from the German FiT system. A predictable rate of price reduction could give PV product manufacturers a strong incentive to continually reduce costs in order to accommodate policy changes.

It is expected that PV deployment in the four countries will continue to grow at a high rate over the next decade. With the expansion of PV power generation, daily or seasonal demand-supply balance will be a problem [48]. The resulting high PV penetration will be a major issue in the limited expansion of PV power generation. The continuous scaling-up of PV deployment would be a great challenge for the government to reduce PV curtailment and maintain grid balance. Policymakers should consider reorienting policies to overcome grid constraints and promote flexibility. The introduction of batteries in PV systems is a favorable solution. In Japan, Germany, and the USA, a series of PV battery subsidy policies have been introduced from the local to central government. The main focus is on investment subsidies. However, there is a lack of long-term planning for battery subsidy policies, which has led to insufficient revenue for residential PV BESS. The introduction of batteries in PV systems is a good solution. In addition, new demand-side management modes such as a VPP (virtual power plant) can effectively achieve peak load reduction on the grid and optimize power resources.

The cost of residential PV systems in Japan is substantially higher than in other countries, which also results in the lowest return on PV system investment in Japan among the four countries. The high system prices have also led to high FiT prices. The high FiT price is probably the main reason Japanese PV products have remained more expensive compared with those in other countries. The high FIT fixed price for PV power generation has made local manufacturers less willing to further reduce the cost of their products, while the high specification requirements for FiT-certified PV products have made it difficult to introduce lower-priced products from abroad into the Japanese PV market. These factors have curbed the reduction of PV system costs in Japan. In the future, the FiT fixed price is bound to continue to decrease, which will inevitably affect the domestic PV industry. Therefore, local PV manufacturers should be more proactive about reducing product costs through R&D and other means, or they should collaborate with foreign manufacturers to introduce lower-priced products through original equipment manufacturers (OEMs). When the PV system investment cost is reduced, the relative FiT fixed price can also be reduced, thus forming a mutually beneficial virtuous circle.

Driven by policies and supportive measure changes in recent years, the residential PV installations will be increased more rapidly. However, the high initial investment cost and long payback periods of distributed PV are barriers to private investors. The effective adoption of a systemic approach to support the deployment of distributed energy, including business model innovation and various renewable energy source integration, would be a great challenge in these four leading countries.

References

1. Sen S, Ganguly S (2017) Opportunities, barriers and issues with renewable energy development – a discussion. Renew Sust Energ Rev 69:1170–1181
2. Sovacool BK, Gilbert A (2013) Feed-in tariffs and other support mechanisms for solar PV promotion. Ref Modul Earth Syst Environ Sci. https://doi.org/10.1016/b978-0-12-409548-9.04797-7
3. Hart D, Birson K (2016) Deployment of solar photovoltaic generation capacity in the United States. Office of Energy Policy and Systems AnalysisU.S. Department of Energy
4. IEA (2019) PVPS Snapshot of global photovoltaic markets 2019
5. U.S.DOE (2006) FY 2006 congressional budget request energy supply
6. IEA (2018) PVPS Annual report 2018
7. Farris C, Industries SS (2003) US PV industries roadmap
8. IEA (2011) National survey report of PV power applications in the United States 2010
9. IEA (2003) National survey report of photovoltaic power applications in the United States 2002
10. IEA (2009) National survey report of PV power applications in the United States 2009
11. IEA (2004) National survey report of PV power applications in the United States 2004
12. US DOE (2015) The SunShot initiative: making solar energy affordable for all Americans. 2008:1–2
13. US DOE (2017) The SunShot 2030 goals: 3¢ per kilowatt hour for PV and 5 ¢ per kilowatt hour for dispatchable CSP. 2–6
14. Dewald U, Fromhold-Eisebith M (2015) Trajectories of sustainability transitions in scale-transcending innovation systems: the case of photovoltaics. Environ Innov Soc Trans 17:110–125
15. Ito Y (2011) The FIT system in Europe and information about the cost for companies and measures to reduce the cost
16. Altenhöfer-Pflaum G (2003) IEA. PVPS. National survey report of PV power applications in Germany 2002. 29
17. Agency IE, On CP, Power P, Conservation N (2007) IEA. PVPS. National survey report of PV power applications in Germany 2007. Power 1–23
18. IEA (2010) PVPS Annual report 2010
19. Wehrmann B (2018) Solar power in Germany – output, business & perspectives. Energy Transit. https://www.cleanenergywire.org/factsheets/solar-power-germany-output-business-perspectives
20. Wirth H (2017) Recent facts about photovoltaics in Germany
21. Kosuke K (2007) Direction of solar photovoltaic technologies. Econ Cult Hist Jpn Spotlight Bimon 26:28–30
22. Hahn E (2014) The Japanese solar PV market and industry -business opportunities for European companies

23. IEA (2014) IEA.PVPS annual report 2014
24. Ito Y (2015) A brief history of measures to support renewable energy: implications for Japan's FIT review obtained from domestic and foreign cases of support measures
25. Ogimoto K, Kaizuka I, Ueda Y, Oozeki T (2013) A good fit: Japan's solar power program and prospects for the new power system. IEEE Power Energy Mag 11:65–74
26. IEA (2012) National survey report of PV power applications in Japan 2011
27. Kimura K (2017) Feed-in tariffs in Japan: Five years of achievements and future challenges. Renew Energy Institute
28. Huang P, Negro SO, Hekkert MP, Bi K (2016) How China became a leader in solar PV: An innovation system analysis. Renew Sust Energ Rev 64:777–789
29. Zhang S, He Y (2013) Analysis on the development and policy of solar PV power in China. Renew Sust Energ Rev 21:393–401
30. Jia F, Sun H, Koh L (2016) Global solar photovoltaic industry: An overview and national competitiveness of Taiwan. J Clean Prod 126:550–562
31. Zhao XG, Wan G, Yang Y (2015) The turning point of solar photovoltaic industry in China: will it come? Renew Sust Energ Rev 41:178–188
32. (2016) Solar PV feed-in tariff. In: IEA/IRENA Renewables Policies Database. https://www.iea.org/policies/5100-solar-pv-feed-in-tariff?country=People%27sRepublic Of China&page=2&qs=China&technology=Solar
33. IEA (2012) PVPS national survey report of PV power applications in China 2011
34. IEC (2018) China: energy efficiency report, 2018
35. Myojo S, Ohashi H (2018) Effects of consumer subsidies for renewable energy on industry growth and social welfare: the case of solar photovoltaic systems in Japan. J Jpn Int Econ 48:55–67
36. Brachert, M, Hornych C (2010) Entrepreneurial opportunity and the formation of the photovoltaic industry in Eastern Germany
37. Zhao ZY, Zhang SY, Hubbard B, Yao X (2013) The emergence of the solar photovoltaic power industry in China. Renew Sust Energ Rev 21:229–236
38. Zhang F, Sims K (2016) Innovation and technology transfer through global value chains: evidence from China's PV industry. Energy Policy 94:191–203
39. Yu HJJ, Popiolek N, Geoffron P (2015) Solar photovoltaic energy policy and globalization: a multiperspective approach with case studies of Germany, Japan, and China. Prog Photovolt Res Appl. https://doi.org/10.1002/pip.2560
40. Kimura O (2006) 30 years of solar energy development in Japan: co-evolution process of technology, policies, and the market. 2006 Berlin Conference on the Human Dimensions of Global Environmental Change
41. Blankenberg A, Dewald U (2013) Public policy and industry dynamics: the evolution of the photovoltaic industry in Germany. 35th DRUID Celebration Conference 2013
42. Quitzow R (2015) Dynamics of a policy-driven market: the co-evolution of technological innovation systems for solar photovoltaics in China and Germany. Environ Innov Soc Trans 17:126–148
43. Zou H, Du H, Ren J, Sovacool BK, Zhang Y, Mao G (2017) Market dynamics, innovation, and transition in China's solar photovoltaic (PV) industry: a critical review. Renew Sust Energ Rev 69:197–206
44. Haley UCV, Schuler DA (2013) Government policy and firm strategy in the solar photovoltaic industry. Environ Manag Regul Uncertain 55:17–38
45. SEIA (2018) The solar investment tax credit (ITC)
46. Nie P, Wang C, Chen Y (2018) Top runner program in China: a theoretical analysis for potential subsidies. Energy Strat Rev 21:157–162
47. Dusonchet L, Telaretti E (2015) Comparative economic analysis of support policies for solar PV in the most representative EU countries. Renew Sust Energ Rev 42:986–998
48. Li Y, Gao W, Ruan Y (2018) Performance investigation of grid-connected residential PV-battery system focusing on enhancing self-consumption and peak shaving in Kyushu, Japan. Renew Energy 127:514–523

Chapter 4
The Implementation and Integration of Renewable Energy and Its Impact on the Public Grid

Tingting Xu and Yanxue Li

4.1 Renewable Energy Integration: Opportunity and Challenge

According to analysis from IRENA [1], a decarbonization of the power sector, in line with the climate objectives outlined in the Paris Agreement, would require an 85% share of renewable energy in total electricity generation by 2050. By that time solar and wind power capacity would account for 60% of the total power generated. The innovations in grid integration strategies have become crucial in increasing the share of VRE in the power system, enhancing the system's flexibility in a cost-effective manner [1].

4.1.1 The Development and Status of Renewable Energy

In October 2020, Japan declared its commitment to become carbon-neutral by 2050. Exploration of renewable energy resources and zero-energy building were selected as two of the key technologies for realizing a carbon-neutral society. In order to tackle tight grid demand-supply balance pressure during peak periods, enormous political and technical efforts were taken to replace the loss of nuclear energy after the Great East Japan Earthquake. Maximizing the renewable source integrations in its power supply fraction has become a key political agenda. Government incentive

T. Xu (✉)
Institute of Mechanical Engineering, Tongji University, Shanghai, China

Y. Li
iSMART, Qingdao University of Technology, Qingdao, China

W. Gao (ed.), *Distributed Energy Resources*, Green Energy and Technology,
https://doi.org/10.1007/978-3-031-21097-6_4

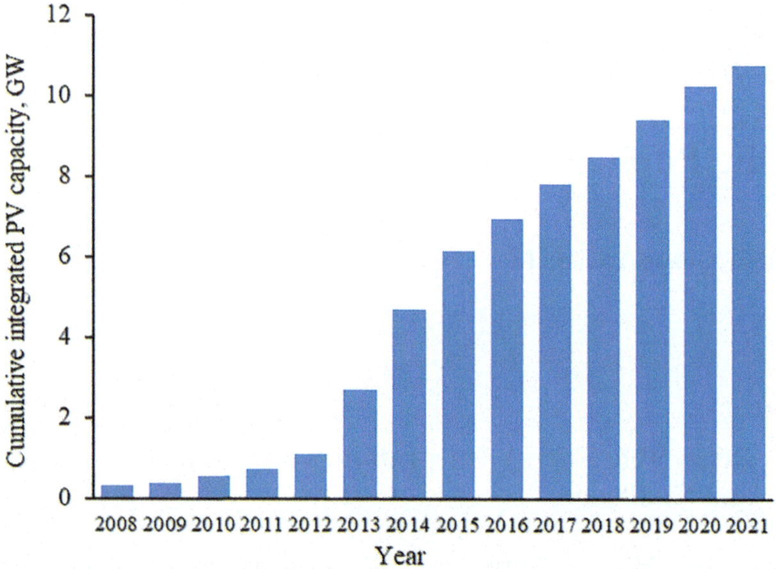

Fig. 4.1 The cumulative capacity of integrated PV in the Kyushu public grid

policies, continuous cost drop in renewable production, improvement in renewable output prediction, and control jointly accelerated the exploration of renewable energy resources. Development of the Japanese public grid moves toward greater amounts of renewable energy. The renewable energy share in the power mix experienced a significant increase over the last years. These renewable technologies had a major contribution to decarbonizing the public electricity grid. PV generation is playing an important role in enhancing national energy self-sufficiency after the feed-in tariff launched in 2012 in Japan.

As shown in Fig. 4.1, integrated PV capacity experienced sharp increases. The cumulative capacity of the grid-connected PV system in Kyushu rose from 1100 MW in 2012 to 10,810 MW in 2022. Intermittent renewable generation is non-dispatchable. The increasing amount of renewable energy also presents a further challenge to security of electrical power supply. The challenge also influences the Japanese FiT scheme. As presented in Fig. 4.2, the FiT program offers an attractive rate at an early stage, for example, 42 yen/kWh for the installed PV system less than 10 kW in capacity. The value of the FiT drops to 17 yen/kWh in 2022 [2]. From 2020, only projects below 250 kW were eligible for the FiT, and the rest had to participate in a market auction.

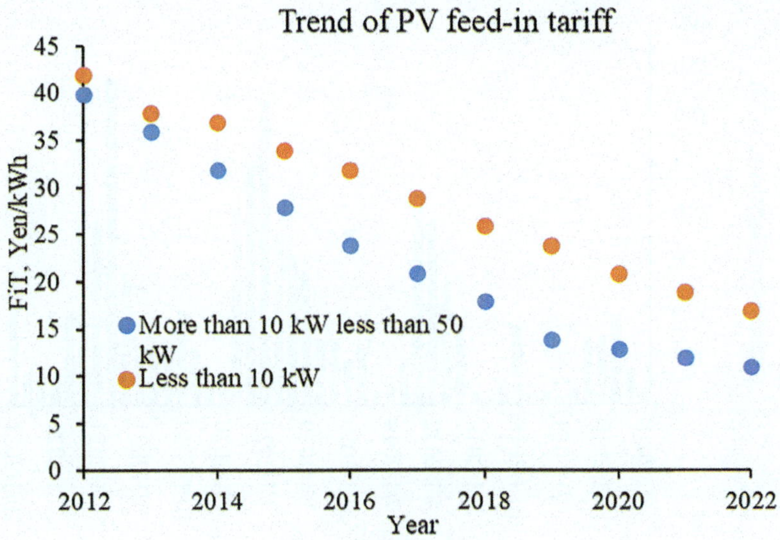

Fig. 4.2 Change of the feed-in tariff for PV systems in Japan

4.1.2 Characteristics of Renewable Energy in the Real Grid

The transition to a low-carbon grid provides challenges and opportunities. As the state incorporates increasing amounts of intermittent renewable energy into the public grid, the grid residual load would be significantly reshaped. For example, California's renewable resources can generate more electricity than is needed during the middle of the day. Due to the ramp-up and ramp-down limitation, the California ISO automatically curtails the renewable generation. Figure 4.3 shows the monthly wind and solar curtailment in California [3].

Figure 4.4 shows the layout of the power system structure and a typical daily (May 4, 2016) supply/demand balancing scenario in Kyushu. Aggregated PV generation plays a dominant role in power supply during daytime, which contributes to the largest daily variabilities compared with other renewable resources. Geothermal and nuclear generators generally run with constant output throughout the day. Thermal generators (natural gas, coal, and oil) as flexible generators will adjust their output, with a ratio of peak capacity to demand generally ranging from 0.3 to 1.0, to meet the changes of renewable output and consumption variations.

Grid flexibility refers to the capability of a power system to maintain balance between generation and load during uncertainty. Reliable renewable integration would be highly dependent on adequate energy flexibility. Renewable energy in Kyushu is generated primarily by solar photovoltaic systems. Massive PV integration suppresses the output of thermal power plants during the daytime, especially in

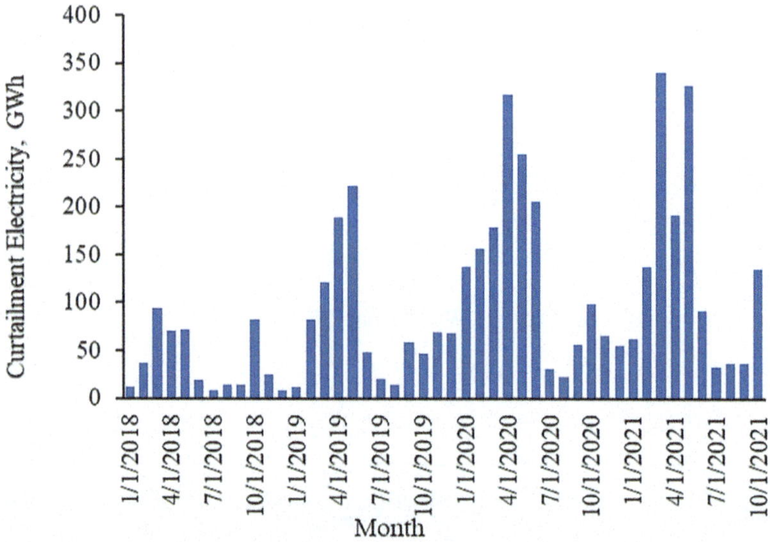

Fig. 4.3 Monthly curtailed wind and PV power in California

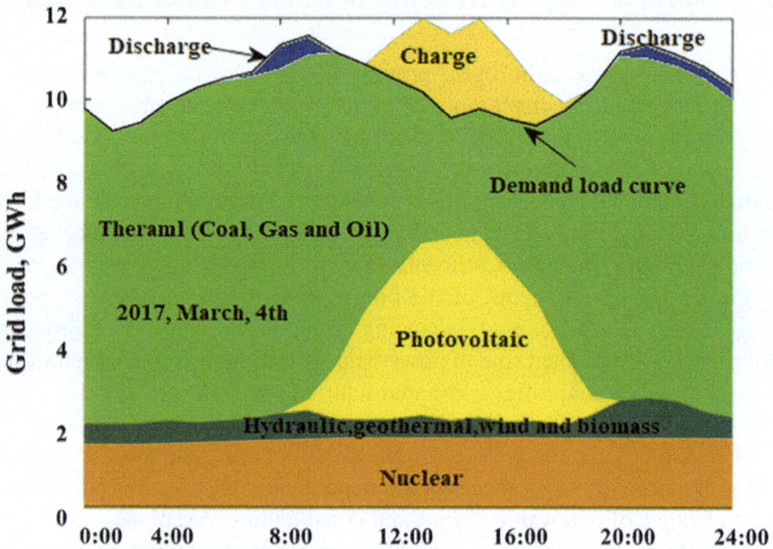

Fig. 4.4 Scenario of the power supply-and-demand balance in Kyushu, Japan

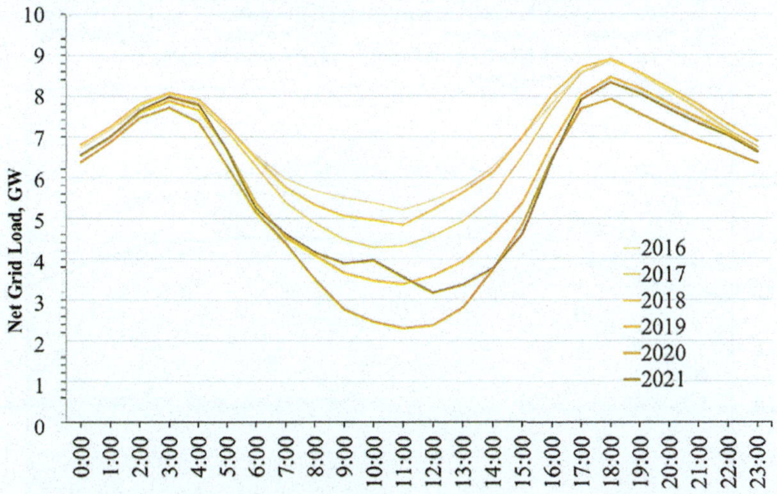

Fig. 4.5 Duck curve scenario in Kyushu, Japan

the middle of the day. As shown in Fig. 4.5, the appearance of the "duck curve" in Kyushu is worsened due to the expansion of PV generation year by year [4]. Overgeneration happens when more electricity is supplied than is needed to satisfy real-time electricity requirements. To stabilize the situation, the utility has to prepare a significant amount of controllable power plant capacity or dispatch storage systems to absorb the fluctuations and balance real-time demand load. Limited to grid-flexible resources, the supply-demand balance is still disturbed even when suppressing the output of the base plant to a minimum level. Therefore, in order to ensure reliability under changing grid conditions, the operator needs resources with ramping flexibility and the ability to start and stop multiple times per day. To ensure supply-and-demand match at all times, controllable resources will need the flexibility to change output levels and start and stop as dictated by real-time grid conditions.

The PV and wind power pretends different trend in monthly power generation profile. In Japan, nuclear power is kept constant in the power grid, due to the difficulties of start-up operations. Figure 4.6 shows the daily power profiles of different power generators. The adjustment of PHS facilities is shown as well. These facilities are used to eliminate the peak load and compensate for electricity when the power demand is low at nighttime. Therefore, with the increasing proportion of renewable energy represented by PV and wind power in the grid, the existing grid has faced some new challenges owing to renewable energy‘s intermittent and uncontrollable characteristics. It is important to evaluate the impact of PV and wind power on the public electricity supply system when they are introduced into the grid.

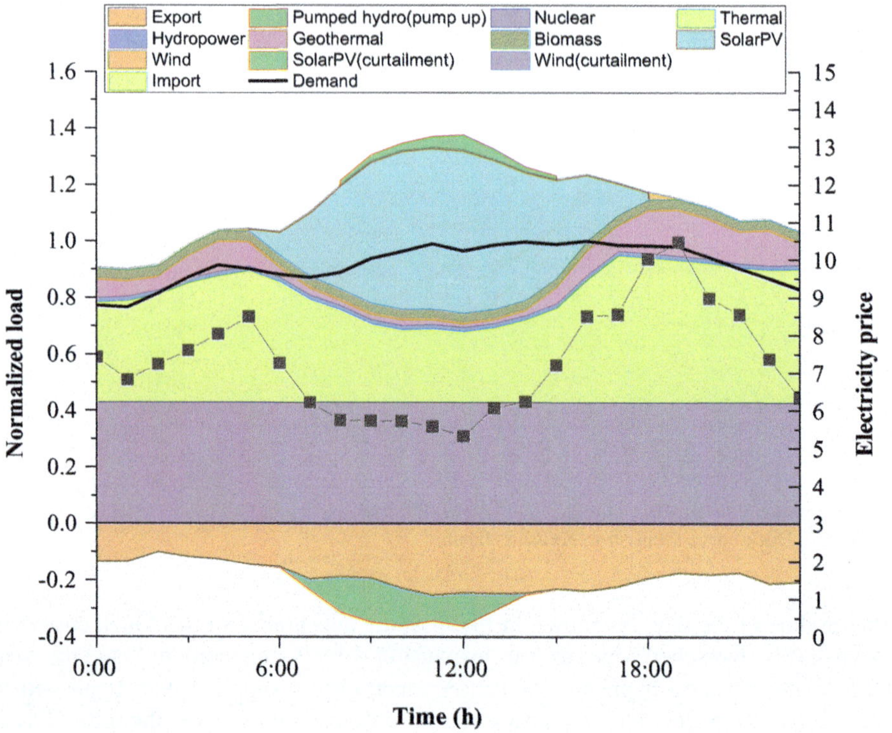

Fig. 4.6 Daily balance of power generation

4.2 The Application of the Electricity Storage System

In view of the surplus of renewable energy power, increasing the power generation proportion of dispatchable units [5], adding energy storage systems [6], and using interregional transmission lines [7] are common solutions. Among them, an energy storage system plays an important role in improving the stability and reliability of the power grid. It has the ability to improve the share of renewable energy sources (RES) in the grid by compensating for the mismatch between the power supply side and demand side [8]. In terms of energy storage technology, different energy storage systems have been compared comprehensively. For a large-scale public power grid, the most common energy storage technology is PHS [9–11]. It has the advantages of large storage capacity and long life. PHS has a history of more than 100 years; however, it still is widely used in the United States [12], Japan [13], China [14], Greece [15], and other countries [16]. Currently, the total installed capacity of PHS is 180 GW worldwide, of which China (32.1 GW), Japan (28.5 GW), and the United States (24.2 GW) account for 50%. Most of the operation strategies are absorbing thermal and nuclear power generation at night and compensating for the peak load demand during the daytime

[17]. In addition, PHS is also applied to hybrid single PV-wind energy systems to increase capacity and address their suppression. In recent years, PHS hybrid PV-wind energy systems have received the attention of many researchers due to the complementary characteristics of their power generation nature [18, 19]. The mature development of pumped storage is accompanied by challenges from the feasibility of site selection and the constraints of the ecological environment [20].

4.2.1 Pumped Storage System

Pumped-hydro storage (PHS) is a way of generating electricity. When the electricity is surplus, the surplus electricity is pumped from the lower reservoir (lower pond) to the upper reservoir (upper pond), as shown in Fig. 4.7. When the demand for electricity increases, water is drawn from the upper pond dam to the lower pond dam. It is a way of generating electricity using the conversion of gravitational potential energy to electrical energy. The utilization rate of existing pumped storage is low. The following are requirements for the development potential of pumped storage in the future:

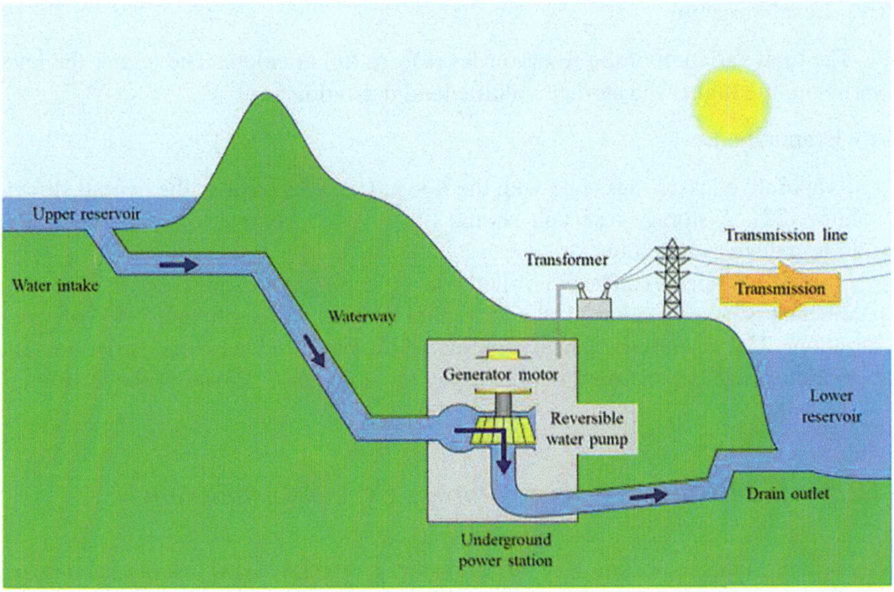

Fig. 4.7 The interconnected configuration of PHS [23]

(i) Decentralized demand for power supply.
(ii) The cost of batteries has dropped sharply, and the share of energy storage systems in the future will change. However, PHS has many advantages in terms of scale, responsiveness, and inertia, so it still occupies the mainstream position in pumping and storing energy.
(iii) Pursue small- and medium-sized decentralized PHS construction sites.

An example of a currently installed PHS is drawn in Fig. 4.7. The aspects considered when planning a storage reservoir and its topographical influence are summarized in the following [21]:

(i) Storage volume

The main objective of a storage reservoir is to store water and energy. The higher the usable storage volume, the better.

(ii) Land requirement

The area occupied by the reservoir, one of the main environmental, social, and economic impacts of reservoir dams. This should be minimized as much as possible.

(iii) Flooded area variation

The reservoir area that changes with the tidal variation as the reservoir is utilized. The flooded area variation has social, environmental, and economic impacts and should be reduced as much as possible.

(iv) Level variation

The total variation of the reservoir level from full to empty. The higher the level variation, the higher the storage volume/land use ratio.

(v) Evaporation

Evaporative losses that scale with the flooded area and reduce the overall storage volume [22]. A storage reservoir should have a high storage volume/flooded area ratio to reduce evaporation.

Only a few aspects can be controlled when planning a storage reservoir. The main parameters are the location of the dam, dam height and length, and reservoir level variation. The resulting storage volume, land use, flooded area variation, and evaporation will depend on the topography, geology, and climate of the location.

4.2.2 *Prediction of Energy Storage System Installation*

According to the composition of pumped storage, the predicted installation demand can be summarized as follows:

(i) The lower pool utilizes the existing multifunctional dam and seeks a high platform near the dam that can be used to build the upper reservoir [24].

Table 4.1 Design requirements and process of the new PHS power station based on the existing dam in Japan

Research item	Content
Dam number and function	Detail the characteristics of dams distributed in Kyushu (except exsiting dams for power generation)
Effective water storage capacity, $V_0(*10^3m^3)$	According to the basic data of each dam
Potential pumped storage, $V_1(*10^3m^3)$	$V_1 = V_0*20\%$
Head conditions, H (m)	$H \geq 200$; head is defined as 200
Installed capacity, S (mw)	$S = V_1/3600/5/9.8*200*\ \eta*0.001$
Capacity of power generation, P (MWh)	$P = S*5$, assuming the discharge period is 5 h, once a day
Total power generation for a year, P_y (MWh/y)	$P_y = P*300$

(ii) Draw the water from the upper reservoir to the lower reservoir to generate electricity and then pump the water from the lower reservoir to the upper reservoir through hydraulic pipes to store electricity.
(iii) The lower pool is a multifunctional dam. Although the multifunctional dam is responsible for disaster prevention, irrigation, and power generation, power generation uses only part of the water to generate and store electricity. It will not hinder the function of the multifunctional dam itself.
(iv) Japan now has more than 2700 decentralized dams. As shown in Table 4.1, we seek the most suitable scale for PHS power generation [25, 26].

The water consumption of PHS power generation is part of the total water storage capacity of the multifunctional dam. In this chapter, it is set to 30%.

The total water storage capacity is the sum of sand pile capacity, water conservation capacity, and flood regulation capacity. Sand pile capacity is the hypothetical value of the sand body capacity accumulated in the dam in a certain year (100 years). The effective water storage is the total water storage minus the sand pile capacity. Conservation capacity is the effective water storage minus the flood regulation capacity. Although water capacity varies with the seasons, it is calculated based on the amount of water in the normal season regardless of floods and droughts. Water storage rate is the percentage of water storage capacity and water conservation capacity. The possible water storage capacity is 20–30% of the effective water storage capacity.

As shown in Fig. 4.8, a pumped storage equipment consists of two reservoirs at different altitudes, and two penstocks are used to connect the pump and generator. According to the terrain and geological conditions of the site, the powerhouse includes electromechanical and control equipment such as pumps, turbines, valves, generators, and transformers. Energy conversion occurs when electrical energy is converted into water's gravitational potential energy during the pumping process and vice versa. The charging process occurs when the RES power is surplus, where water from the lower reservoir is pumped to the upper reservoir. RES power generation can be performed to compensate for insufficient power supply or reduce the use of thermal power generation.

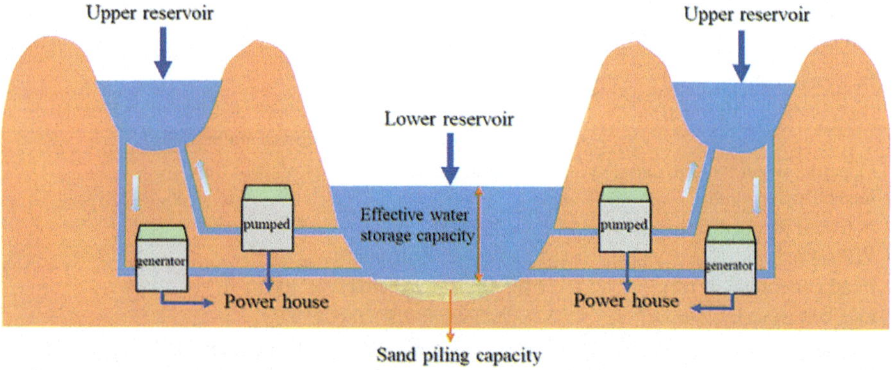

Fig. 4.8 Basic layout of the PHS with dams with main parameters

In terms of the possible pumping capacity of the dam, considering flood control, irrigation, water conservation, and other factors of the dam itself, it is set to 30% of the effective water storage capacity of the dam (the total water storage capacity of the dam minus the sand pile capacity at the bottom of the reservoir). It is worth noting that huge dams with a water storage capacity of more than 100 million should be eliminated to protect the rare creatures in the reservoir and reduce the difficulty of development.

The head of the upper and lower reservoirs shall be greater than or equal to 200 m. A buffer zone with a radius of 1.5 km shall be set around each reservoir to detect all upper reservoirs within the buffer zone that meet the marked distance. The upper reservoir is set to be 100 m in length and width and 10 m in height.

The energy storage capacity of PHS, $PHS_{storage}$, can be calculated as follows:

$$PHS_{storage}(\text{kWh}) = \rho\left(\frac{\text{kg}}{\text{m}^3}\right) * V(\text{m}^3) * g * \left(\frac{\text{m}}{\text{s}^2}\right) * h(\text{m}) \tag{4.1}$$

where V is the volume of the upper reservoir, g is gravitational acceleration, and h stands for the head between the upper and lower reservoirs. The storage capacity is the sum of the maximum power output for 5 h. Therefore, the rated power generation, C, is explained as

$$C(\text{kW}) = S(\text{kWh})/5(\text{h}) \tag{4.2}$$

4.2.3 *Utilization of Energy Storage Systems*

When considering the application of PHS in EnergyPLAN, PHS converts renewable electrical energy into mechanical energy and vice versa. It plays a major role in reducing power surplus caused by the volatility of renewable energy, reducing the

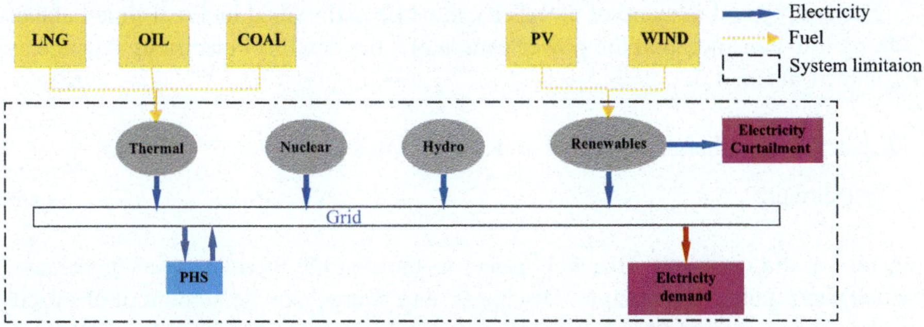

Fig. 4.9 Facility composition of the electricity coordination of supply and demand

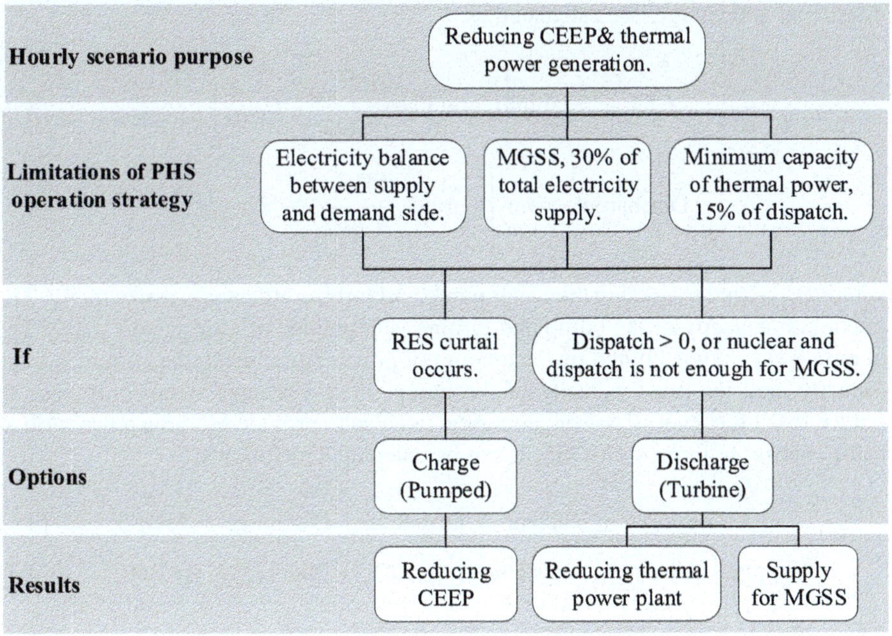

Fig. 4.10 Energy management flowchart with the strategies of PHS operation

use of fossil fuel energy, and balancing power supply and demand. Because the single pump-turbine configuration must be switched during braking and reversing to reach the charge and discharge modes, it increases the response time. However, the double-penstock configuration can reduce demand response time by pumping and generating power at the same time. In this chapter, the double-penstock PHS configuration is selected to cater to the main purpose of maximizing the penetration of RES and reducing thermal power generation. The schematic diagram of a PHS, using a mix of renewable energy, nuclear, and thermal sources, is shown in Fig. 4.9. These sources are used for power supply. The energy management flowchart with the strategies of PHS operation is shown in Fig. 4.10.

The supply and demand of electricity must be maintained in a real-time balance. Taking into consideration the power generators, the details of electricity balance are explained in Eq. 4.3.

$$\text{Dispatch}^i + \text{Discharge}^i - \text{Charge}^i + \text{RES}^i + \text{Nuclear}^i + \text{Hydro}^i - \text{Curtail}^i = \text{Demand}^i \tag{4.3}$$

where i is the time unit. Dispatch^i refers to the flexible thermal power generators, which can regulate their output. Discharge^i and charge^i are the amount of electricity generated and pumped by PHS, respectively. To reduce the CEEP from RES, the pump will be operated when the electricity curtail occurs. While generating electricity by turbines to reduce thermal power generation or supply for the MGSS, the discharge^i and charge^i are calculated as

$$\text{Charge}^i = \min\left[\text{Curtail}^i, \frac{\text{PHS}^i_{\text{storage}}}{\eta_{\text{pump}}}, C_{\text{pump}}\right] \tag{4.4}$$

$$\text{Discharge}^i = \min\left[\text{Curtail}^i, \frac{\text{PHS}^i_{\text{storage}}}{\eta_{\text{pump}}}, C_{\text{pump}}\right] \tag{4.5}$$

where, η_{pump} and η_{turbine} are the water pump and turbine efficiency, respectively. The maximum capacity of the pump and turbine is explained as C_{pump} and C_{turbine}. The pumping capacity is limited by the amount of power curtailment, the current storage capacity, and the rated capacity of the pump. PHS discharge is not only used to reduce fossil fuel energy power generation but also to provide MGSS a power grid as basic energy. In terms of MGSS, it can be calculated as follows:

$$30\% * \text{Supply}^i \le \text{Nuclear}^i + \text{Dispatch}^i + \text{Discharge}^i = \text{MGSS}^i \tag{4.6}$$

$$\text{Supply}^i = \text{Dispatch}^i + \text{Discharge}^i + \text{RES}^i + \text{Nuclear}^i + \text{Hydro}^i \tag{4.7}$$

where supply^i is the total electricity generation by all power sectors.

4.2.4 Results and Analysis of the Impact of PHS on the Power Grid

28.5 GW is the current installed capacity of PHS in Japan, which has 40 pumped power plants. Hokkaido and Kyushu account for 800 and 2300 MW of installed capacity of PHS, respectively. The potential for development of medium- and small-sized PHS promotes the possibility to grow to 20 GW capacity.

Table 4.2 Supply side

Scenario name	Thermal power	Kyushu (MW)	Hokkaido (MW)
T1	Thermal 1	10,490	4529
T2	Thermal 2	6315	2758
T3	Thermal 3	3275	70

Table 4.3 Energy storage

Scenario name	Kyushu (MW)	Hokkaido (MW)
With PHS	20,350	20,800
Without PHS	0	0

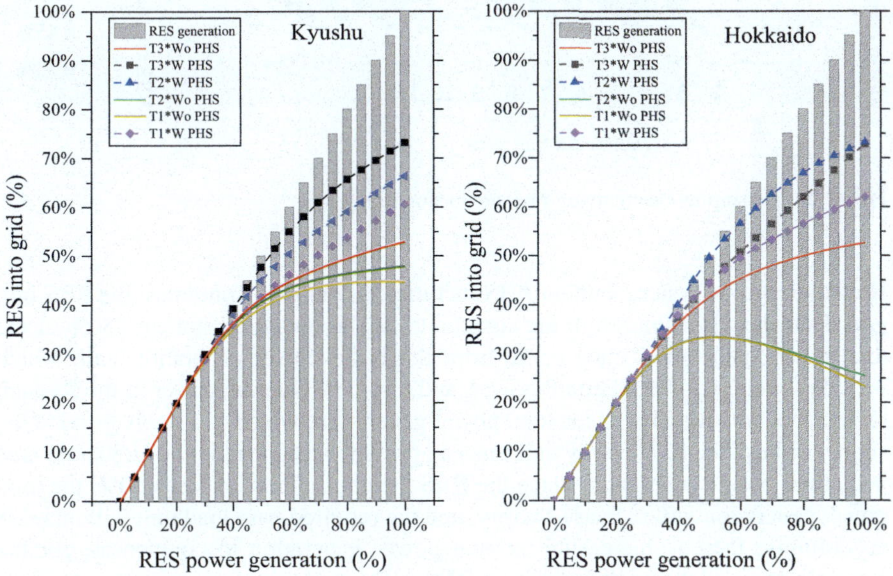

Fig. 4.11 RES output integrated into the Kyushu and Hokkaido grids

The installed capacity of thermal power scenarios in 2020, 2030, and 2040 is abbreviated as T1, T2, and T3. Under this circumstance, the scenarios are designed by combining thermal power and the existence of PHS facilities (with PHS) or using thermal power alone (without PHS (without PHS)) (as shown in Tables 4.2 and 4.3, respectively). Scenario design of electricity supply–energy storage is shown as follows:

On the one hand, the penetration of RES into the grid in different scenarios is shown in Fig. 4.11. It supports the conclusion that PHS can increase the penetration rate of RES into the grid. However, as the share of RES increases, the actual RES in the grid tends to be saturated. PHS plays a significant role when the RES penetration share is greater than 30% in Kyushu and 25% in Hokkaido. In the scenario of

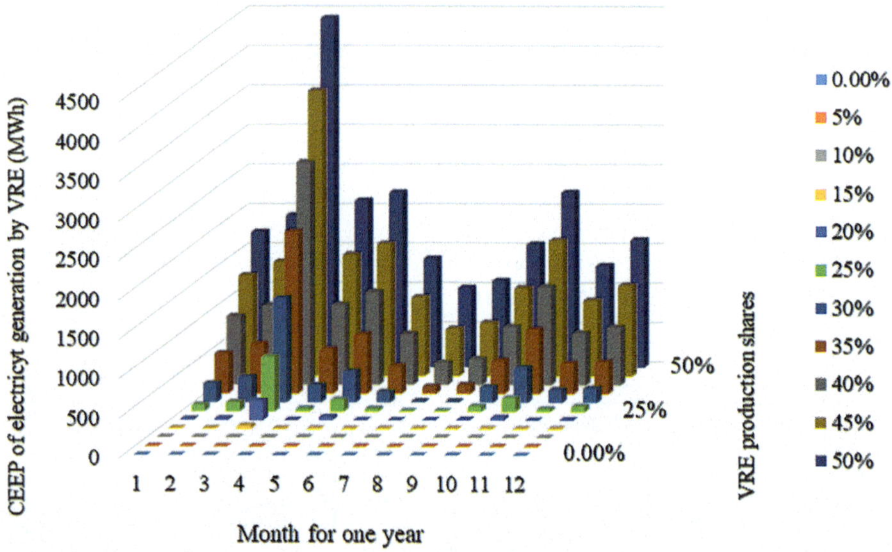

Fig. 4.12 The surplus electricity of VRE production in Kyushu

Hokkaido power system without PHS facilities, as the RES continues, the RES into grid decreases. It results from the constraints of the operating strategy. 30% of the total power generation must come from the basic power generation unit, which includes nuclear power, thermal power, and turbines. Nuclear power in the Kyushu region accounts for 25% of the total power generation, which can stably provide the supply of basic load. The only unit that can provide basic load in the Hokkaido area is thermal power generation. Once the RES penetration rate exceeds 50%, the total power generation will increase sharply, and the required base load unit will increase accordingly, thereby increasing thermal power generation to compensate for the basic electricity demand and reducing RES in the grid.

Figure 4.12 shows the monthly CEEP of VRE production. The data resource is 10,490 installed capacity of thermal power plants. The situation of excess power mainly occurred in the transitional season at April and November.

On the other hand, to reflect the influence of the share of RES and the existence of PHS facilities, taking scenario 1 in the Hokkaido region as an example, the chromaticity diagram of thermal power generation is shown in Fig. 4.13, when the RES share is 0%, 50%, and 100%, respectively. Horizontally comparing the impact of PHS on the operation of thermal power generation, when the penetration of RES is 0%, thermal power generation does not show obvious difference due to the addition of PHS units. When the penetration of RES is 50% and 100%, the usage rate of thermal power generation is reduced by PHS, especially during the daytime. These conditions result from the operation strategy of PHS, which is mainly used to increase the penetration rate of RES and reduce the operation of thermal power

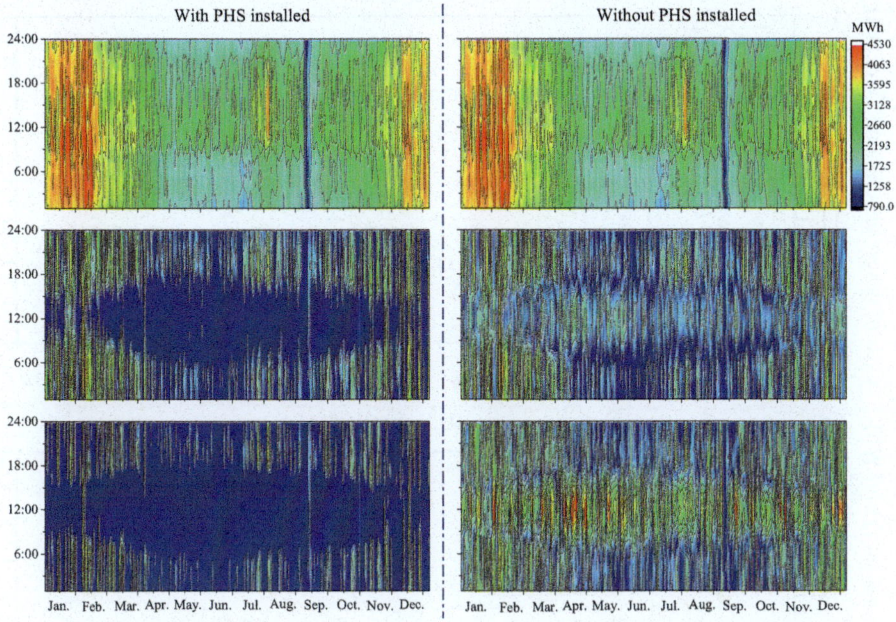

Fig. 4.13 Operation of thermal power plants in monthly intervals in the Hokkaido power grid

generation. In case RES excess occurs, the pump operates to reduce power suppression. At the same time, PHS generates power to mitigate thermal power generation or to improve power grid stability due to the penstock in PHS equipment. The impact of initial installed capacity of thermal power on the amount of actual thermal power generation are analyzed in Fig. 4.13. When PHS equipment is available, the higher the penetration of RES, the lower the utilization rate of thermal power generation. When there is no PHS equipment, the utilization rate of thermal power generation with 50% RES penetration is lower than that without RES. However, in case the RES penetration is 100%, the amount of thermal power generation is greater than 50% RES production, which is due to the limitation of basic load operating conditions. 30% of the total power generation capacity must come from basic load supply units, such as thermal power generation, PHS facilities, and thermal power plants. Once the PHS capacity is utilized to the maximum, the stability of the power grid can only be provided by increasing thermal power generation. This phenomenon can be used to evidence that the permeability of RES in Fig. 4.11 increases first and then decreases with the increase of RES.

Figure 4.14 summarizes the fluctuation of average power generation cost with the increase of RES share in different scenarios, which is affected by many factors. It includes the RES curtailment, the power generation proportion of PHS, and the installation of nuclear power. The analysis can be concluded as follows. Firstly, the average power generation cost in Kyushu and Hokkaido does not exceed 16 and

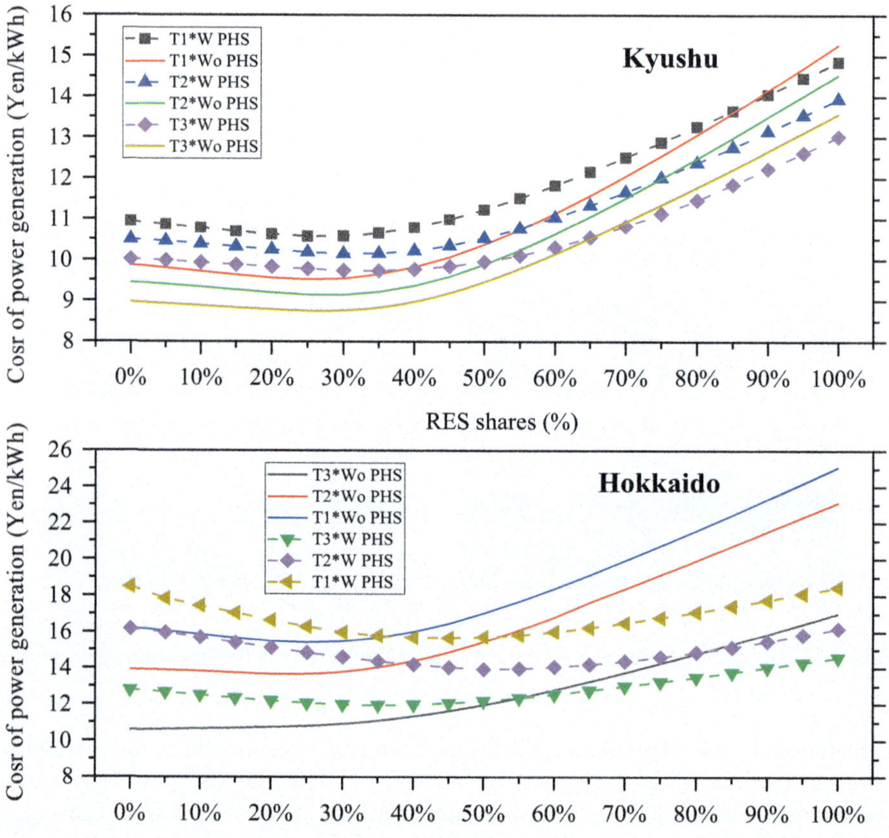

Fig. 4.14 Cost of power generation in Kyushu and Hokkaido region

26 yen/kWh, respectively. Nuclear power in Kyushu can reduce power generation costs. Thermal power accounts for a larger share in Hokkaido, increasing power generation costs. Secondly, in Kyushu and Hokkaido, the average power generation cost with PHS installed is firstly higher and then lower than that of non-energy storage systems. PHS can significantly reduce the curtailment of RES when its penetration exceeds 25% while decreasing the use of fossil fuel energy. Therefore, the power generation cost will be gradually lower than that of non-energy storage systems.

However, the final power generation cost shows an upward trend regardless of PHS facilities. The higher the share of RES power generation, the greater the amount of power suppression. When PHS and grid load demand cannot be absorbed, the average cost of electricity therefore continues to increase. Thirdly, the cost of power generation under the same RES share decreases with the initial installation of thermal power (as shown in Fig. 4.14). At the same time, the intersection of it and the added energy storage system is also forward. This intersection refers to the crossing of the average power generation cost curve with or without the energy storage equipment for the same thermal power installation capacity. The intersection of the Kyushu area shown up after RES was 68%, while that of the Hokkaido area appeared at 35%. This is related to the reduction of the PHS operation rate of nuclear power in Kyushu. The maximum PHS power generation in Kyushu and Hokkaido is 20% and 37%, respectively.

4.3 Suggestions for Future Renewable Energy Development

Major changes in the energy sector in recent years mean that the need for smart, flexible energy is increasing. Low-carbon transformation in the power sector is expected to play a significant role in developing a renewable energy–dominated public grid. Energy production and energy consumption will become more coordinated. The VRE intermittent output has a great influence on grid supply-demand balance, especially under a high renewable energy penetration level. Increasing the renewable energy penetration level brings promising environmental benefit and energy self-sufficiency security at the district level. However, variable renewable energy differs from the conventional power supply technologies. Variable renewable energy shows low peak capacity credit. Due to the grid flexibility constraint and low correlation between solar generation and load profiles, we can observe that the PV production can greatly shape the grid residual load and PV integration mainly decreases the output from medium-based plants. The pumped storage system is used in the energy management system in order to promote renewable energy integrations. The existing small- and medium-sized dams in Japan are exploited to expand the capacity of PHS installed. The result shows that PHS can reduce RES suppression and thermal power operation while providing a share of grid stability. The reduction of the initial installation of thermal power improves the penetration of RES and the operation rate of PHS. The average power generation cost decreases with the reduction of thermal power installation capacity. Besides that, the power generation cost with the addition of PHS is first higher and then lower than the scenario without energy storage equipment. Under the same RES share, the maximum cost in Kyushu is 16 yen/kWh, and that in Hokkaido is 26 yen/kWh. This is attributed to the nuclear power in Kyushu, which accounts for 25% of power generation. The penetration of VRE in the future public grid and its impact on other power generation units were covered.

Energy supply security contributes to a sustainable future and improves social welfare by reducing carbon emissions. With the development of sustainable energy, PV energy and wind energy are expected to become important flexible resources of the power supply system in the future. At present, the variability and intermittence of renewable energy are the main obstacles to its large-scale grid connection. The utilization rate of renewable energy can be maximized through peak shaving and valley filling of the power storage system. Pumped storage has been used to balance power supply and demand for nearly a century. However, its development potential is limited due to the resource constraints of large reservoirs, especially in Japan.

Therefore, the installation capacity can be increased by using the resources of decentralized small- and medium-sized pumped storage power stations. In the future, public utility decision makers can combine a variety of energy storage methods and regional power supply and demand characteristics to realize the maximum grid connection of renewable energy.

References

1. IRENA (2019) Innovation landscape for a renewable-powered future: solutions to integrate variable renewables. Available at: https://www.irena.org/publications/2019/Feb/Innovation-landscape-for-a-renewable-powered-future
2. METI (2022) Japan feed-in tariff scheme for renewable resources
3. ISO C (2022) Managing oversupply. Wind and solar curtailment totals by month
4. Kyuden (2022) Public grid electricity supply-demand balance scenario in Japan
5. Tingting X, Gao W, Li Y, Qian F (2021) Impact of the COVID-19 pandemic on the reduction of electricity demand and the integration of renewable energy into the power grid. J Renew Sustain Energy 13:026304. https://doi.org/10.1063/5.0045825
6. Dujardin J et al (2017) Interplay between photovoltaic, wind energy and storage hydropower in a fully renewable Switzerland. Energy 135:513–525
7. Komiyama R, Fujii Y (2019) Optimal integration assessment of solar PV in Japan's electric power grid. Renew Energy 139(August):1012–1028
8. Honda A, Tezuka K et al (2020) Analysis of production and storage of secondary energy derived from renewable energy. J Jpn Soc Energy Resour 41(5):186–195
9. Gotoh R, Tezuka T (2020) Study on power supply system for large scale renewable energy introduction under different strategies of existing power plant replacement. J Jpn Soc Energy Resour 40(2):38–50
10. Javed MS et al (2020) Solar and wind power generation systems with pumped hydro storage: review and future perspectives. Renew Energy 148:176–192
11. Ichimura S (2020) Utilization of cross-regional interconnector and pumped hydro energy storage for further introduction of solar PV in Japan. Glob Energy Interconnect 3(1):68–75
12. Beaudin M et al (2010) Energy storage for mitigating the variability of renewable electricity sources: an updated review. Energy Sustain Dev 14(4):302–314
13. Ichimura S, Kimura S (2019) Present status of pumped hydro storage operations to mitigate renewable energy fluctuations in Japan. Glob Energy Interconnect 2(5):423–428
14. Kong Y, Kong Z et al (2017) Pumped storage power stations in China: the past, the present, and the future. Renewable and Sustainable Energy Reviews 71:720–731
15. Anagnostopoulos JS, Papantonis DE (2012) Study of pumped storage schemes to support high RES penetration in the electric power system of Greece. Energy 45(1):416–423
16. Rehman S, Al-Hadhrami LM, Alam MM (2015) Pumped hydro energy storage system: a technological review. Renew Sust Energ Rev 44:586–598
17. Heath, B, Figueroa-Acevedo, AL (2018) Potential capacity contribution of renewables at higher penetration levels on MISO system. 2018 IEEE international conference on probabilistic methods applied to power systems (PMAPS). IEEE
18. Ma T et al (2014) Technical feasibility study on a standalone hybrid solar-wind system with pumped hydro storage for a remote island in Hong Kong. Renew Energy 69:7–15
19. Connollya D, Lund H, Mathiesen BV, Picana E, Leahy M (2012) The technical and economic implications of integrating fluctuating renewable energy using energy storage. Renew Energy 43:47–60

20. Kucukali S, Al Bayatı O, Hakan Maraş H (2021) Finding the most suitable existing irrigation dams for small hydropower development in Turkey: a GIS-fuzzy logic tool. Renew Energy 172: 633–650
21. Hunt JD, Byers E, Riahi K, Langan S (2018) Comparison between seasonal pumped-storage and conventional reservoir dams from the water, energy and land nexus perspective. Energy Convers Manag 166(15):385–401
22. Scherer L, Pfister S (2016) Global water footprint assessment of hydropower. Renew Energy 99:711–720. https://doi.org/10.1016/j.renene.2016.07.021
23. Wikipedia. PHS power generation. Available at: https://ja.wikipedia.org/wiki/%E6%8F%9A%E6%B0%B4%E7%99%BA%E9%9B%BB [In Japanese]
24. Power Development Numahara Power Station, Ministry of Land, Infrastructure, Transport and Tourism. National land image information (color aerial photograph) (photographed on November 18 and 22, 1976). Available at: https://ja.wikipedia.org/wiki/%E6%8F%9A%E6%B0%B4%E7%99%BA%E9%9B%BB#/media/%E3%83%95%E3%82%A1%E3%82%A4%E3%83%AB:Numappara_power_station_survey.jpg
25. National Research and Development, Japan Science and Technology Agency. Low Carbon Social Strategy Center. Potential capacity and cost of pumped-storage power in Japan (Vol. 2). Available at: https://www.jst.go.jp/lcs/pdf/fy2019-pp-03.pdf [In Japanese]
26. Proposal chapter for policy making and governmental action toward low carbon societies. Potential capacity and cost of pumped-storage power in Japan. Available at: https://www.jst.go.jp/lcs/pdf/fy2018-pp-08.pdf [In Japanese]

Chapter 5
Design and Optimization of Distributed Energy Systems

Hongbo Ren, Qiong Wu, and Xingzhi Shi

5.1 Concept of Design and Optimization of DER System

Plan and design of DER system is seen as the path toward an economic and ecological sustainable energy system while taking into account limited technical, financial, and natural resources.

The DER system analyzed in this chapter consists of interconnected power generations, storage, and several energy carriers: electricity, gas, biomass, etc [1]. Furthermore, the different types of end-use services (e.g., electricity, cooling, heating, and hot water) can be also integrated into the distributed energy system [2]. Figure 5.1 shows the general image of the formation of alternative DER systems. With the predefined energy carriers and generation technologies, a set of system alternatives can be deduced, where each system alternative typically consists of several physical components with predefined connections to the rest of the energy system. The same components can be included in several competing system alternatives, making the different alternatives mutually exclusive from an economic point of view.

Assuming the numbers of energy carriers and technologies are p and q, respectively, then we will have $p \times q$ possible equipment. If no equipment is selected, let the variable $y_0 = C^0_{p \times q} = 1$; similarly, if only one equipment is employed for the system, then $y_1 = C^1_{p \times q} = p \times q$. The total number of possible systems that can be constituted is identified by the value of S in

H. Ren (✉) · Q. Wu
College of Energy and Mechanical Engineering, Shanghai University of Electric Power, Shanghai, China

X. Shi
Zhejiang Eco-environmental Low-Carbon Development Center, Hangzhou, China

W. Gao (ed.), *Distributed Energy Resources*, Green Energy and Technology,
https://doi.org/10.1007/978-3-031-21097-6_5

Fig. 5.1 General image of the formation of alternative DER systems

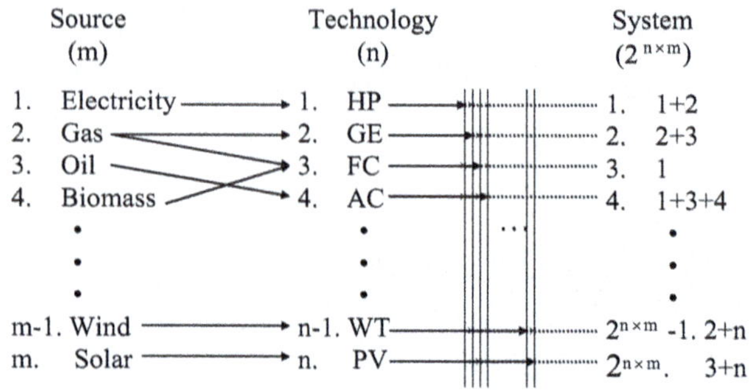

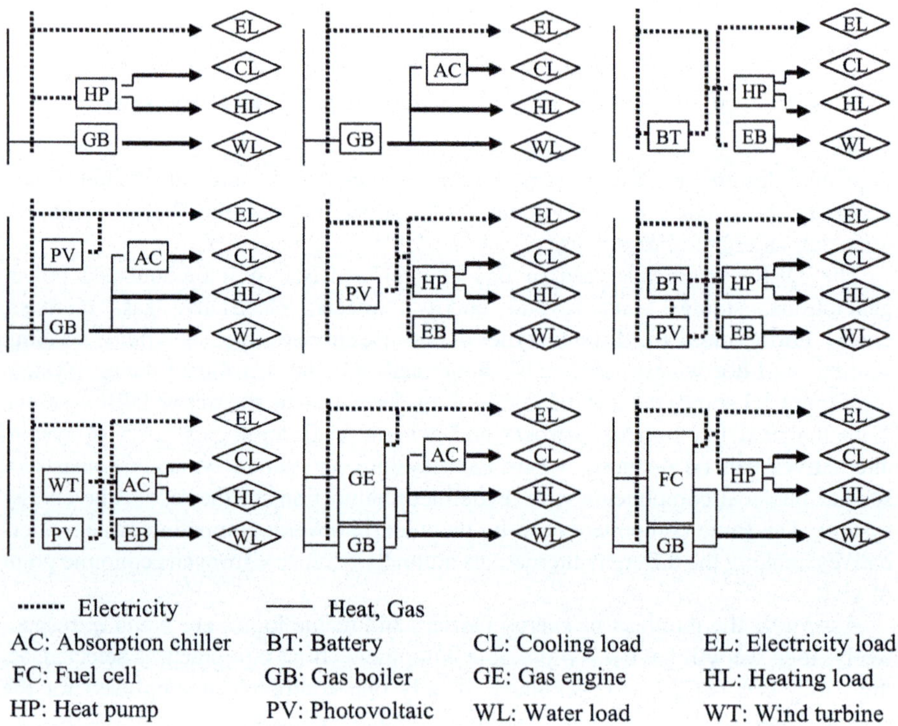

Fig. 5.2 Sample layout of DER system design

$$S = \sum_{i=0}^{p\times q} y_i = \sum_{i=0}^{p\times q} C_{p\times q}^{i} = 2^{p\times q} \tag{5.1}$$

Each element in the set represents a unique combination of technology alternatives. Figure 5.2 shows the principle layout of some sample designs, including various technology components. It is obvious that selecting a suitable system from the numerous alternatives will be an enormous work. Even with the help of computer, it will result in long computational time if many technology alternatives have been considered.

Fig. 5.3 General image of operation schedule

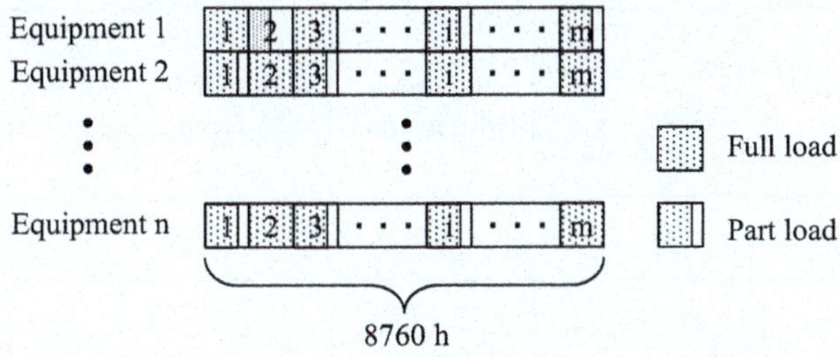

Operation planning is another vital task for the introduction of DER system [3]. Generally, it is to find the optimal operation of a given system that minimizes the costs of satisfying a predefined energy demand within the studied location. The system boundaries are implicitly defined by import and export of energy to the modeled system. The planning horizon is usually 1 year, and the typical time step is 1 h. Figure 5.3 illustrates the general image of operation planning. As to a real site, both electricity and thermal demands fluctuate seasonally and hourly, so it is necessary to take account of the system's annual operational strategies according to the variations of load demands. In addition, the operation of DER system is subjected not only to the variation of load demands but also to the fuel prices and energy policies as well.

As discussed above, in order to realize high economical and energy-saving potentials of the DER system, it is necessary to determine its structure rationally by selecting some kinds of equipment from many alternative ones so that they match energy requirements for an objective user. Furthermore, it is also important to determine rationally the number and capacities of each kind of equipment selected and the system's annual operating strategies corresponding to hourly variations in energy demands.

Facing the abovementioned problems, a systematic optimization procedure is needed. The role of optimization is to reveal the best (under certain criteria and constraints) design and the best operational point of the DER system automatically, with no need for the designer to study and evaluate one by one the multitude of possible variations [4]. It is a time-dependent optimization and without pre-specification, so the degree of freedom is very large and the problem is very complex. In order to solve such a complex work in an effective manner, a large amount of information is required, and energy modeling has been and still is the most basic approach in aiding the plan of evaluation of DER system.

According to the concept described above, in the plan and design of a DER system, there is a hierarchical relationship among determination of the system structure, the numbers and capacities of equipment, and the system's operating strategies [5]. Namely, a certain system structure is created by selecting some kinds of equipment from their candidates; under this system structure, the number and capacities are specified for each kind of equipment from their alternatives; and under these specifications, the system's operating strategies are determined so as to satisfy energy requirements.

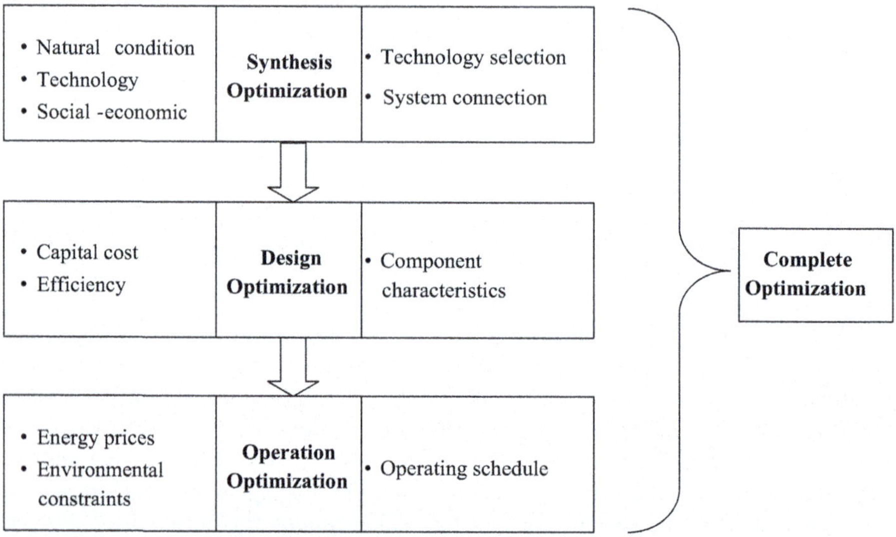

Fig. 5.4 Structure of the levels of DER system optimization

Corresponding to the hierarchical process of plan and design of a DER system, the optimization model of the DER system can also be considered at three levels, as shown in Fig. 5.4.

1. Synthesis optimization. It decides the components that appear in a system and their connections. After the synthesis of a system has been successfully composed, the flow diagram of the system can be drawn. It is a macroanalysis, which sometimes takes into the technological, socioeconomic, and even natural conditions of the specific area.
2. Design optimization. It is used to imply the technical characteristics of the components and the properties of the substances entering and exiting each component at the nominal load of the system.
3. Operation optimization. For a given system (i.e., one in which the synthesis and design are known) under specified conditions, the optimal operating schedule is requested, as it is defined by the operating properties of components and various economic and environmental conditions.

However, the optimization of DER system is usually a complete optimization; each level cannot be considered in complete isolation from the others. This is because the operational strategy will sometimes affect the selection of specific equipment. For example, sometimes we prefer to select two 50 kW generators instead of one 100 kW generator because of the relative low efficiency at part loads. So, it is necessary to integrate the synthesis of the system, the design characteristics of the components, and the operating strategy so as to lead to an overall optimum. It can be stated mathematically as follows:

$$\underset{u,v,w}{\text{Min}} \quad f(u, v, w) \tag{5.2}$$

where u, v, w are independent variables for synthesis, design, and operation optimizations, respectively.

For a given synthesis of the system, i.e., for given u, the problem becomes one of design and operation optimization:

$$\underset{v,w}{\text{Min}} \quad f_d(v, w) \tag{5.3}$$

Furthermore, if the system is completely specified (both u and v are given), then an operation optimization problem is formulated:

$$\underset{w}{\text{Min}} \quad f_{\text{op}}(w) \tag{5.4}$$

5.2 Single-Objective Optimization Model

5.2.1 Description of the Model

Economy is the core goal considered in the design of a DER system. From the vast set of different optimization methods, the DER optimization is formulated here as a MILP (mixed-integer linear programming) model. It should on one side be kept as simple as possible, to facilitate the work of the optimization algorithm and to shorten the total computing time. On the other side, it has to guarantee a certain level of accuracy. Figure 5.5 is a flowchart illustrating the structure of the model.

It has been developed to find the optimal combination of distributed generation units and corresponding operation strategies, with an objective of minimizing annual total cost including investment cost and running cost. Decision variables are composed of design and operational ones. Since the units can be only with discrete numbers, integer variables are necessary. The input date for the model can be divided into the following categories: customer information (load profiles, climate conditions, natural resources, etc.), technical information (power and thermal efficiencies, etc.), as well as financial and policy information (capital costs, energy tariffs, taxes, etc.). Whiling considering the balance between the supply and demand of energy resources, by analyzing the equipment availability, supply share, and costs, the operation characteristics of the complex system are examined. The calculation is executed for every hour in a whole year (8760 h). The results obtained from this process are the optimal combination of on-site generation and heat recovery,

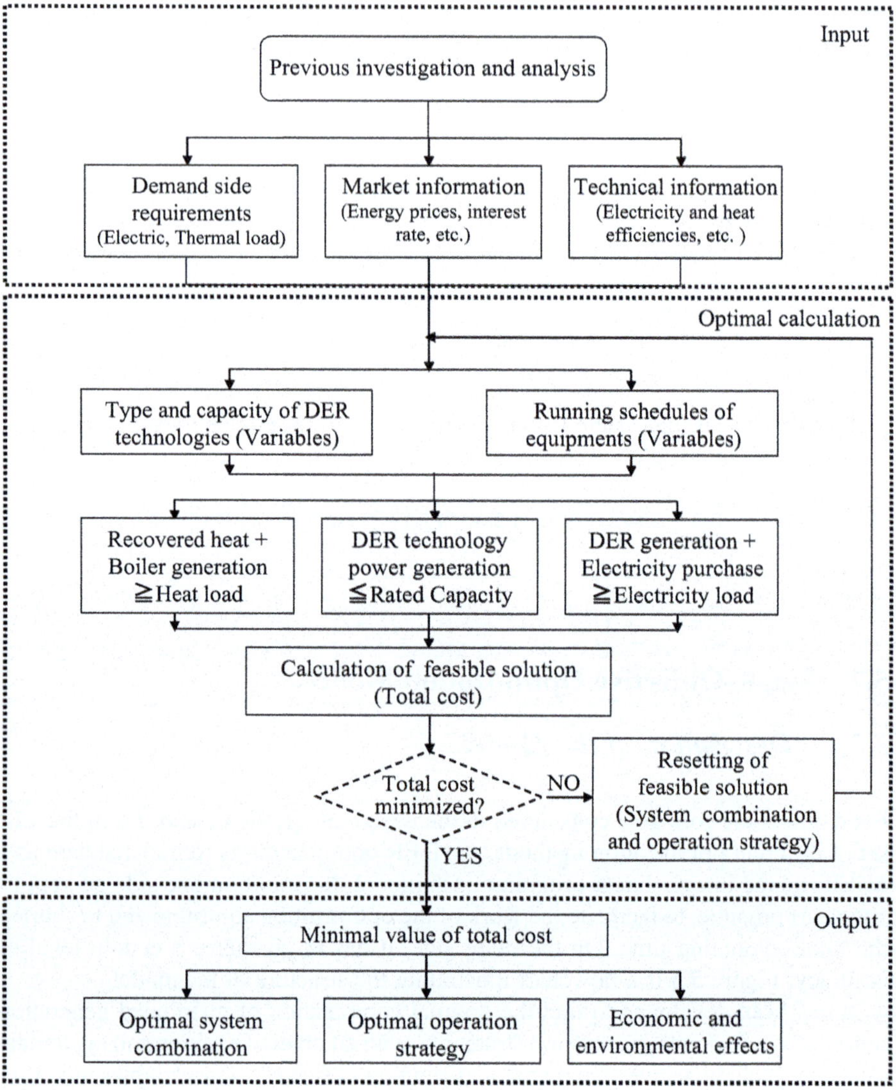

Fig. 5.5 Flowchart of the design and optimization model

capacity of each technology to be installed, an elementary operating schedule of how the equipment should be used, and summary results for each scenario, such as total electricity bill, electricity generation and purchase in each hour, etc. Furthermore, according to the determined system combination and operation, the economic, energetic, and environmental effects can be also deduced.

While constructing the model, in order to make it more rational, some important assumptions have been considered [6].

1. Customer decisions are taken based only on direct economic criteria. Other effects of DER adoption can be converted and assessed in an economic manner. For example, by introducing the carbon tax, the environmental benefits can be integrated into the economic comparison.
2. All data are known with complete certainty, i.e., energy loads and fuel prices for the duration of the test year are all given. The candidate technologies are comparable from a life cycle cost viewpoint.
3. When electricity buyback is taken into consideration, the customer is not allowed to sell and buy at the same time.
4. The DER system is grid-connected; when more electricity is consumed than generated, the DER system will buy from the macro-grid at the default tariff rate.
5. There is a fixed relationship between the amount of recoverable heat and electricity generated by each DER unit based on the manufacturer's technical specifications.
6. Neither reliability and power quality benefits, nor economies of scale in investment and running costs for multiple units of the same technology are taken into account.

In addition, in order to make the model easy to be used and spread, the following issues have been paid enough attention to while developing the model:

1. The procedure is flexible and generic, allowing different scale areas in any location and climate, with different demand needs and supply possibilities to be analyzed.
2. Various types of energy demands including electricity, cooling, heating, and hot water are able to be considered.
3. The temporal matching of supply and demand is reached in order to analyze variations throughout the day and between seasons.
4. Intermittent suppliers, alternative fueled technologies for CHP generation, energy storage technologies, as well as other uses for excess electricity are available for consideration.
5. It is possible to analyze and compare a number of potential combinations quickly and easily.

The model has three possible applications: first, it can be used to guide choices of equipment for specific sites or provide general solutions for example sites and propose good choices for sites with similar circumstances; second, it can additionally provide the basis for the operations of installed on-site generation; and third, it can be used to assess the market potential of technologies by anticipating which kinds of customer might find various technologies attractive.

5.2.2 Decision Variables

Decision variables are composed of integer and continuous ones. The integer variables express the number of equipment, the on-off status of the operation, as well as the existence of energy storage devices. The continuous variables express the input and output energy flows of the system components.

5.2.3 Objective Function

The objective function of the model is to minimize the overall cost of supplying energy service to a specific customer by using distributed generation to meet part or all of its electricity and thermal requirements, as shown in Eq. 5.5. It is evaluated as the sum of energy (both electricity and fuel) costs, annual capital, operational and maintenance (O&M) costs, as well as carbon tax cost and cost associated with start and stop of equipment, subtracting the revenue from selling excess electricity.

$$\mathrm{Min}\{C_{\mathrm{Total}} = C_{\mathrm{Inv}} + C_{\mathrm{Elec}} + C_{\mathrm{Fuel}} + C_{\mathrm{OM}} + C_{\mathrm{Ctax}} + C_{\mathrm{Ss}} - C_{\mathrm{Sal}}\} \tag{5.5}$$

Annual capital cost is described in Eq. 5.6. It is calculated by spreading the initial cost across the lifetime of the equipment while accounting for the time value of money (realized through a specific interest rate).

$$C_{\mathrm{Inv}} = \sum_i NInv(i) \cdot [FCostf(i) + FMaxp(i) \cdot (FCostv(i) - FSubs(i))] \cdot \frac{IRate}{\left(1 - \frac{1}{(1+IRate)^{FLTim(i)}}\right)} \tag{5.6}$$

The outside electricity purchase cost is described in Eq. 5.7. It is composed of the demand charge and total energy cost. The demand charge is calculated with charge rate multiplied by the peak electricity demand in every month. The energy cost is calculated with cumulative amount of electricity purchase multiplied by the utility electricity rate.

$$C_{Elec} = \sum_m EDchar \cdot \max\left(\sum_u PElec(m,d,h,u)\right) + \sum_m \sum_d \sum_h \sum_u PElec(m,d,h,u) \cdot Eprice(m,d,h) \tag{5.7}$$

The fuel cost is composed of monthly base service fee, flux charge, and volumetric cost, as shown in Eq. 5.8. The flux charge is calculated with charge rate multiplied by the peak fuel demand in every month. The volumetric cost consists of two separate parts, DER use and non-DER use. It is calculated with cumulative fuel consumption for each period of DER equipment and direct fuel use multiplied by the tariff rate.

$$
\begin{aligned}
C_{\text{Fuel}} = {} & \sum_{f}\sum_{m} FBase(f,m) \\
& + \sum_{f}\sum_{m} \max\left[\begin{array}{l} \sum_{u} PFuel(f,m,d,h,u) + \\ \sum_{i}\left(\frac{\sum_{u} EGen(i,m,d,h,u) + ESal(i,m,d,h) + IEne(i,j,m,d,h)}{Eff(i)}\right)\end{array}\right] \\
& \cdot FDchar(f,m) \\
& + \sum_{f}\left[\begin{array}{l} \sum_{m}\sum_{d}\sum_{h}\sum_{u} PFuel(f,m,d,h,u) + \\ \sum_{i}\left(\frac{\sum_{m}\sum_{d}\sum_{h}\sum_{u} EGen(i,m,d,h,u) + ESal(i,m,d,h) + IEne(i,j,m,d,h)}{Eff(i,f)}\right)\end{array}\right] \\
& \cdot Fprice(f) j \in \{elec\}
\end{aligned}
\tag{5.8}
$$

The O&M cost is composed of fixed and variable costs, as illustrated in Eq. 5.9. The fixed O&M cost is calculated with the installed DER equipment capacity multiplied by a unit cost coefficient. The variable O&M cost is calculated with cumulative power generation for each period of DER equipment multiplied by a unit cost coefficient.

$$
\begin{aligned}
C_{\text{OM}} = {} & \sum_{i}\sum_{m}\sum_{d}\sum_{h}\left(\sum_{u} EGen(i,m,d,h,u) + ESal(i,m,d,h)\right. \\
& \left. + IEne(i,j,m,d,h)\right) \cdot OMv(i) + \sum_{i} NInv(i) \cdot FMaxp(i) \cdot OMf(i)
\end{aligned}
\tag{5.9}
$$

The carbon tax cost is indicated in Eq. 5.10. It is described as the cost for carbon emissions from utility electricity, on-site power generation, as well as direct fuel consumption.

$$C_{\mathrm{Ctax}}=\sum_{f}\left(\begin{array}{l}\sum_{m}\sum_{d}\sum_{h}\sum_{u}PFuel(f,m,d,h,u)+\\ \sum_{i}\frac{\sum_{m}\sum_{d}\sum_{h}\left(\sum_{u}EGen(i,m,d,h,u)+ESal(i,m,d,h)+IEne(i,j,m,d,h)\right)}{Eff(i,f)}\end{array}\right)$$
$$\cdot CTax\cdot FCInt(f)+\sum_{m}\sum_{d}\sum_{h}\sum_{u}PElec(m,d,h,u)\cdot CTax\cdot ECInt\quad j\in\{elec\} \tag{5.10}$$

Additional cost is assumed to be associated with the start and stop of the equipment, as is illustrated in Eq. 5.11.

$$C_{\mathrm{Ss}}=\sum_{i}\sum_{m}\sum_{d}\sum_{h}(FStart(i,m,d,h)-FStop(i,m,d,h))\cdot\omega \tag{5.11}$$

An income will be received when electricity buyback is available by the utility grid, as shown in Eq. 5.12.

$$C_{\mathrm{Sal}}=\sum_{i}\sum_{m}\sum_{d}\sum_{h}ESal(i,m,d,h)\cdot Sprice(m,d,h) \tag{5.12}$$

5.2.4 *Main Constraints*

The main constraints in this model are energy balance and supply-demand relationships, as well as the performance characteristics of the system components.

5.2.4.1 Energy Balance and Supply-Demand Relationships

A key constraint is that energy demand in each time period must be met in one of three ways: purchase of energy from utilities, operation of a technology or set of technologies selected by the model, or a combination of these options. It is formulated with the energy flow rate, as is described in Eq. 5.13.

$$\begin{aligned}Cload(m,d,h,u)=&\sum_{i}EGen(i,m,d,h,u)+PElec(m,d,h,u)\\&+\sum_{f}\beta(f,u)\cdot PFuel(f,m,d,h,u)+\sum_{i}(\gamma(i,u)\cdot RHeat(i,m,d,h,u))\\&+\sum_{j}\delta(j,u)\cdot OEne(j,m,d,h,u)\ \forall m,d,h,u\end{aligned} \tag{5.13}$$

Furthermore, according to the agreement with the electric utilities, constraint is set to prohibit the customer from buying and selling energy at the same time, as shown in Eq. 5.14. Also, the electricity purchase has to be not lower than a specific share of the total requirements, as illustrated in Eq. 5.15.

$$ESal(i,m,d,h) = 0 \quad \text{if} \quad \sum_u \sum_i EGen(i,m,d,h,u) < \sum_u Cload(m,d,h,u)$$
$$\forall i,m,d,h \quad \text{if} \quad u \in \{\text{electricity}\} \tag{5.14}$$

$$\sum_m \sum_d \sum_h \sum_u PElec(m,d,h,u) \geq \theta$$
$$\cdot \sum_m \sum_d \sum_h \sum_u \begin{pmatrix} PElec(m,d,h,u) + \sum_i EGen(i,m,d,h,u) + \\ \delta(j,u) \cdot OEne(j,m,d,h,u) \end{pmatrix} \tag{5.15}$$
$$j \in \{\text{elec}\}$$

5.2.4.2 Performance Characteristics of System Components

As the second group constraint equations, the performance characteristics of the system components are formulated as a relationship between the input and output energy flow.

1. General DER generation unit

A main performance constraint is that any installed DER technologies must be operated between the minimum and maximum operating capacity of the units, as illustrated in Eq. 5.16. Equation 5.17 indicates the number of equipment under operation at each time interval.

$$Operate(i,m,d,h) \cdot FMinp(i) \leq \sum_u EGen(i,m,d,h,u) + ESal(i,m,d,h)$$
$$+ IEne(i,j,m,d,h) \leq Operate(i,m,d,h) \cdot FMaxp(i) \forall j \in \{elec\},\ i,m,d,h \tag{5.16}$$

$$Operate(i,m,d,h) \in \{0,1,2,\cdots,NInv(i)\} \tag{5.17}$$

2. PV generation unit

PV is assumed to produce electricity in proportion to the capacity of the installed system and the amount of solar irradiation, as illustrated in Eqs. 5.18 and 5.19.

$$\sum_u EGen(i,m,d,h,u) + ESal(i,m,d,h) + IEne(i,j,m,d,h) \leq NInv(i) \cdot PVp(i,m,d,h) \quad \forall i \in \{PV\}, j \in \{elec\}, m, d, h \tag{5.18}$$

$$PVp(i,m,d,h) = \min\{FMaxp(i), R(m,d,h) \cdot Eff(i)\} \quad \forall i \in \{PV\}, m, d, h \tag{5.19}$$

3. Wind turbine unit

The electricity produced by wind turbine is determined by local wind speed and equipment characteristics, as shown in the following equations.

$$\sum_u EGen(i,m,d,h,u) + ESal(i,m,d,h) + IEne(i,j,m,d,h) \leq NInv(i) \cdot Wp(i,m,d,h) \quad \forall i \in \{wind\}, j \in \{elec\}, m, d, h \tag{5.20}$$

$$Wp(i,m,d,h) = FMaxp(i) \cdot \frac{V^k(m,d,h) - V^k_{C(i)}}{V^k_{R(i)} - V^k_{C(i)}} \quad \forall i \in \{wind\}, V_{C(i)} \leq V(m,d,h) \leq V_{R(i)} \tag{5.21}$$

$$Wp(i,m,d,h) = FMaxp(i) \quad \forall i \in \{wind\}, V_{R(i)} \leq V(m,d,h) \leq V_{F(i)} \tag{5.22}$$

$$Wp(i,m,d,h) = 0 \quad \forall i \in \{wind\}, V(m,d,h) < V_{C(i)} \cup V(m,d,h) > V_{F(i)} \tag{5.23}$$

4. Heat recovery unit

Constraint is also set to how much heat can be recovered for both immediate usage and diversion to storage from each type of DER technology, as described in Eq. 5.24.

$$\sum_u RHeat(i,m,d,h,u) + IEne(i,j,m,d,h) \leq \alpha(i) \cdot \sum_u EGen(i,m,d,h,u) \quad \forall i, j \in \{heat\}, m, d, h \tag{5.24}$$

5. Energy storage unit

Additional constraints are needed to ensure the operation of energy storage units (both electricity and heat). Equation 5.25 shows the energy inventory balance constraint. It states that the total amount of energy stored at the beginning of each time interval is equal to the non-dissipated energy stored at the beginning of the previous time interval plus net energy flow (energy stored into the storage unit minus stored energy which is released to meet end-use loads).

$$\begin{aligned} SEne(j,m,d,h+1) = \varepsilon(j) \cdot SEne(j,m,d,h) + \sum\nolimits_i IEne(i,j,m,d,h) \\ - \sum\nolimits_u OEne(j,m,d,h,u) \quad \forall j,m,d,h \end{aligned} \tag{5.25}$$

Equation 5.26 prevents the quantity of energy stored from exceeding the maximum storage capacity, while below the minimum level.

$$ESMin \leq SEne(m,d,h) \leq ESMax \quad \forall m,d,h \tag{5.26}$$

6. Start and stop of equipment

The operational experience shows that frequent start and stop will speed up the corruption of equipment. In this chapter, the cost coefficient is associated to the start and stop of equipment. By using the following equations, the start and stop times are expected to be minimized.

$$D(i,m,d,h+1) - D(i,m,d,h) = FStart(i,m,d,h) - FStop(i,m,d,h) \quad \forall i,m,d,h \tag{5.27}$$

$$\begin{aligned} \sum\nolimits_u EGen(i,m,d,h,u) + IEne(i,j,m,d,h) + ESal(i,m,d,h) \\ - M \cdot D(i,m,d,h) \leq 0 \quad \forall i,j = \{elec\},m,d,h \end{aligned} \tag{5.28}$$

5.2.5 Solution Method

Generally, there are two options for implementing the above solution techniques: to transform the model into an algorithm and code it in a high-level language and to use the professional optimization software. Both options have their own advantages and disadvantages: the former may be more efficient in terms of the speed of solution but lends itself to relatively less user-friendliness and less flexibility in modifying input data and interacting with other software. In this chapter, the MILP model is programmed as one coherent system of equations using the mathematical modeling language LINGO. Modeling the overall system performance with a series of equations has advantages in conjunction with recycled streams because there is no need to determine a calculation order. Moreover, additional system components can be easily integrated into the model by connecting them with the upstream and downstream components. In this way, the optimization model will be flexible in modifying input data to perform sensitivity analysis and easy to interact with most window-driven software for a continuous flow of data input, processing, display, and analyses.

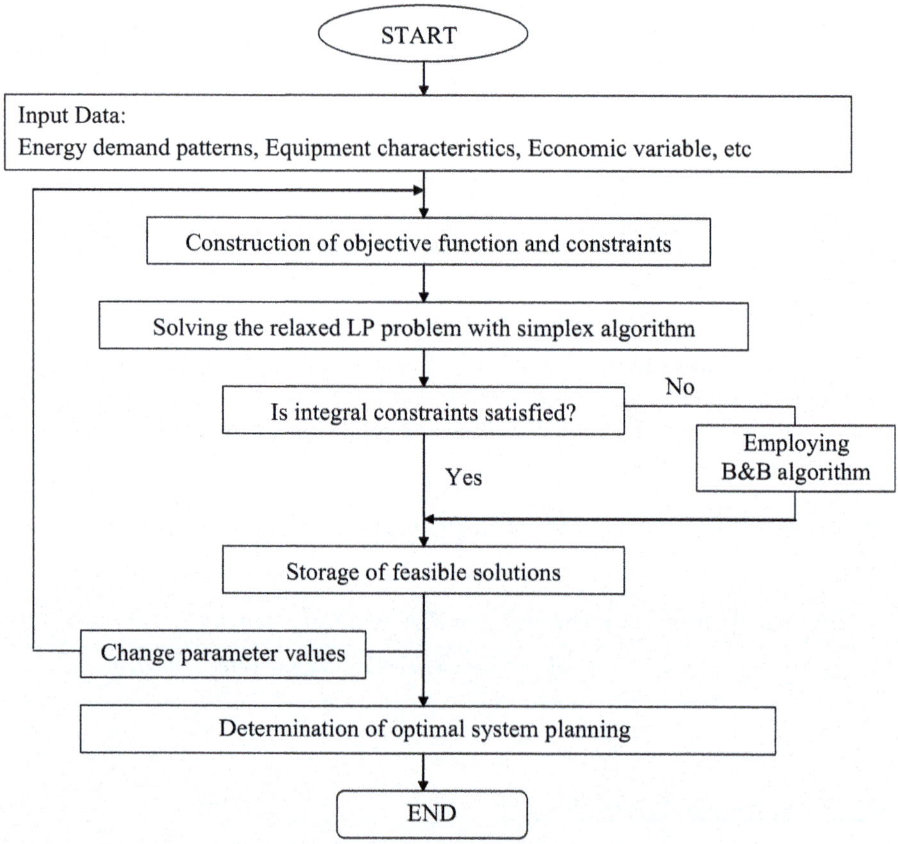

Fig. 5.6 Flowchart of the solution process for the optimization model

Due to the great differences in the order of magnitude of the variables, the entire model has to be scaled to ensure an equal treatment by the solver regarding the compliance of all equations. In addition, to facilitate the work of the solver and to improve the quality of the results, upper and lower bounds for all variables are needed. The optimization is controlled by the MILP solver B&B which accesses the LINGO model, performs the optimization, and stores the best solutions and a run in a data file for a subsequent analysis. Figure 5.6 shows the flowchart of the solution process, which can be concluded as following steps: firstly, the necessary input data is delivered to the model; secondly, the model is activated to construct the objective function and constraints with data realization; thirdly, the simplex algorithm is employed to solve the relaxed LP problem; fourthly, if the achieved results can satisfy the integer constraints, it is put into the storage as a feasible solution; otherwise, the B&B algorithm will be employed; then, the above steps are recycled until the end of the simulation; finally, the results of optimal system planning are deduced.

5.3 Multi-objective Optimization Model

5.3.1 Multi-objective Optimization Problem

The process of decision-making determines the choice of a possible course of action among a wide variety of available alternatives. The difficult point in decision-making is the multiplicity of the criteria set for judging the alternatives. The decision-maker needs to attain more than one objective in achieving the final goal set while satisfying constraints dictated by the economics, resources, and environment.

As a complex decision-making problem, energy system planning inherently involves multiple, conflicting, and incommensurate objectives. Therefore, mathematical models become more realistic if distinct evaluation aspects, such as cost and environmental concerns, are explicitly considered by giving them an explicit role as objective functions rather than aggregating them in a single economic indicator objective function. Multi-objective models can provide decision support to decision-makers by rationalizing the comparison among different alternative solutions, thus enabling the decision-maker to grasp the inherent conflicts and trade-offs among the distinct objectives for selecting a satisfactory compromise solution from the set of non-dominated solutions. In multi-objective models, the concept of optimal solution in single-objective problems (unique, in general) gives place to the concept of non-dominated solutions (feasible solutions for which no improvement in any objective function is possible without sacrificing at least one of the other objective functions). Figure 5.7 illustrates the general flowchart of a multi-objective model for energy system planning.

Fig. 5.7 Flowchart of multi-objective optimization

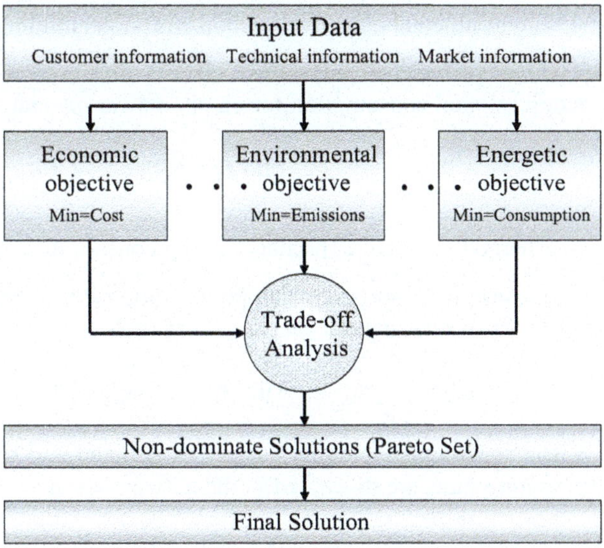

5.3.2 Mathematical Formulation of the Problem

In this section, a multi-objective mixed-integer linear programming (MOMILP) methodology is applied to determine the supply share for some chosen technologies for a district where various distributed energy resources can be found and could be exploited to satisfy part of the energy needs. These resources are examined from all aspects using specific mathematical model. Considering the existing constraints, a series of solutions is derived providing decision-makers the flexibility to choose the appropriate solution with respect to the given situation.

In this section, the MOMILP model considers three objective functions, which quantify the annual energy cost, the environmental impact, and the primary energy consumption, all to be minimized. The environmental impact objective functions attempt to capture the increasing awareness of environmental externalities resulting from energy generation.

By keeping the objectives separate, trade-offs among different objectives are clearly illustrated, and more informed design decisions can be made. In this case, it allows to find and to rank the best integrated generation technology solutions from the superstructure, which are cost-effective, less polluting, and energy conservative. The solutions returned by the algorithm are an approximation to the optimal—such a solution cannot be made less polluting without being more costly and energy saving, vice versa.

1. Economic objective (energy cost minimization)

As illustrated in Eq. 5.29, total energy cost is shown in the following:

$$\text{Min } f_{\text{C}} = C_{\text{Elec}} + C_{\text{Gas}} + C_{\text{Inv}} + C_{\text{OM}} + C_{\text{CTax}} + C_{\text{Ss}} - C_{\text{Sal}} \tag{5.29}$$

2. Environmental objective (carbon emissions minimization)

The environmental impact associates with the total CO_2 emissions from utility electricity, fuel consumption for on-site power generation, and thermal use.

$$\text{Min } f_{\text{E}} = \text{CE}_{\text{Elec}} + \text{CE}_{\text{Gas}} \tag{5.30}$$

3. Energetic objective (primary energy consumption minimization)

The primary energy consumption composed of energy consumption for purchased electricity, natural gas, and other fuels.

$$\text{Min } f_{\text{P}} = \text{EC}_{\text{Elec}} + \text{EC}_{\text{Gas}} \tag{5.31}$$

The constraints including balance of demand and supply and the availability of generating units are illustrated in the above section.

5.3.3 Solution Method

There are a lot of methods for solving multi-objective optimization problems, such as compromise programming, global criterion method, goal programming, and so on. In this chapter, the developed MOMILP model is programmed in the LINGO software, and the compromise programming method has been employed to solve it.

To apply compromise programming, the decision model is modified so as to include only one objective. The aim in this method is to minimize the distance of the criterion values from their optimum values. Considering this, the decision problem is formulated as follows:

$$\text{Min } z = \phi \tag{5.32}$$

Besides the constraints illustrated above, the following constraints should be taken into consideration.

$$\phi \geq (f_{\mathrm{C}} - f_{\mathrm{Cmin}}) \cdot (w_{\mathrm{C}} / f_{\mathrm{Cmin}}) \tag{5.33}$$

$$\phi \geq (f_{\mathrm{E}} - f_{\mathrm{Emin}}) \cdot (w_{\mathrm{E}} / f_{\mathrm{Emin}}) \tag{5.34}$$

$$w_{\mathrm{C}} + w_{\mathrm{E}} = 1 \tag{5.35}$$

where ϕ corresponds to the Chebyshev distance, f_{Cmin} and f_{Emin} are the optimum values of the two objectives when optimized independently, and w_{C} and w_{E} are corresponding weight coefficients reflecting the relative importance of the two criteria. The use of weight coefficients allows the planners to express their preferences regarding the criteria and must satisfy the constraint as shown in Eq. 5.35.

5.4 Structural Optimization Including Network Layout

5.4.1 Description of the Structural Optimization Problem

As illustrated above, a DER system is a system comprising a set of energy suppliers and consumers, district heating pipelines, heat storage facilities, and so on. The distributed energy production plants are built on locations defined in the regional plan. For the energy production technology, several options are available.

Compared with the electricity transmission, the transport and storage of heat are much more complicated and cost-intensive. The costs comprise the pipeline investment costs and the circulation water pumping costs, the costs for heat losses along the pipelines, as well as the exchanger equipment costs. Therefore, after the suitable technologies are recognized, the structure of the DER system is of vital importance. In this section, the above plan and evaluation model is extended to include the design of district heating network. In the model, production and consumption of electric

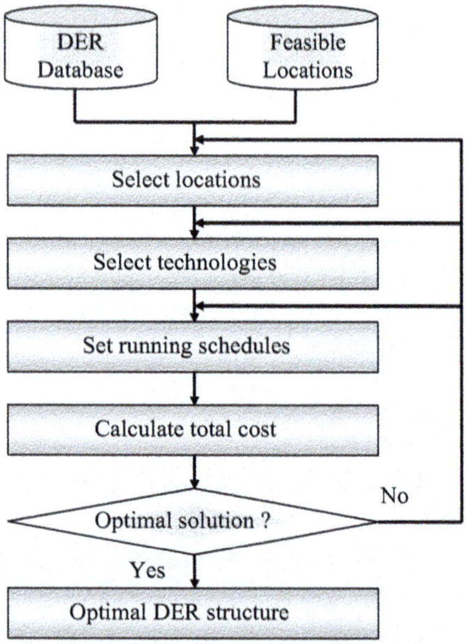

Fig. 5.8 Flowchart of DER structure optimization model

power and heat, power transmissions, transport of fuels to the production plants, transport of water in the district heating pipelines, and storage of heat are taken into account. The problem is formulated as a MILP problem where the objective is to minimize the overall cost of DER system, i.e., the sum of the running costs for the included operations and the annualized investment costs of the included equipment. The solution gives the DER structure, i.e., which production units, heat transport lines, and storages should be built as well as their locations, together with design parameters for plants and pipelines. Figure 5.8 shows the flowchart of the structure optimization model.

A geographical region comprising a set of energy suppliers that can supply heat and/or electric power and a set of energy consumers with given heat and power demands is considered. The region is shown schematically in Fig. 5.9. It is considered as self-content in respect of the total heating demand, while for the electric power, the demand can be satisfied with power produced within the region and/or with power purchased from the utility grid. It is purchased with given tariffs that may vary, for example, between day- and nighttime and between different seasons.

Heat is delivered to the consumers via a district heating network, which is defined in the model as a superstructure covering the connection options from suppliers to consumers. The line distances between suppliers and consumers are calculated prior to the optimization.

A supplier can be situated inside the region in a number of predetermined locations. There can be many types of heat or power production units or combinations of these. CHP is one option, boiler plant that produces district heat is another option, and wind power, PV, etc. are other options. Sites can have any number of any given types of production units, or a site can be given restrictions, for example, a coast site may have been reserved only for a wind power plant.

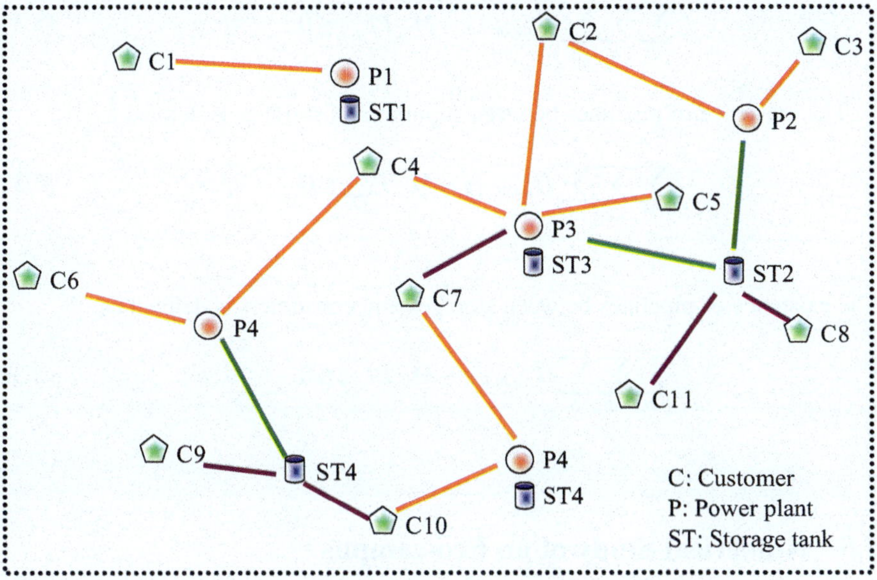

Fig. 5.9 Schematic view of a DER system with district heating network routes

Consumers can produce their own heat in competition to the district heat. If a heat pump is installed at a consumer site, there will be additional electricity consumption at that site. A consumer can also produce his own electricity by PV panels or the like, but the investment costs are for the time being still high compared with the main grid power or power from CHP plants.

5.4.2 *Mathematical Formulation of the Problem*

The objective function is to minimize the total cost for the regional energy system, as illustrated in Eq. 5.5. However, here, the pipeline investment cost should be included, which is defined as sum of the investment costs of different pipelines, as shown in Eq. 5.36.

$$C_{inv,P} = \left(\sum_p \sum_q Pf_{p,q} \cdot c_p \cdot l_{p,q} + \sum_p \sum_r Pf_{p,r} \cdot c_p \cdot l_{p,r} + \sum_r \sum_q Pf_{r,q} \cdot c_p \cdot l_{r,q} \right) \cdot YP \tag{5.36}$$

Besides the constraints illustrated in Sect. 5.2.4, the existence of various pipelines should be defined.

The existence of pipelines between supplier and consumers is defined as

$$\sum_{m}\sum_{d}\sum_{h}H_{p,q,m,d,h} - M \cdot Pf_{p,q} \leq 0 \qquad \forall p, q \tag{5.37}$$

The existence of pipelines between supplier and storages is defined as

$$\sum_{m}\sum_{d}\sum_{h}H_{p,r,m,d,h} - M \cdot Pf_{i,k} \leq 0 \qquad \forall p, r \tag{5.38}$$

The existence of pipelines between storages and consumers is defined as

$$\sum_{m}\sum_{d}\sum_{h}H_{r,q,m,d,h} - M \cdot Pf_{r,q} \leq 0 \qquad \forall r, q \tag{5.39}$$

5.5 Numerical Study of an Eco-campus

A numerical study of the model usage is presented below. It is an eco-campus in Kitakyushu, Japan [7–10]. In order to provide a comfortable living and working conditions with little environmental load, some DER technologies, such as gas engine, fuel cell, PV, etc., are considered for adoption. In the following, the results of single-objective optimization, multi-objective optimization, and structural optimization are presented and discussed [11–14].

5.5.1 Single-Objective Optimization Results

5.5.1.1 Setting of Scenarios

Here, only the economic objective function is considered. In order to investigate various aspects that affect the optimal adoption of DER system from the technical point of view, the following four scenarios have been assumed for analysis.

Scenario 1: Conventional system. It is the base scenario which indicates the conventional energy supply system. The electricity load is satisfied by utility grid and thermal load by gas boiler. No DER technology is considered.

Scenario 2: DER system without CHP technologies. All distributed generators except CHP technologies are accounted for. The electricity load can be served through either utility electricity or DER operation or a combination of both. However, the thermal load can only be supplied by the gas boiler as the conventional system. This option is considered to be suitable for locations without enough heat demands.

Table 5.1 Optimal system combinations

Scenarios	Installed capacity (kW)	System combination
Scenario 1	0	–
Scenario 2	3000	GE (3000)
Scenario 3	2683	H-GE (300), H-GE (2383)
Scenario 4	2683	H-GE (300), H-GE (2383)

Scenario 3: DER system with all distributed generators. Besides the technologies assumed in Scenario 2, the CHP technologies are also paid enough attention to. Under this consideration, the thermal load can be satisfied by the recovered heat from CHP plants, and the deficiency is supplemented by the gas boiler. It is expected to be a good alternative for sites (e.g., hospital, hotel, etc.) with enough thermal loads accompanying the electricity requirements.

Scenario 4: DER system with both distributed generators and storage technologies. It is an integrated consideration, in which both energy generation and storage technologies (thermal storage only) are considered. Under this consideration, by including the thermal storage tank, it is able to keep an inventory of heat for use in subsequent periods, and sites are able to lower costs even further by relying on storage. It may be a good consideration for sites with much thermal loads, which are not accompanied by the electricity demands.

5.5.1.2 Optimal System Combination

Table 5.1 shows the optimal system combination for various scenarios illustrated above. Generally, it can be found that the total installed DER equipment capacity is below the peak electricity demand (about 4000 kW) for the three scenarios considering DER technologies. It means that grid connecting may be always necessary from the economic viewpoint, although energy can be generated on-site.

According to Table 5.1, for Scenario 1, no DER technology is considered, and all energy requirements (both electricity and heat) are served through direct electricity and gas purchases. In the DER without heat recovery scenario (Scenario 2), a 3000 kW gas engine is selected. For Scenario 3, heat recovery is taken into consideration. The total installed capacity is reduced, with a value of 2683 kW. Two equipment are selected, a 300 kW gas engine and a 2383 kW gas engine. Both equipment can be used with heat recovery for heating but not for absorption cooling. Furthermore, as to Scenario 4 which considers not only heat recovery but also energy (heat only for this chapter) storage, the selected equipment are the same as Scenario 3. It means that the inclusion of heat storage has marginal effect on the introduction of DER technologies in this illustrative example. This is partly because of the relatively small heat to power ratio of the examined site.

In addition, in can be found that no fuel cell or PV is adopted. It is mainly because of their high capital cost. Therefore, from the economic view point, gas engine is the most popular DER technology currently. However, as more and more attention is paid to the global environment, and due the rapid technology development, it is believed that renewable energy technologies will become dominant in the near future.

5.5.1.3 Optimal Operation Strategies

Optimal operation schedule is one of the most important outputs of the plan and evaluation model. It is very important to decision-makers for the design of a DER system.

Taking the summer as an example, Fig. 5.10 illustrates the electricity balance for various scenarios. The cooling load is here considered as an electricity load and is complemented by electric chillers.

According to the figure, for Scenario 2, most electricity is supplied by DER equipment. In addition, it can be found that the gas engine operates only during the

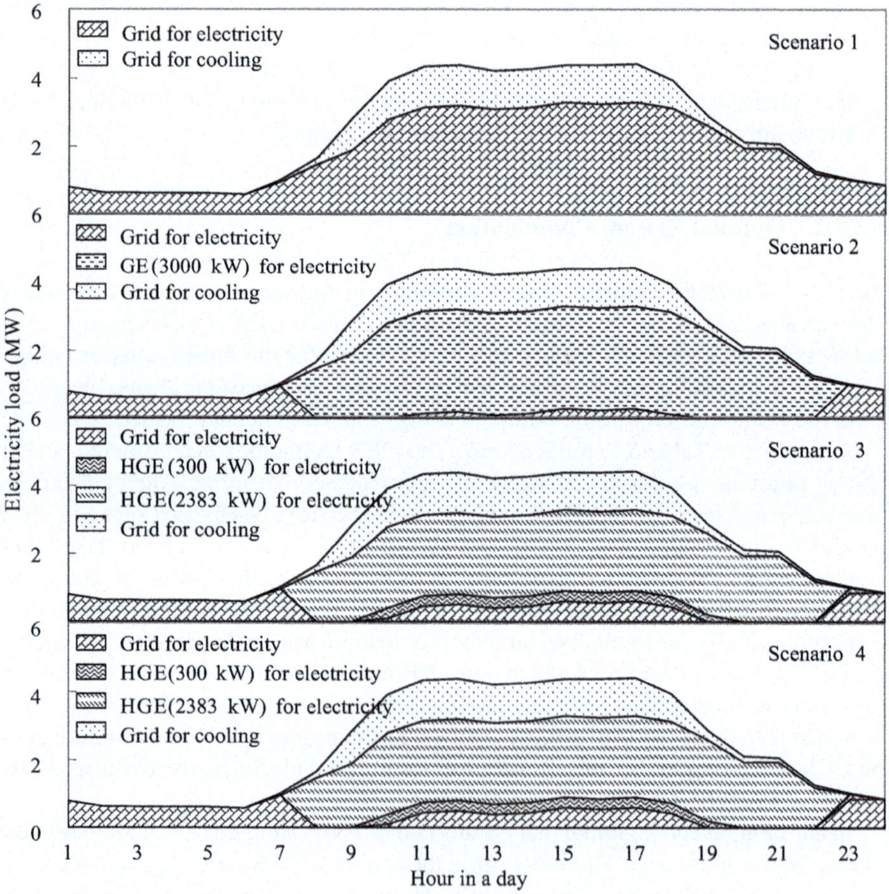

Fig. 5.10 Electricity balance for various scenarios in the summer

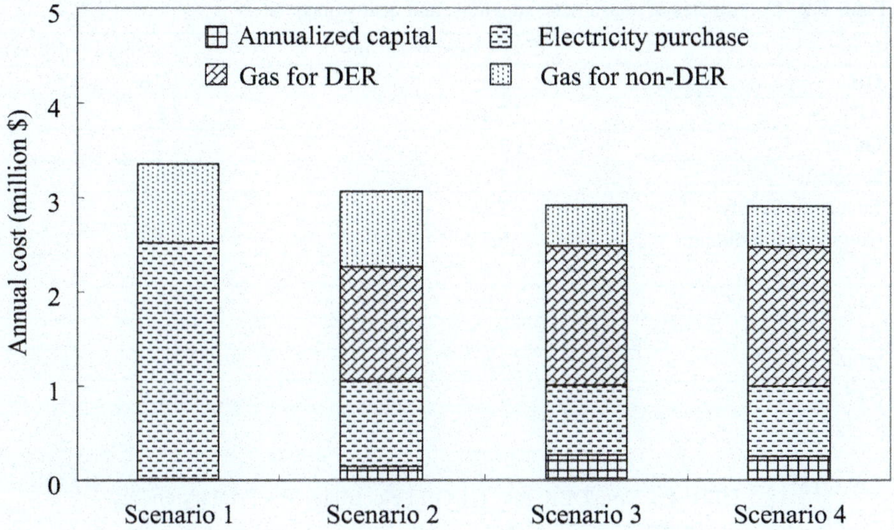

Fig. 5.11 Annual energy cost for various scenarios

daytime when tariff rate is relatively high. In the night, the customer prefers to purchase from the grid rather than generate on-site. Scenario 3 and Scenario 4 have similar running schedules. Both DER equipment operate during the daytime, and the 2383 kW gas engine has a longer operating time than the 300 kW gas engine.

5.5.1.4 Economic Analysis

Figure 5.11 illustrates the economic related results for various scenarios. The conventional system (Scenario 1) has a total cost of 3.9 million dollars. The introduction of DER equipment results in reduced electricity purchase and increased gas purchase. The overall cost is decreased through on-site generation. Compared with the conventional system (Scenario 1), the total costs of Scenario 2, Scenario 3, and Scenario 4 are reduced by 9.6%, 13.4%, and 13.4%, respectively.

Furthermore, it can be found that the adoption of CHP technology results in increased investment cost, although the total installed capacity is decreased (see Table 5.2). This is because of the high capital cost of the CHP technologies compared to the ones without heat recovery. However, it can be found that the total cost is reduced by about 0.2 million dollars. This is because the inclusion of heat recovery extends the operation of DER technologies, which leads to reduced grid electricity purchase and gas consumption for non-DER use. In addition, according to the figure, the consideration of heat storage (Scenario 4) does not lead to obvious economic merits. This is because of the similar system combination and operation schedule as illustrated above.

Table 5.2 Coordinates of consumers, suppliers, and heat storage sites

Item	x-Coord ($\times 10^2$ m)	y-Coord ($\times 10^2$ m)
DH	1.2	7.6
APT	7.4	8
Office	2	2
Hospital	5	6
Education facility	5.1	1
Commercial building	0.3	4.5
S1	4.8	7.2
S2	2.9	3.9
S3	7	2
ST1	4.8	7.2
ST2	4	1.9

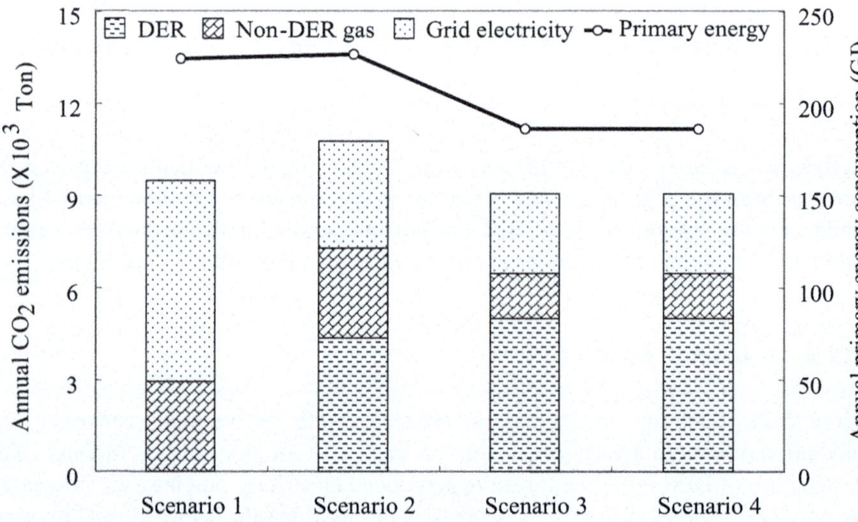

Fig. 5.12 Energetic and environmental characteristics

5.5.1.5 Environmental and Energetic Analysis

As shown in Fig. 5.12, carbon emissions and primary energy consumption resulting from various scenarios are analyzed. Compared with the conventional system (Scenario 1), all scenarios (except Scenario 2) result in reduced carbon emissions (although marginal) and primary energy consumptions. Annual CO_2 emissions of Scenario 2 are increased by 13.6%, and the primary energy consumption is also increased by about 1.1%. This is because without heat recovery, the efficiency of gas engine is relatively low, compared to either utility grid or gas boiler. It means that the introduction of DER system does not definitely lead to better environmental effects if only economic benefits are taken into consideration. In addition, it can be found that the inclusion of heat recovery (Scenario 3 and Scenario 4) results in considerable reduction amount of primary energy consumption. It is due to the higher total efficiency of the CHP plants. Therefore, it can be concluded that the installation of DER equipment using fossil fuels has more energy conservation merits than the environmental merits. In addition, it should be indicated that while evaluating the CO_2 emission reduction, the determination of reference system is of vital importance, which can change the results more or less.

5.5.1.6 Sensitivity Analysis

Sensitivity analyses on the model runs are performed to understand the influence of key parameters on the decision to install DER technologies and their resulting cost and environmental effectiveness. In the following, the sensitivity analyses are executed on the scale of energy demand, electricity and gas tariffs, as well as the carbon tax rate. All the technologies including heat recovery and storage are considered.

Energy Demand Scale Sensitivity

Energy demand is one of the uncertainty factors at the demand sites. In the following, the energy demands (both electricity and thermal) are assumed to have an increase until twice of current value, and the corresponding results are analyzed.

According to the results shown in Fig. 5.13, the installed DER capacity does not show a linear increase to the energy demand scales. For example, as the energy

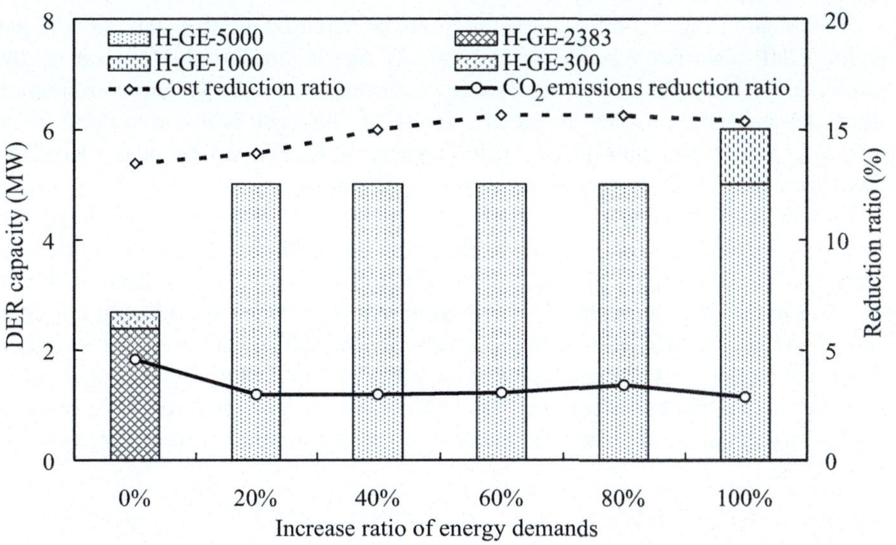

Fig. 5.13 Optimal system combination and corresponding economic and environmental effectiveness for various scales of energy demands

demand is increased by 20%, the total DER capacity has an increase from 2683 to 5000 kW. However, as the demand is further increased to 140% and even 180% of current value, no additional DER equipment is adopted. When the energy demand is twice of current value, besides the 5000 kW gas engine CHP plant, a 1000 kW one is also included.

Furthermore, from this figure, it can be found that the economic merit is always larger than the environmental one at various energy demand scales. This is partly because of the economic objective which has been assumed in this section. Compared with the base scenario, the cost reduction ratio is increased by about 2.2% as the energy demand is increased by 60%. As the further increase of energy demand, the cost reduction ratio has slight reduction, which means economies of scale cannot be actually realized while introducing DER system. As to the environmental merit, the reduction ratio of CO_2 emissions illustrates an uncertain trend as the increase of demand scale. However, the variation range is relatively small, which is between 2.8% and 4.5%.

Electricity Tariff Sensitivity

As discussed above, besides the energy demands, the operation of DER system is also subjected to the energy prices, which will affect the combination and capacity of DER system correspondingly. Along with the increased consumption of energy resources especially fossil fuels, the energy prices (e.g., electricity, gas, etc.) are expected to have a continued increase in the following years. Under this consideration, in this section, the sensitivity of electricity tariff, which is the most important factor affecting the total energy cost, is analyzed by increasing the rate until twice of current value.

As shown in Fig. 5.14, when the tariff rate is increased by 20%, the 300 kW gas engine CHP plant is replaced, and a 1000 kW one is introduced. However, as the tariff rate is further increased until twice of current value, the system combination illustrates no variation. This is because electricity buyback is not accounted for in this case study. The introduced DER capacity is limited to the peak electricity demand, which avoids the excess electricity generation.

Furthermore, looking into the figure, it can be found that the economic performance of the DER system is greatly affected by the tariff rate. As the tariff rate is as high as twice of current value, the cost reduction ratio is increased from 13.4% to 40.0%. On the other hand, the CO_2 emission reduction ratios almost keep the same level at different tariff rates, with the values around 5.0%. However, the increased tariff rate prolongs the operation time of DER system, which results in reduced power purchase and corresponding less CO_2 emissions. The reduced CO_2 emissions may be partly replaced by the increased city gas consumption due to more on-site generation.

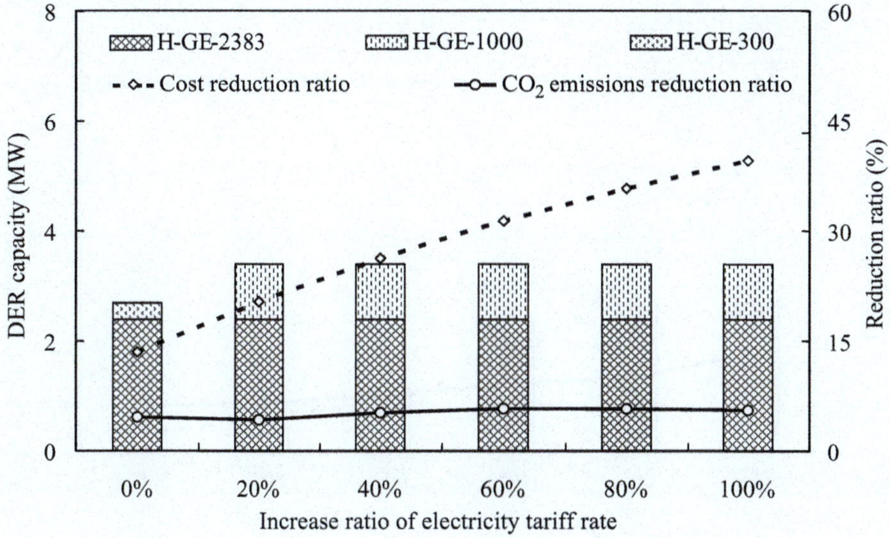

Fig. 5.14 Optimal system combination and corresponding economic and environmental effectiveness at various electricity tariff rates

City Gas Price Sensitivity

As one of the most popular fuels for DER technologies, city gas is considered to be a transition fuel before reaching a clean fuel society. In this section, most of the DER technologies are assumed to be fueled by city gas, and the economic system combination is considered to be partly affected by the gas price. According to the results shown in Fig. 5.15, the increase of gas price leads to reduced total DER capacity, which is 500 kW as the gas price is twice of current value. The optimal system combination also illustrates quite difference at different gas prices. For example, when the gas price is increased by 60%, two gas engine CHP plants (1000 and 100 kW) are adopted with a total capacity of 1100 kW. In addition, as the price is further increased to 180% of current value, two 300 kW gas engine CHP plants are the best combination.

Furthermore, according to the figure, the increase of gas price has a negative effect on the economics of DER system. As the gas price is twice of current value, total cost reduction ratio decreases from 13.4% to as low as 0.9%. The environmental merit is also reduced due to less DER adoption, with a reduced CO_2 emission reduction ratio from 4.5% to 2.1%.

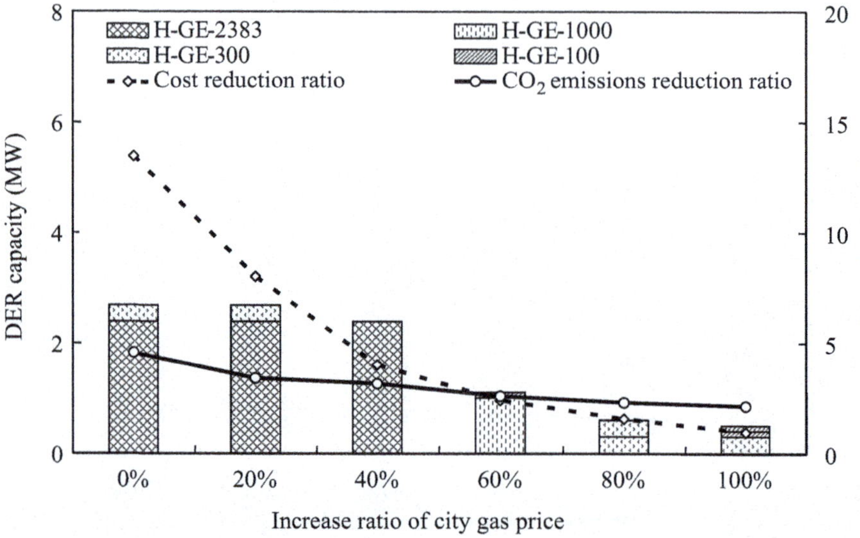

Fig. 5.15 Optimal system combination and corresponding economic and environmental effectiveness at various city gas prices

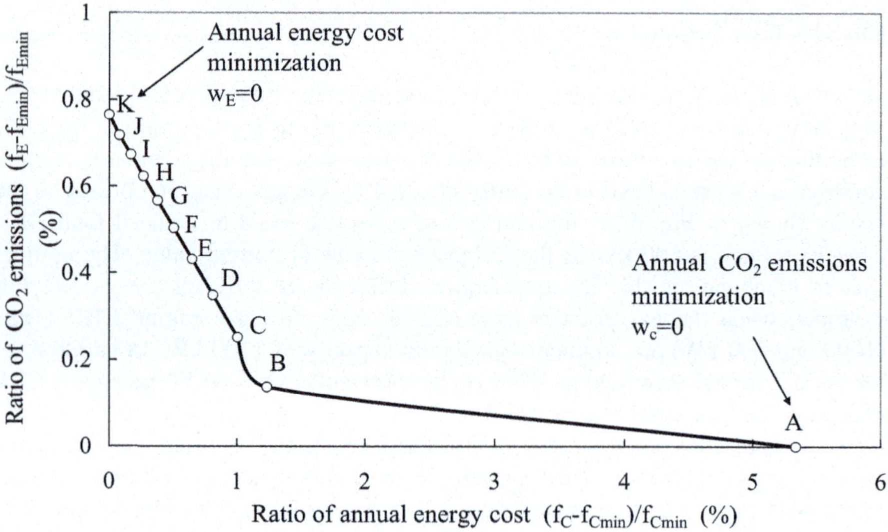

Fig. 5.16 Trade-off relationship between economic and environmental characteristics

5.5.2 *Multi-objective Optimization Results*

Here, both economic and environmental objectives are considered for the operation optimization of the DER system. The MOMILP solution for the operational optimization of the DER system produces a Pareto optimal front, as shown in Fig. 5.16, generated by 11 possible non-inferior solution points (or optimal operational strategies) for the competing objectives including energy cost and CO_2 emissions. The solution points A to K correspond to the weight coefficient of the economic objective at the range of 0 to 1. As observed, a high reduction in the energy cost is gained if the selection of solutions passes from the right extreme (CO_2 emissions minimization) to point B than when it changes from point B to the left extreme (energy cost minimization). This means that passing from solution A to B provides a small increment in the CO_2 emissions but produces a high decrease in the energy cost.

In order to understand the operational strategy that affects the economic and environmental characteristics, Fig. 5.17 illustrates the optimal operating results at various trade-off points. In general, due to the increase in economic objective satisfaction degree (from A to K), the share of electricity requirement from utility grid has an increase from 41.8% to 59.1%. Correspondingly, the share of thermal load from recovered heat is decreased from 42.8% to 26.5%. In addition, looking into the figure, it can be found that as the degree of economic objective satisfaction increases from 0 to 0.5 (from A to F), the share of thermal load from recovered heat has a decrease of 11.8%. However, when the satisfaction degree rises from 0.5 to 1 (from F to K), the recovered heat share is reduced by only 3.2%. This means that as more attention is paid to the environmental performance (near point A), the operation becomes more sensitive to the change of satisfaction degree of the objective.

Furthermore, according to the results obtained and presented in Fig. 5.17, the operating hours of gas engine reduce from 8760 to lower than 5000 as the objective is changed from emissions minimization to cost minimization. As to the fuel cell, when the satisfaction degree of economic objective increases from 0 to 0.1 (from A to B), the running hours have a sharp decrease from 6866 to 5125. However, due to the further increase in the satisfaction degree, the running hours keep at a steady level. It means that compared with fuel cell, gas engine is more sensitive to the change of optimization objectives. This is because on the one hand, as the environmental objective is paid the main attention, gas engine has longer running time than fuel cell due to its higher total efficiency which is the determining factor of the environmental performance; on the other hand, as the economic objective is focused, the running hours of gas engine are reduced below that of fuel cell due to its lower power efficiency which is recognized as the main factor affecting the economic performance.

5.5.3 Structure Optimization Results

5.5.3.1 Description of the Case Study

An optimal structural design method aims to determine the structures of energy supply systems in consideration of the multi-period operation. As illustrated in Sect. 5.4, the optimization problem can be formulated as a MILP problem with binary variables for selection and on/off status of operation of equipment and continuous

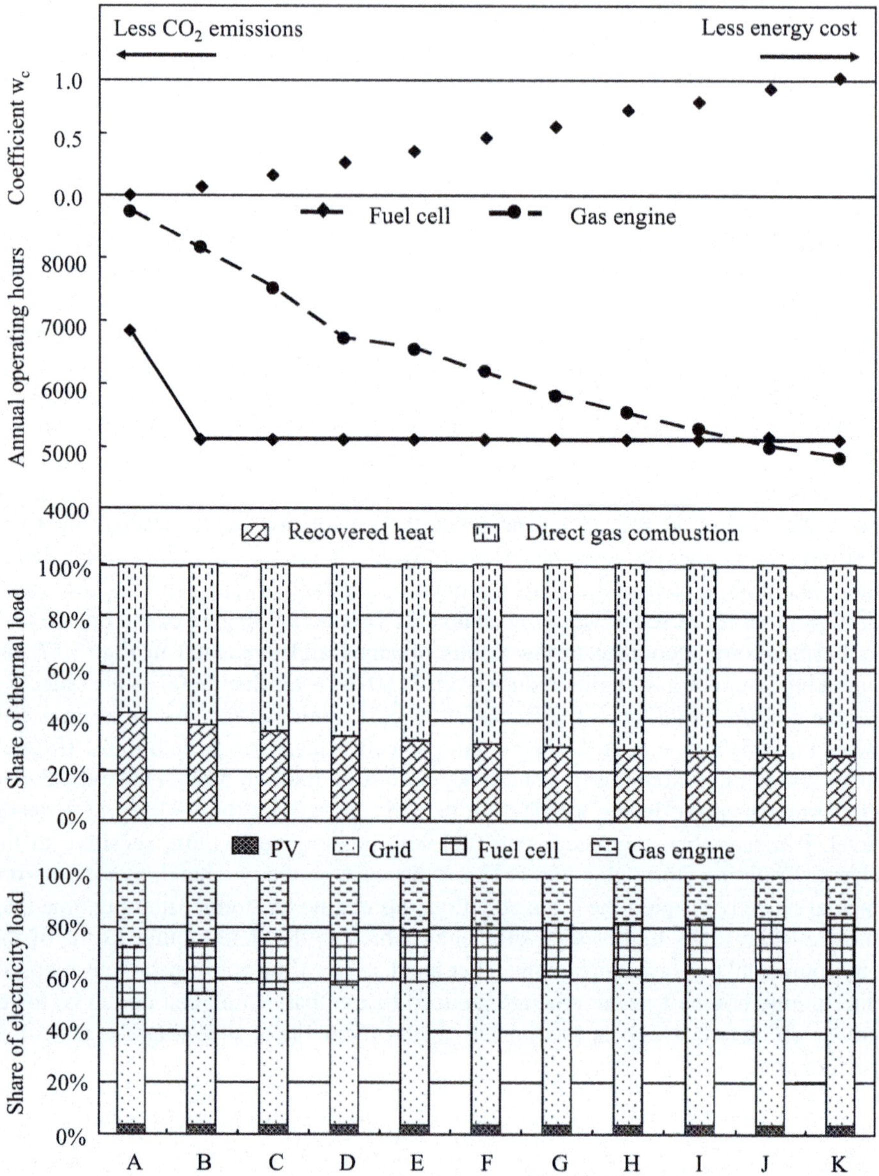

Fig. 5.17 Optimal operating results at various trade-off points

variables for capacities and load allocation of equipment. In addition, the planning of the path for the heat piping network is taken into consideration.

An illustrative example of the model usage is presented below. The case is an eco-campus which has been introduced in Sect. 5.2 in detail. The graphic image of the district is shown in Fig. 5.18. Three available power and heat production sites are

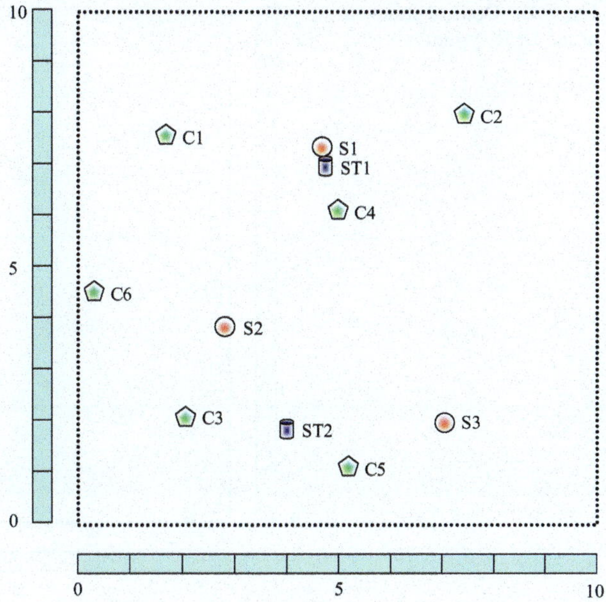

Fig. 5.18 Graphic image of the research object

determined in the regional plan for six main consumer sites. Heat storage can be built in two sites.

At each production site, two alternative production units are allowed: a CHP plant and a boiler plant with only heat production. Any combination of these two is allowed at a site. A site can also be left unbuilt. Heat losses from the pipelines are not considered in the example.

Table 5.2 presents the coordinates for the consumer, supplier, and heat storage sites. Coordinates are defined starting from the southwest corner of the region.

5.5.3.2 Optimal Energy System Structure

Figure 5.19 shows the obtained structure for the energy system. The district heating lines are shown. Two production plants, S1 and S2, and one heat storage, ST1, are obtained in the solution. At the site S1, a 440 kW CHP plant and a 964 kW gas boiler are selected. At the site S2, a 14 kW CHP plant and a 6 kW gas boiler are determined. Furthermore, a storage tank of 600 kWh is used in the ST1 site. In can be found that from the economic point of view, the site S1 is selected as the main source for supplying the heat requirements in the district. The customer C2 (APT) and C6 (hospital) are supplied directly by the source at site S1. The customer C1 (DH), C4 (office), and C5 (education facility) are satisfied by the heat from the storage tank. The capacity of production plant at site S2 is relatively low. It is used for all the heat demand of customer C3 (commercial building) and part of the heat load of customer C1 (DH).

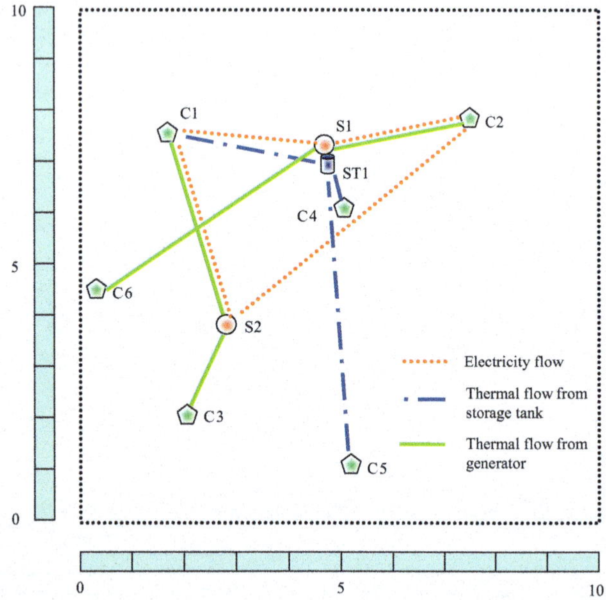

Fig. 5.19 Optimal structure of the DER system

5.6 Nomenclature

Name	Definition
C	Cost, $
CLoad	Customer load, kW
CTax	Carbon tax, $/kg-C
d	Day in a month
D	Indicator variable for the state of equipment operation
ECInt	Carbon intensity of electricity, kg-C/kWh
EDchar	Regulated demand charge of electricity, $/kW
Eff	Efficiency of DER technology, %
EGen	Power generation, kW
Eprice	Tariff rate for electricity purchase, $/kWh
ESal	Buyback electricity, kWh
ESMax	Maximum dispatch level of energy storage, kWh
ESMin	Minimum dispatch level of energy storage, kWh
f	Index of fuel type
FCInt	Carbon intensity of fuel, kg-C/kJ
FCostf	Fixed part in capital cost of DER equipment, $
FCostv	Variable part in capital cost of DER equipment, $/kW
FDchar	Regulated demand charge of fuel, $/kJ
FLTime	Lifetime period of equipment, year
FMaxp	Nameplate power rating of equipment, kW

Name	Definition
FMinp	Minimum load rating of equipment, kW
Fprice	Fuel volumetric charge, \$/kJ
FStart	Indicator variable for the start of DER technology
FStop	Indicator variable for the stop of DER technology
FSubs	Subsidies for DER equipment, \$/kW
h	Hour in a day
i	Index of DER technology
IEne	Energy stored into the storage unit, kW
IRate	Interest rate, %
j	Index of storage type, electricity or heat
k	Power generation coefficient of wind turbine
m	Month in a year
M	A very large number
NInv	Number of adopted DER equipment
OEne	Energy output from storage unit, kW
OMf	Fixed operation and maintenance cost, \$/kW
OMv	Variable operation and maintenance cost, \$/kWh
Operate	Number of units under operation
p	Number of energy carriers
PElec	Purchased electricity from the grid, kW
PFuel	Fuel consumption, kJ
PVp	Power generation of PV technology, kW
q	Number of total DER technologies
R	Irradiation data, kW/m^2
RHeat	Recovered heat from DER equipment, kW
S	Number of possible systems
SEne	Amount of stored energy, kW
Sprice	Electricity selling price, \$/kWh
u	End uses, including electricity, cooling, heating, and hot water
V	Wind speed, m/s
Wp	Power generation of wind technology, kW
Greek symbols	
α	Heat recovery efficiency, %
β	Heat efficiency from direct fuel consumption, %
γ	Utilization efficiency of recovered heat, %
δ	Utilization efficiency of stored energy, %
ε	Storage coefficient, %
θ	Minimum percentage of electricity purchase, %
ω	Unit start and stop cost, \$
Subscripts	
C	Cut-in
Ctax	Carbon tax
Elec	Grid electricity

Name	Definition
F	Cutoff
Fuel	Fuel consumption
Inv	Investment
OM	Operation and maintenance
R	Rated value
Sal	Sales
Ss	Start and stop
Total	Total cost

References

1. Mancarella P, Chicco G (2009) Global and local emission impact assessment of distributed cogeneration systems with partial-load models. Appl Energy 86(10):2096–2106
2. Ruan Y, Liu Q, Zhou W, Firestone R, Gao W, Watanabe T (2009) Optimal option of distributed generation technologies for various commercial buildings. Appl Energy 86(9):1641–1653
3. Medrano M, Brouwer J, McDonell V, Mauzey J, Samuelsen S (2008) Integration of distributed generation systems into generic types of commercial buildings in California. Energy Build 40(4):537–548
4. Dicorato M, Forte G, Trovato M (2008) Environmental-constrained energy planning using energy-efficiency and distributed-generation facilities. Renew Energy 33(6):1297–1313
5. Chen H, Ooka R, Iwamura K, Huang H, Yoshizawa N, Miisho K, Yoshida S, Namatame S, Sakakura A, Tanaka S (2008) Study on sustainable redevelopment of a densely built-up area in Tokyo by introducing a distributed local energy supply system. Energy Build 40(5):782–792
6. Zhou N, Marnay C, Firestone R, Gao W, Nishida M (2006) An analysis of the DER adoption climate in Japan using optimization results for prototype buildings with U.S. comparisons. Energy Buildings 38(12):1423–1433
7. Kashiwagi T (2002) Natural gas cogeneration plan/design manual 2002. Japan Industrial Publishing Co. Ltd, Tokyo
8. Ojima Lab (2002) Consumption unit of electricity, heating, cooling and hot water. Waseda University, Tokyo
9. Kyushu Electric Power Co., Inc. Available at: http://www.kyuden.co.jp/index.html
10. Saibu Gas Co., Ltd. Available at: http://www.saibugas.co.jp
11. Energy and Environmental Analysis, Inc (2003) Market potential for advanced thermally activated BCHP in five national account sectors. Energy and Environmental Analysis, Inc, Arlington
12. Marnay C, Venkataramanan G, Stadler M, Siddiqui A, Firestone R, Chandran B (2007) Operation of microgrids in commercial buildings, LBNL_62315. LBNL, Berkeley
13. Wahlund B, Yan J, Westermark M (2004) Increasing biomass utilisation in energy systems: a comparative study of CO2 reduction and cost for different bioenergy processing options. Biomass Bioenergy 26(6):531–544
14. Möllersten K, Yan J, Westermark M (2003) Potential and cost-effectiveness of CO2 reductions through energy measures in Swedish pulp and paper mills. Energy 28(7):691–710

Chapter 6
Integrated Plan and Evaluation of Distributed Energy System by Area Energy Network in Smart Community Toward Low Carbon Society

Liyang Fan

6.1 Distributed Energy with Area Energy Network

6.1.1 The Trend of Distributed Energy and Area Energy Network

Distributed energy systems (DES) have been drawing increasing attention as a substitute for grid in the low carbon society development [1, 2]. Compared with the traditional centralized energy supply system, the distributed energy generations are easy for renewable energy use and can avoid the loss in energy delivery as well. As the integration of distributed energy generation become major concerns, one problem occurred that the conventional energy supply model, the unidirectional top-down grid could hardly be multipurpose to it [3]. It can only support the energy flow from the energy station to static users. A much smarter energy supply system will be desirable to support multidirectional energy flows that can dynamically switch between the user and local energy providers. It needs for more observable, accessible, and controllable network infrastructures. The future energy system, termed as area energy system with smart city technologies, is the emerging system.

Area energy network refers to the idea of mutually accommodating electricity and heat among multiple buildings. As Fig. 6.1 displayed, in conventional system, buildings use electricity and heat individually. However, in an energy system based on area energy network, electricity and heat are distributed among multiple buildings to level out usage patterns of electric energy and thermal energy.

L. Fan (✉)
Nikken Sekkei Research Institute, Tokyo, Japan
e-mail: fan.liyang@nikken.jp

W. Gao (ed.), *Distributed Energy Resources*, Green Energy and Technology,
https://doi.org/10.1007/978-3-031-21097-6_6

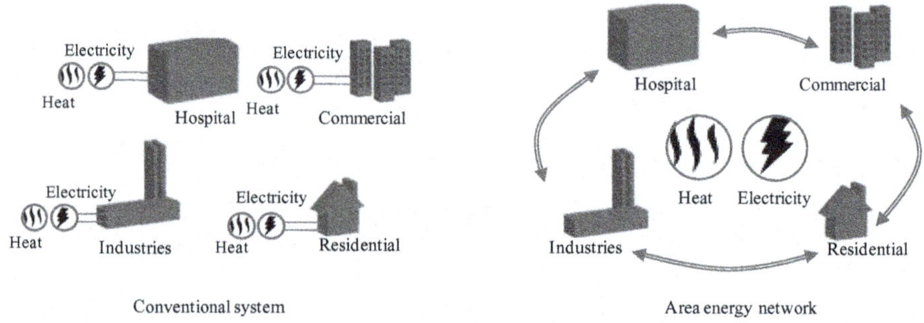

Fig. 6.1 The conventional system and area energy network

6.1.2 General Structure of Area Energy Network

The area energy network within microgrid and distributed generation paradigm should be recognized as the future power paradigm. The model discussed in this chapter is not an isolated system but connected with the conventional system. The system can both import and export surplus electricity at times of low local demand. Area energy network is a network of energy chains, starting from the primary energy supply and ending in the end-user sectors. The main parts of area energy network are primary energy supply sector, energy transformation sector, energy distribution sector, and end-user sector.

Primary Energy Supply Sector
Primary energy supply sector includes the nonrenewable energy (fossil fuel, etc.), renewable energy (solar, biomass, etc.), and untapped energy (industrial by-pass, water energy, etc.).

Energy Transformation Sector
The small-scale distributed technologies will be dominated in the area energy network model with distributed energy generation. The energy transformation sectors as CHP plant make use of the energy from primary energy supply sector and transfer them to the energy in the local network, the energy distribution sector in area energy network.

Energy Distribution Sector
The energy distribution sector is an essential part in the area energy network. It connected the distributed energy generations to form a network with the electricity and heat sharing. The optimization of the distribution sector with the intelligent control can level out the load fluctuation and cut the capital investment, rewarding both in environmental aspect and economic aspect.

End-User Sector

The integrated energy system can supply customers with different kinds of end-use services. In area energy network, the end users usually formed with more than one building. Load profiles represent the basis for generation and distribution system sizing and economic dispatch optimization [4]. Design and plan in the end-user sector is the base for the area energy network.

6.1.3 Classification of Area Energy Network with Heat

Various research about the area energy network is mostly focused on electricity supply chain, among which the heat supply system can hardly be mentioned. This chapter focuses more on the area energy network from the aspect of heat, as heat is also an important aspect for improving the efficiency of area energy network.

From the viewpoint of heat utilization, the area energy network in Japan is classified into three types, according to the scale: the district heating, centralized plant, and the interchanging energy system.

1. **District Heating (Cooling)**

District energy systems (DES), displayed in Fig. 6.2, centralize the production of heating or cooling for a neighborhood or community. It is a large energy plant, supplying energy to a board district. The supply regulations are based on the heat supply business law. The energy is supplied by heat supply companies that are set by the law, and supply obligation is based on the law, defining the conditions by the supply rules.

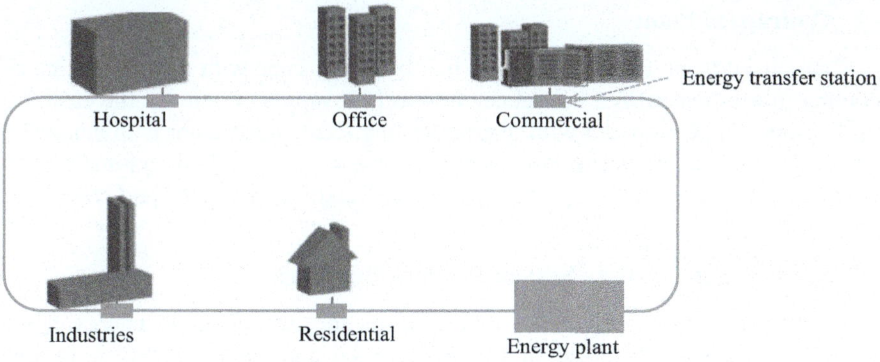

Fig. 6.2 The concept of district heating (cooling) system

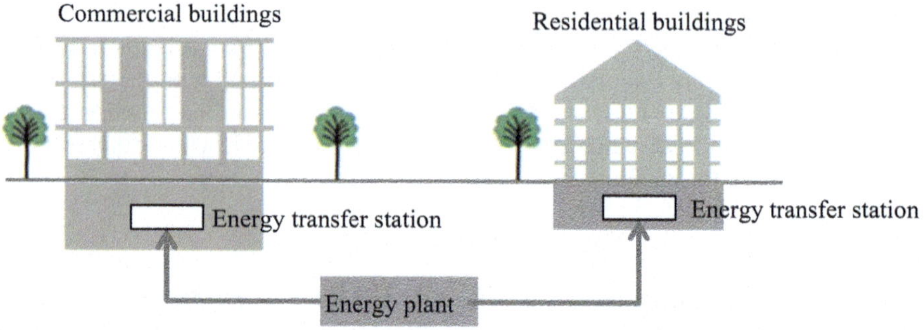

Fig. 6.3 The concept of centralized plant

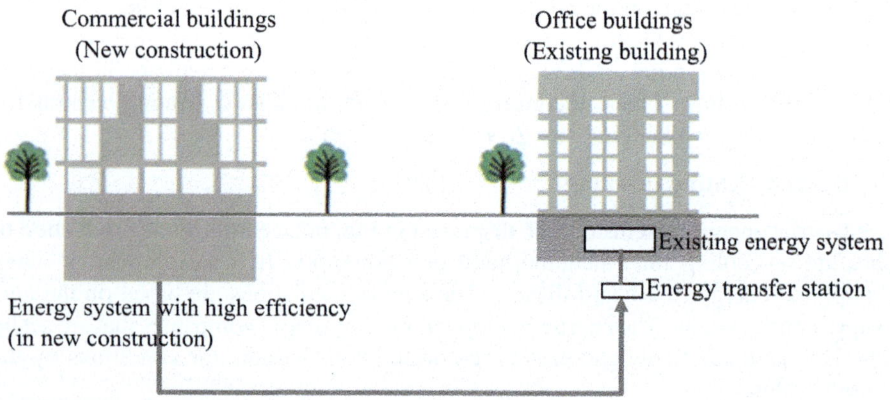

Fig. 6.4 Interchanging energy system between buildings

2. **Centralized Plant**

Centralized plant, displayed in Fig. 6.3, is almost same with district heating and cooling. It uses centralized heat generation as heat supply system for the customers in the same site. Compared with district heating, centralized plant is in small scale and excluded from the heat supply law. This type is usually called regional heat and cooling, widely used in the residential area, universities, hospitals, and commercial areas.

3. **Interchanging Energy System Between Buildings**

Compared with the other two types before, the interchanging energy system between buildings (displayed in Fig. 6.4) is the area energy network in small scale, with the energy interchanging between buildings. In this kind of project, the nearby buildings are connected with pipes and share the heat. The characteristic of this type is that the buildings in this area network have their own heat generation system as the base heat supply. In other words, the buildings in this type can also

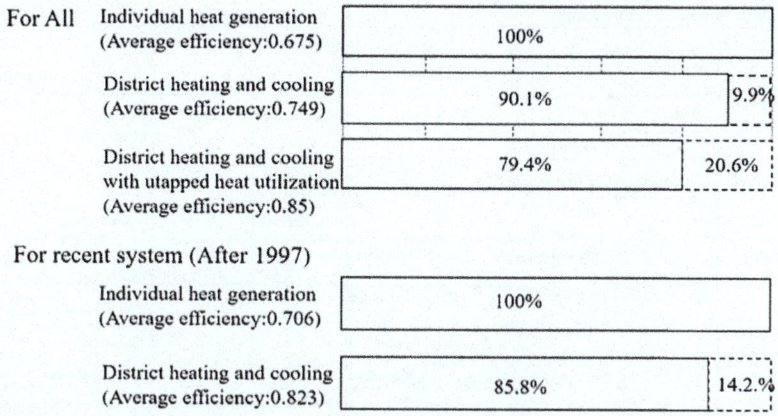

Fig. 6.5 The energy saving effect of area energy network

work individually. However, the efficiency of the heat supply system in the existing buildings can be improved by highly efficient equipment introduction in the new constructions and the energy interchanging management.

6.1.4 Energy Conservation Effect of Area Energy Network

The area energy network aggregates the energy load in the region. In that case, it can adopt more efficient energy generation plant and can manage the system more efficiently to support part load by appropriate equipment division. The optimization of the system construction and controlling can improve the energy saving effect.

The survey [5] in the year 2008, displayed in Fig. 6.5, suggested that the district heat supply system can save 9.9% primary energy compared with the individually heat supply system. Recent systems built after the year 1997 have higher energy efficiency and can save 14.2% primary energy. Further, the distributed heat energy system can utilize the untapped heat resource and realize the 20.6% primary energy conservation.

6.1.5 Planning and Design of Area Energy Network Procedure for Its Introduction

Procedures for the introduction of area energy network are different according to its type, district characteristic, and system form. Usually, the procedure is from the basic concept and the plan of the business, the system design, construction, operation and maintenance, evaluation, and system renovation.

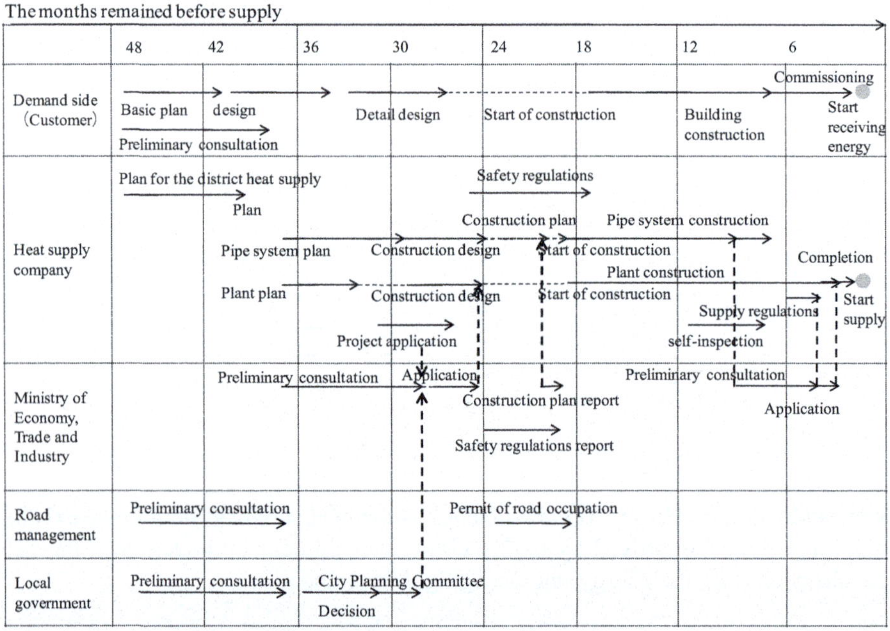

Fig. 6.6 Time schedule for area energy network construction

The area energy network is proposed firstly based on the city development, as the construction or renewal of the urban blocks. Then the developers will analyze the surrounding heat resource and the situation of the heat supply, deciding the type of the project. Figure 6.6 displayed the flow and time schedule for area energy network construction.

The area energy network needs detailed planning during every stage for the survey on the untapped energy use, energy demand side prediction, and plant system setting. Further, it needs more information about the characteristic of the district and the urban plan information to make the project more practicable. Figure 6.7 displayed the considerations in the area energy network planning decisions.

6.2 A Model for Area Energy Network by Offline Heat Transport System and Distributed Energy Systems

Based on the normal area energy network with hit utilization, the heat supply should also convey the concept of distributed energy generation (DEG), demand energy storage (DES), and demand side response (DSR), under the concept of smart city. It can make use of onsite exhaust heat, such as recovery heat of CHP plant and nearby factory exhaust heat (FEH) [6]. It should be a dynamic controllable as well, which can smooth out the heat fluctuation.

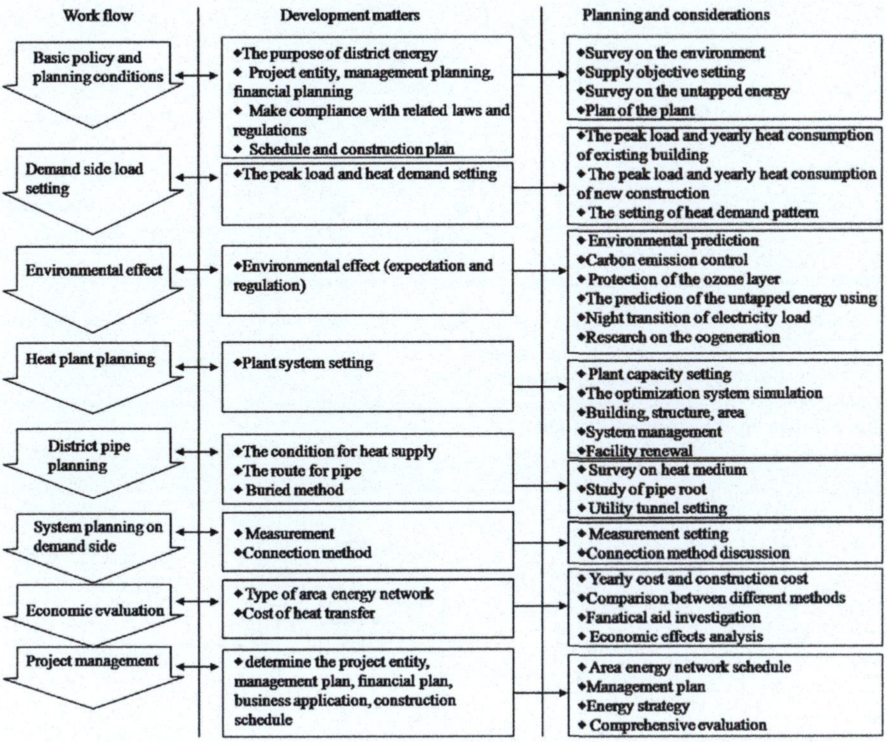

Fig. 6.7 Considerations in the area energy network planning decisions

In this part, we develop an area energy network by offline heat transport system and distributed energy systems, with the concept of DEG, DES, and DSR.

6.2.1 The Demand-Response Network Model

The demand-response network (DRN) model for smart community is a treelike hierarchical model that comprises the community energy management system (CEMS), energy station (ES), and building energy management system (BEMS).

Figure 6.8 demonstrated the hierarchical model. The end users (managed by BEMS) reside at the bottom of the hierarchy. They will be prioritized and organized into groups. Every group is managed by one ES. The ES is at the lowest rank unit for the energy strategy decision that controls the introduction of DEG, DES, and DSR. The ES collects information of the energy generation and consumption in the group and sends signals to the CEMS. The CEMS connected with each other and formed city energy network, which is organized in a topological structure. ES is assumed to

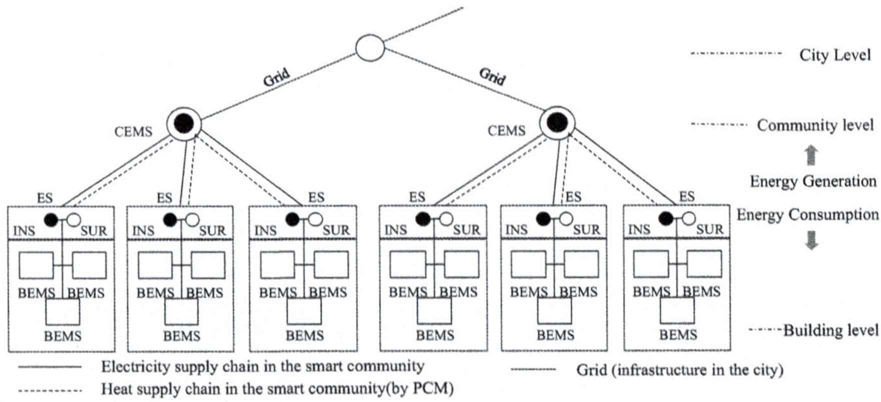

Fig. 6.8 The treelike hierarchy of DRS

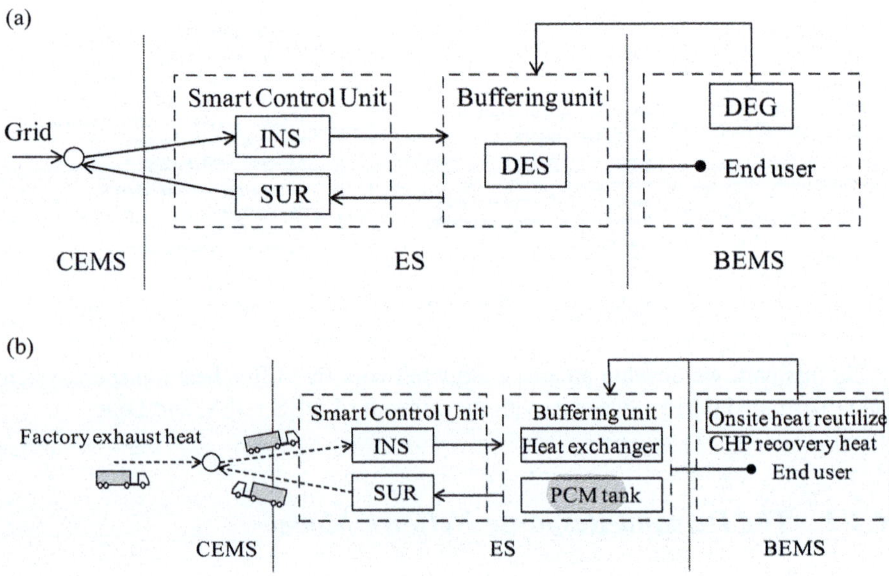

Fig. 6.9 DRN system energy supply chain (**a**) Electricity supply chain (**b**) Heat supply chain

have two modes, the energy surplus mode (SUR) and the energy insufficient mode (INS). The ES can dynamically switch its mode depending on the energy generation and consumption in the group. The mode signal will send to CEMS who collects and distributes the energy.

The proposed DRN system in this research is different from the smart grid that it not only has electricity supply chain but also has heat supply chain. Figure 6.9a illustrated the electricity supply chain. The DES in DRN system only acts as a backup and the buffer unit. The energy produced by DEG is supplied to the end users from the buffer unit. When the energy generated by DEG is more than the energy consumption in the group, the ES will be in SUR mode and become an energy supplier to other ES. Oppositely, the ES will be in INS mode when the energy generated by DEG is less than the energy consumption and becomes an energy consumer. Figure 6.9b illustrates the heat supply chain. Similar with the electricity supply chain, it has a buffering unit that comprises the PCM tank and the heat exchanger. Under the INS mode, the CEMS will transport heat to the ES by trucks with PCM tank. Under the SUR mode, the tank in the buffer unit will collect the surplus heat and be transported to other ES when it received the order from the CEMS. The mode signals in the heat supply and that in the electricity supply are self-governed.

6.2.2 District Energy Using Concept and Operation Hypothesis

As the DRN system illustrated before, the building in the community will be divided into groups. Every group is managed by ES, the basic unit to make energy strategies. Figure 6.10 describes the district energy using concept.

1. Introduction of the renewable energy: all the buildings will be introduced with PV system. The electricity generated by PV system will be preferentially used by the building themselves and the remaining electricity will be sent back to the grid.
2. Introduction of the CHP system: the CEMS will characterize buildings by their demand types. The buildings have both high electricity consumption and heat consumption (such as commercial buildings and public buildings) which will be introduced with the CHP system, named as CHP group. The capacity of the CHP system is set as electricity peak load of the group. The buildings without CHP

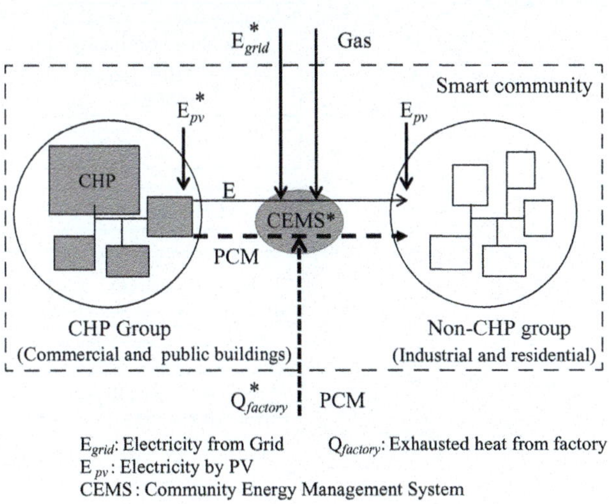

Fig. 6.10 District energy using concept

system is considered as non-CHP (NCHP) group. The electricity produced by CHP plant will satisfy themselves first and then send the remaining electricity to NCHP system. The CHP group will generate all their own demand beside PV. Therefore, as the DRN described before, the electricity of these groups is only in SUR mode. The NCHP group will be in INS mode if PV cannot afford their electricity consumption.

The CEMS will manage the model signal and control and dispatch the electricity. It will preferentially use the DEG, thus maximum output of CHP plant. The electricity produced by PV can sell back to grid, but the electricity produced by CHP plant cannot. In that case, when the electricity generated by the CHP is more than the district electricity demand, the CEMS will restrict the CHP output. It will preferentially choose the CHP plant with higher efficiency and lower down the CHP plant with low efficiency. If the efficiencies of the CHP plants are the same, CEMS will cut down the CHP plants in same rate.

3. Reutilization of the onsite exhaust heat and the FEH: the recovery heat of the CHP system will be preferentially used by the group first. However, if the recovery heat is more than the heat demand, the heat supply mode of the ES will turn to SUR. This part of heat surplus will be collected by CEMS.

Further, the CEMS will select out the FEH resource based on the characteristic of the PCM system, which collects the FEH and utilizes it in the community.

The onsite CHP exhaust heat and the FEH stored in the PCM system will be preferentially used. The CEMS will distribute the heat according to the SUR signal from the ES. It will be sent to the ES which has higher amount of heat insufficiency.

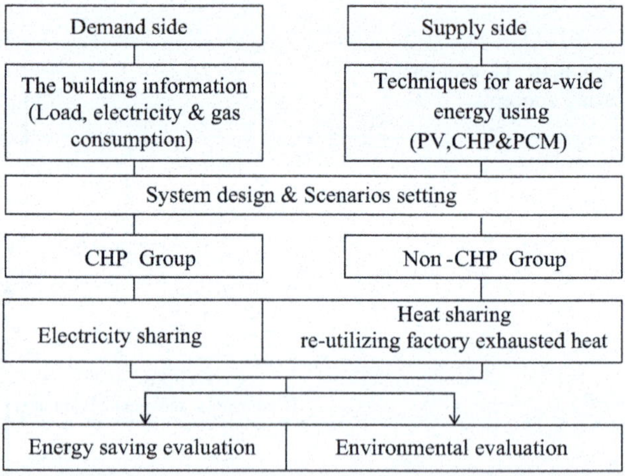

Fig. 6.11 The simulation flow

6.3 Energy Balance Management and Simulation Modeling

6.3.1 Simulation Flow

The energy balance management and the simulation flow are conducted as in Fig. 6.11. The simulation of the DRN system is also a bottom-up model. Firstly, based on the district zoning, the research will estimate the building energy consumption and describe profiles by groups. As the treelike hierarchy described in the second part, buildings in one group will be managed by one ES. Secondly, the CEMS will characterize the groups by its energy consumption character and introduce proper DEG in every ES: some are with CHP system, but some are not. The simulation separated them into CHP group and the NCHP group. Thirdly, the research executed the simulation. During the simulation, ES will dynamically switch between the INS mode and the SUR mode by estimating the energy consumption and the generation. Finally, the research will calculate the primary energy consumption and evaluate the environmental effect of every technology as well as the whole community.

6.3.2 Estimation of District Energy Consumption

The energy consumption of the whole community $\left(E_{demand}^{community}\right)$ is calculated as follows (6.1):

$$E_{demand}^{community} = E^1 + E^2 \ldots\ldots + E^n = \sum_n \sum_m \sum_d \sum_i \left(ELEC_{mdh}^n + HEAT_{mdh}^n\right) \quad (6.1)$$

$ELEC_{mdh}^n$ is hourly electricity load, calculated as

$$ELEC_{mdh}^n = \sum_k e_{mdh}^n \times s_k \quad (6.2)$$

$HEAT_{mdh}^n$ is hourly heat load, calculated as (6.3)

$$HEAT_{mdh}^n = \sum_k h_{mdh}^n \times s_k \quad (6.3)$$

n is the group number.
m is month; d is date; h is hour.
E1...En is the energy consumption of every group.
e_{mdh}^n and h_{mdh}^h are the energy consumption unit in Kyushu area, Japan [7].
k is the building function.
S_k is the building area for one function (k).

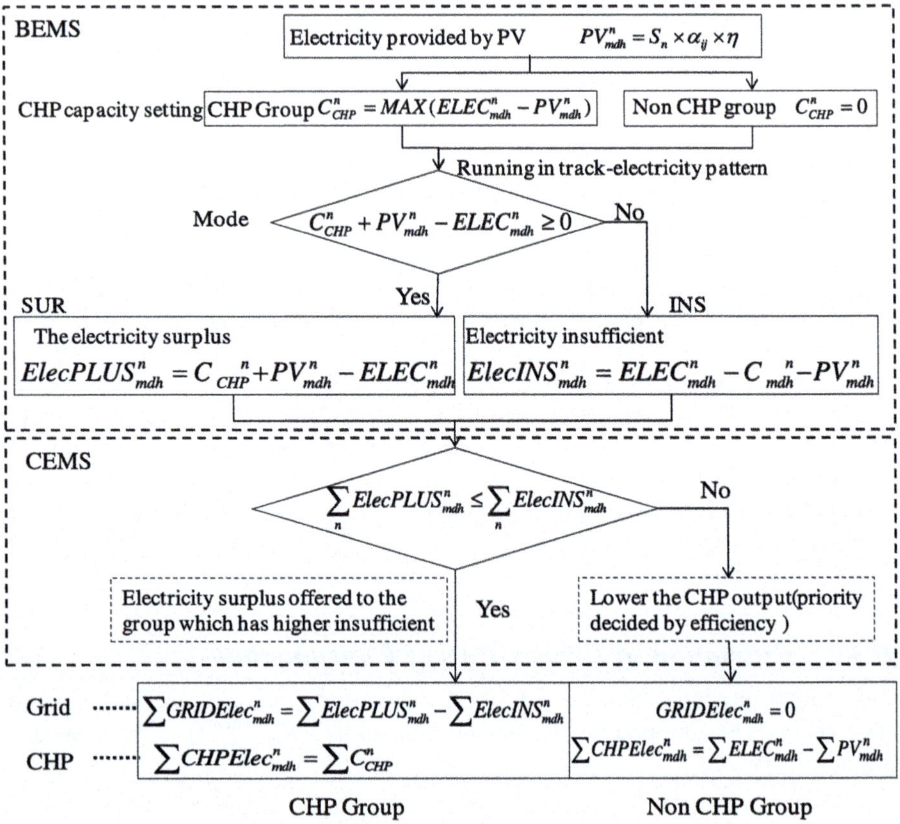

Fig. 6.12 The electricity balanced model

6.3.3 The Electricity Balanced Management

Figure 6.12 illustrates the simulation model for the electricity balance.

The buildings will preferentially use the electricity produced by PV.

The electricity produced by PV system in one group $\left(PV^n_{mdh}\right)$ is calculated as follows (6.4):

$$PV^n_{mdh} = S_n \times \alpha_{mdh} \times \eta \tag{6.4}$$

S_n is the area for PV penal in a group (n).
α_{mdh} is the hourly sun radiation rate [8].
η is the efficiency of the PV penal [8].

The CHP capacity $\left(C^{n}_{CHP}\right)$ is decided as (6.5).

$$C^{n}_{CHP}=\begin{cases}\mathrm{MAX}\left(ELEC^{n}_{mdh}-PV^{n}_{mdh}\right) & (CHPgroup)\\ 0 & (NCHPgroup)\end{cases} \tag{6.5}$$

The ES will decide the mode by the prediction of electricity load profile of the CHP system, PV system, and electricity demand.

When $C^{n}_{CHP}+PV^{n}_{mdh}-E^{n}_{mdh}\geq 0$, the group is in SUR mode. The expected surplus electricity $\left(ElecPLUS^{n}_{mdh}\right)$ is calculated as follows (6.6):

$$ElecPLUS^{n}_{mdh}=C^{n}_{mdh}+PV^{n}_{mdh}-E^{n}_{mdh} \tag{6.6}$$

On the contrary, when the group is in INS model, the expected electricity insufficiency $\left(ElecINS^{n}_{mdh}\right)$ is calculated as follows (6.7):

$$ElecINS^{n}_{mdh}=E^{n}_{mdh}-PV^{n}_{mdh}-C^{n}_{mdh} \tag{6.7}$$

If $\sum_{n}ElecPLUS^{n}_{mdh}\leq\sum_{n}ElecINS^{n}_{mdh}$, CEMS would lower down the total CHP output (prior use of the equipment with higher efficiency). Under this situation, there was no electricity supplement from the grid. The electricity generated by CHP plant $\left(CHPElec^{n}_{mdh}\right)$ is calculated as follows (6.8):

$$\sum CHPElec^{n}_{mdh}=\sum ELEC^{n}_{mdh}-\sum PV^{n}_{mdh} \tag{6.8}$$

If $\sum_{n}ElecPLUS^{n}_{mdh}\leq\sum_{n}ElecINS^{n}_{mdh}$, the surplus electricity from CHP group will be offered to the NCHP group. Under this situation, the electricity from the grid $\left(GRIDElec^{n}_{mdh}\right)$ is calculated as follows (6.9):

$$\sum GRIDElec^{n}_{mdh}=\sum ElecPLUS^{n}_{mdh}-\sum ElecINS^{n}_{mdh} \tag{6.9}$$

Electricity offered by CHP is calculated as follows (6.10):

$$\sum CHPElec^{n}_{mdh}=\sum C^{n}_{CHP} \tag{6.10}$$

6.4 The Design and Modeling for the PCM System

6.4.1 The Offline Heat Transportation System with PCM

The offline heat transport system, suggested in Fig. 6.13, is the rejected heat transport system without pipeline that can utilize the rejected heat at less than 200 °C. It is a track with a PCM container that can collect storage exhausted thermal energy from the factories and store and then transport to the heat demand area.

The system is firstly developed by the aviators and introduced in Japan from the year 2003 and then became widely used these years. Compared with the traditional pipe system, the offline system can collect heat from all the factories in possible distance rather than one. Furthermore, it can avoid large initiate cost of pipes.

When the heat is used for heating or cooling, the system has three times heat capacity than water; thus, this system can offer a reliable heat supply. In the container, the oil can make a good contact with the PCM material that the heat can transit fast and efficiently to PCM. When the surface enlarged, the speed becomes much faster. The maximum can reach to 0.6 MW [9].

The system firstly collected heat from the supply side, which is then stored and then transported to the demand side. Even heat demanding and discharging of the demand side always fluctuate; the PCM container has a buffer tank which can promise stable supply. There are two energy supply patterns as in Fig. 6.14.

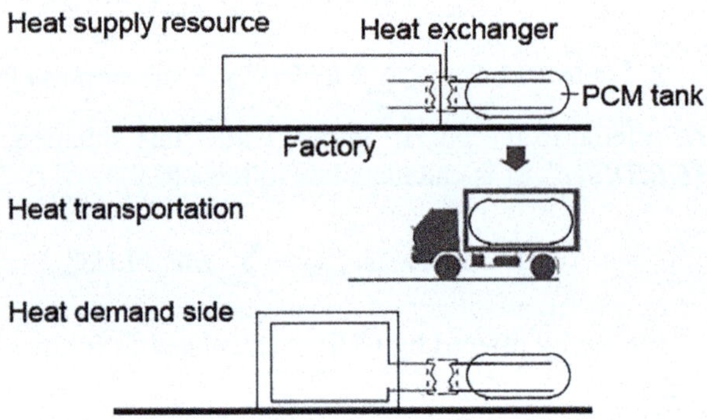

Fig. 6.13 The offline heat transport system

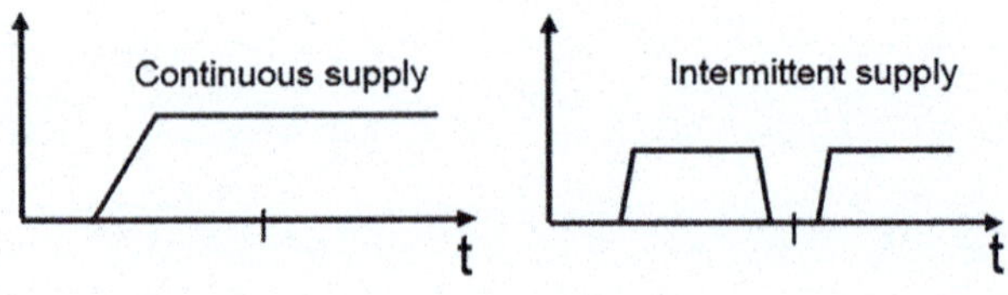

Fig. 6.14 The energy supply pattern of demand side

6.4.2 *PCM for Collecting the FEH*

According to the system parameter, economically the system can utilize heat within 135 km and 20 km round trip [10]. The CEMS will economically select out the possible utilized resource and make a plan for the PCM system. The collecting schedule of the PCM trucks should match with the factory working hour. It will become more complicated as the factory heat resource increases. Considering the various factors for making the plan, the research assumed that the FEH collected would be transported to the demand side and used in the following day.

The number of the tanks for collecting FEH used in 1 day (x) is decided by the capacity ([9]). It should satisfy the following (6.11):

$$Q_{pcm} \cdot x \geq \sum Q_{fac} \qquad x \in (1, p) \tag{6.11}$$

Q_{pcm} is the capacity for the PCM tank.
$Q_{fac}$is the daily factory exhaust of the selected resources.

The exhaust heat that can be used in the demand side is limited by the energy lost during the heat storage, transport, and exchange. CEMS will estimate it can select out the proper resource. The amount of heat ($HEAT_{recFAC}$) that can be used in the demand side is as follows (6.12):

$$HEAT_{recFAC} = \mu \cdot Q_{fac} \tag{6.12}$$

μ is the overall efficiency of the PCM system, set as 0.91 according to the existing research [8] (Table 6.1).

6.4.2.1 PCM for the Heat Delivery Between the Groups

ES will use the estimated consumption pattern for the consistent prediction and send the mode signal to CEMS.

Table 6.1 The type and parameters of the PCM tank

Type	Melting Point/ °C	Heat temperature/°C	Tank capacity/ MWh	Usage		
				Hot Water	Heating	Cooling
Type1	58	85(70)	0.8~1.1	○	○	—
Type2	78	100(90)	—※	○	○	—
Type3	116	150(130)	—※	○	○	○
Type4	118	150(130)	1.1~1.4	○	○	○

※Type2 and Type3 are used outside Japanese ○ The function it has — The function it doesn't has

For every group, CHP recovery heat $\left(CHPREC^{n}_{mdh}\right)$ is as follows (6.13):

$$CHPREC^{n}_{mdh} = \begin{cases} \frac{\eta_h}{\eta_e} ElecCHP^{n}_{mdh} & (CHPgroup) \\ 0 & (NCHPgroup) \end{cases} \tag{6.13}$$

$\eta_e$is the electricity generating efficiency of CHP plant.
η_h is the heat recovery efficiency of CHP plant.

If $CHPREC^{n}_{mdh} - HEAT^{n}_{mdh} \geq 0$, the ES is in SUR mode and the expected value of heat surplus $\left(Heat_0SUR^{n}_{mdh}\right)$ is as follows (6.14):

$$Heat_0SUR^{n}_{mdh} = CHPREC^{n}_{mdh} - HEAT^{n}_{mdh} \tag{6.14}$$

If $CHPREC^{n}_{mdh} - HEAT^{n}_{mdh} < 0$, the ES is in INS mode and the expected value of heat shortage $\left(Heat_0INS^{n}_{mdh}\right)$ is as follows (6.15):

$$Heat_0INS^{n}_{mdh} = HEAT^{n}_{mdh} - CHPREC^{n}_{mdh} \tag{6.15}$$

Every day, the PCM system will carry the FEH and input into the community from the first peak time in the morning, set as h_0 . During the day, the system will preferentially use the heat stored in the PCM and release it before the next day. Therefore, every day at the time h_0 , the heat amount stored in the PCM system is reset.

The amount of stored heat energy in the PCM that can be supplied to the ES in SUR mode at h time in one group $\left(PCMREC^{n}_{mdh}\right)$ is as follows (6.16):

$$\sum^{n} PCMREC^{n}_{mdh} = \begin{cases} \sum^{n} PCMREC^{n}_{md(h-1)} + \sum^{n} Heat_0SUR^{n}_{mdh} - \sum^{n} Heat_0INS^{n}_{mdh} & (h \neq h_0) \\ HEAT_{recFAC} & (h = h_0) \end{cases} \tag{6.16}$$

The total amount of PCM truck (p) should satisfy condition (6.17)

$$MAX\left(\sum^{n} PCMREC^{n}_{mdh}\right) \leq Q_{pcm} \cdot p \tag{6.17}$$

$MAX(\cdot)$ is a function to determine the maximum value of the stored heat in PCM system by the expected value.

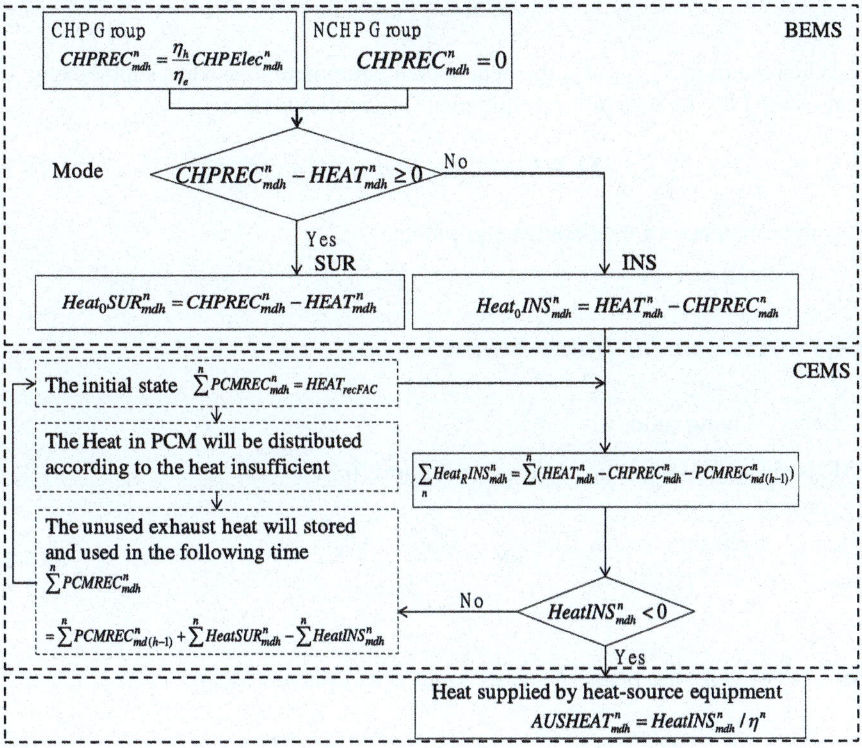

Fig. 6.15 Heat supply calculation flow

6.4.3 The Heat Balanced Management

Figure 6.15 illustrates the heat balanced management. The collected heat in the PCM system including the recovery heat of CHP system and FEH is used for heating, cooling, and hot water in the community. It is also managed by CEMS following total quantity priority that is supplied to the group, which had larger amount of heat shortage, $\mathrm{MAX}\left(Heat_0INS^n_{mdh}\right)$..

With the use of the waste heat collected by the PCM system, the heat shortage $\left(Heat_RINS^n_{mdh}\right)$ is as follows (6.18):

$$\sum_n Heat_RINS^n_{mdh} = \sum^n \left(HEAT^n_{mdh} - CHPREC^n_{mdh} - PCMREC^n_{md\,(h-1)}\right) \quad (6.18)$$

When $Heat_RINS^n_{mdh} \leq 0$, the heat demand can be satisfied with the onsite exhaust heat reutilization that the heat-source equipment $\left(AUSHEAT^n_{mdh}\right)$ is not required (6.19):

$$AUSHEAT^{n}_{mdh} = 0 \tag{6.19}$$

When $Heat_{R}INS^{n}_{mdh} > 0$, the heat-source equipment is used as supplement. The heat offered by the heat-source equipment is as follows (6.20):

$$AUSHEAT^{n}_{mdh} = Heat_{R}INS^{n}_{mdh}/\eta^{n} \tag{6.20}$$

η^{n} is the efficiency of heat-source equipment.

6.4.4 Assessment Index Setting

1. Energy saving ratio.

ESR is energy saving ratio, defined as follows (6.21):

$$\mathrm{ESR} = \frac{Q^{Conv}_{input} - Q^{CHP}_{input}}{Q^{Conv}_{input}} \tag{6.21}$$

For CHP system, the primary energy input is as follows (6.22):

$$Q^{CHP}_{input} = E^{CHP}_{Utility} \times \varepsilon_{Grid} + \left(V^{CHP} + V^{Boiler}\right) \times \varepsilon_{gas} \tag{6.22}$$

$E^{CHP}_{Utility}$ is the electricity input in CHP system; V^{CHP}, V^{Boiler} is the gas input to the CHP plant and boiler.

For conventional system, the primary energy input is as follows (6.23):

$$Q^{Conv}_{input} = E^{Conv}_{Utility} \times \varepsilon_{Grid} + V^{Conv} \times \varepsilon_{gas} \tag{6.23}$$

$E^{Conv}_{Utility}$ is the electricity input in conventional system.
V^{Conv} is the gas input to conventional system for hot water.
ε_{Grid} is primary energy consumption unit of grid in Japan (11.4 MJ/kWh)).
ε_{gas} is primary energy consumption unit of city gas in Japan (45 MJ).

2. CO_2 reduction ratio.

$\eta_{\Delta CO_2}$ is CO_2 reduction ratio, defined as follows (6.24):

$$\eta_{\Delta CO_2} = \frac{EX^{Conv}_{CO_2} - EX^{CHP}_{CO_2}}{EX^{Conv}_{CO_2}} \tag{6.24}$$

$EX^{CHP}_{CO_2}$ is CO_2 emission for CHP system, calculated as follows (6.25):

$$EX_{CO_2}^{CHP} = ex_{CO_2}^{gas} \times \left(V^{CHP} + V^{Boiler}\right) \times \varepsilon_{gas} + ex_{CO_2}^{Pow} \times E_{Utility}^{CHP} \times \varepsilon_{Grid} \quad (6.25)$$

$EX_{CO_2}^{Conv}$ is CO_2 emission for conventional system, calculated as follows (6.26):

$$EX_{CO_2}^{Conv} = ex_{CO_2}^{gas} \times V^{Conv} \times \varepsilon_{gas} + ex_{CO_2}^{Pow} \times E_{Utility}^{Conv} \times \varepsilon_{Grid} \quad (6.26)$$

$ex_{CO_2}^{gas}$ is the CO_2 emission unit for gas in Japan (13.8 g-C/MJ).
$ex_{CO_2}^{Grid}$ is the CO_2 emission unit for grid in Japan (153 g-C/kwh).

A numerical study of this model is presented in a smart community in Kitakyushu, Japan. In order to provide a smart community model with low carbon concept, the latest DEG technologies, such as gas engine, fuel cell, hydrogen fuel cell, PV, PCM system, untapped FEH, etc., are considered in the model. By analysis on various cases, the study will suggest the environmental effect of every technology as well as the overall potential of the smart community in Japan.

6.5 Case Study

6.5.1 *Energy Load*

Kitakyushu lied in the northern part of Kyushu, the westernmost of the four main islands in the Japanese archipelago. It used to be one of Japan's four leading industrial regions and contributed greatly to the rapid economic growth of Japan [11].

The smart community creation project is newly launched in Yahata Higashi district, where it used to be the factory district of the steel company. The government invested 16.3 billion yen over the 5-year period from 2010 to 2014. It has already cut 30% of the CO_2 emission compared with the other place in the city. However, the target for the smart community was to cut 50% of the existing emission, but still 20% need to be cut.

The urban structure has been changed in the past few years under the concept of "Environmentally Growing Town" and "Creation of a Shared Community." Commerce, entertainment, museum, and residential buildings were introduced into this area, which made a "compact district" with mixed function.

Detailed knowledge about energy end-use loads is important for the energy system design and optimization. In this study, the hourly load demand for electricity, cooling, heating, and hot water has been calculated according to the energy consumption unit (the system in Japan that displays energy consumption intensities) of various buildings in Kyushu, Japan [7]. As the method described before, the whole community is divided into four groups. Figure 6.16 displays the image of community and district zoning. The building information, yearly energy consumption, and peak load of the district are shown in Table 6.2.

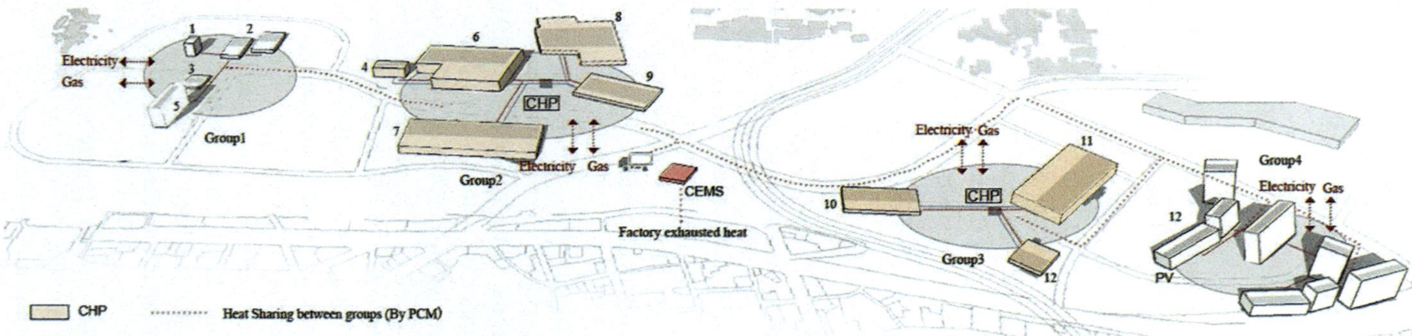

Fig. 6.16 Image of community and district zoning

Table 6.2 Yearly energy consumption and the peak load

Group		Group1	Group2	Group3	Group4	Community
Function		Office&hotel	Commerce	Education	Residence	
Building area(m^2)		30555	17545	28692.3	72212	149004.3
Roof area(m^2)		24358	5420	13342.9	7146	50266.9
Yearly energy consumption	Electricity (GWh)	2.77	7.85	1.50	0.98	13.61
	Electricity* (GWh)	2.51	5.13	0.68	1.49	9.07
	Cooling (TJ)	6.89	24.38	5.05	2.89	39.21
	Heating (TJ)	2.79	3.54	5.75	4.14	16.21
	Hot water (TJ)	3.37	2.85	0.00	8.91	15.13
Peak (kW)	Electricity	598	2260.8	471	1652.4	3313
	Electricity*	532.8	2048.4	344	1634.4	3037
	Cooling	1484.00	5155.50	1920.20	1719.55	9085.60
	Heating	458.76	1808.89	2750.25	566.86	5134.89
	Hot water	275.42	535.85	0.00	1360.87	1622.77

Electricity*: The electricity demand excluded PV

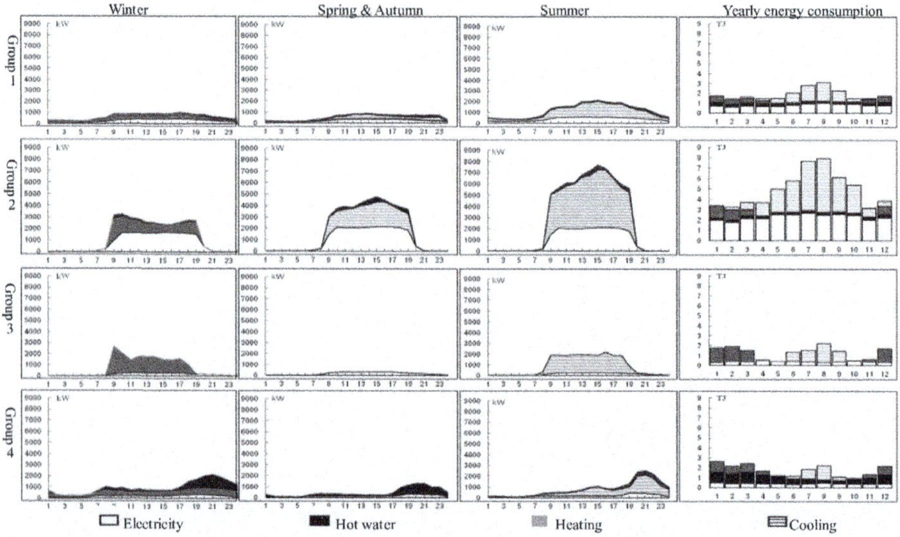

Fig. 6.17 The district energy consumption

Figure 6.17 described the detailed hourly load profiles for every group in summer (August), winter (January), spring, and autumn (May). The energy consumption profile firmly related with the building function.

I) Group 4 is the residential area; thus, the peak of the energy consumption comes during the night. Group 1 also has considerably higher energy consumption compared with group 2 and group 3, because there is a hotel in the group.
II) The commercial group (group 2) has a higher energy consumption during the day, but with almost no energy consumption during the night.
III) The hot water load is higher in residential group (group 4), but lower in group 2 and group 3.

6.5.2 *FEH Load*

1. **Investigation and estimation of the factory exhaust heat**

Kitakyushu is one of the most important industrial cities having a large area of industry. Thus, the CO_2 emission from the factories is more serious than the average level of the whole country. In that case, the city will try to reuse the potential energy from the factory exhaust heat, aiming to construct a low carbon society combining industry, school, government, and residential environment together.

To estimate the exhaust heat of the whole city, the research put forward a questionnaire to all the factories and industries in Kitakyushu. The investigation includes 170 factories and 213 industries. It covers the equipment, temperature, working hours, and so on. All the factories that been reported by the Kitakyushu were included in the estimation in this research.

6.5.2.1 Methods

In Japan, around 90% of the exhaust heat is discharged as gas. Therefore, in this research, the estimation of the exhaust thermal reserves is estimated by gas.

(1) Maximum exhaust heat.

From the result of the questionnaire, the factory measured values can offer the maximum gas discharge amount and the discharge temperature. The maximum exhaust heat energy can be calculated as below:

$$E_{\max}^{RH} = V_{\max}^{RG} \times T_{\max}^{RG} \times C_g$$

$E_{\max}^{RH}$ (KJ/h) is maximum hourly exhausted thermal energy.
$E_{\max}^{RH}$ (Nm^3/h) is maximum hourly gas discharge amount.
$E_{\max}^{RH}$ (°C) is gas discharge temperature.
Cg (KJ/N m3·°C) is the specific heat capacity of gas (1.356 KJ/N m^3).

(2) The ratio between the maximum exhaust heat and the normal exhaust heat.

The facilities with exhaust heat will measure the hourly exhaust gas including the discharge amount and the temperature. It takes out 2 or 6 years a time depending on the scale of the facility. This research uses the last record of measured value and set the mean value as the normal value.

Conversion coefficient between the maximum value and the normal value can be defined like the following:

$$\alpha_V = \frac{V_{\max}^{RG}}{\overline{V}_{real}}$$

$$\alpha_T = \frac{T_{\max_x}^{RG}}{\overline{T}_{real}}$$

α_V is the conversion coefficient between the maximum exhaust gas amount and the normal exhaust gas amount.
$\overline{V}_{real}$ is the mean value of the last measured data of discharged gas amount per hour.
α_T is the conversion coefficient between the maximum exhaust gas temperature and the normal exhaust gas temperature.
$\overline{T}_{real}$ is the mean value of the last measured data of discharged gas temperature.

The result of α_V and α_T are listed out in Figs. 6.18 and 6.19. The mean values are set as the conversion coefficient

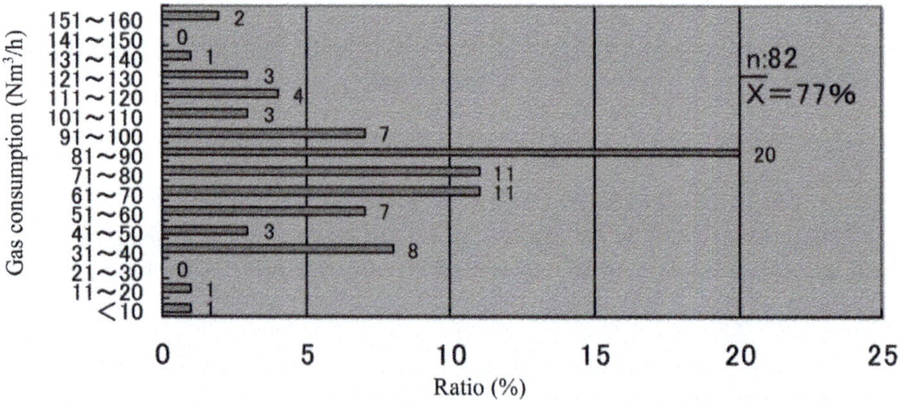

Fig. 6.18 Conversion coefficient between the maximum and the normal exhaust gas amount

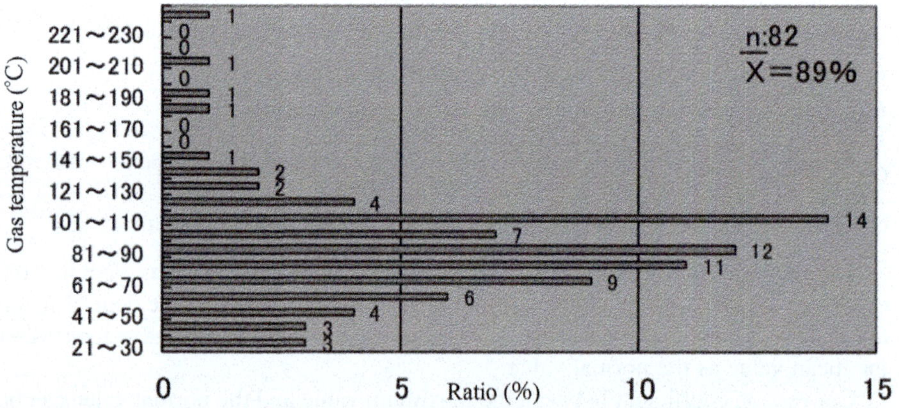

Fig. 6.19 Conversion coefficient between the maximum and the normal exhaust gas temperature

$$\alpha_V \text{ polt } (\overline{\alpha}_V = 0.77)$$

$$\alpha_T \text{ polt } (\overline{\alpha}_T = 0.89)$$

Therefore, the exhaust heat energy can be estimated as follows:

$$\begin{aligned} E^{RH} &= V^{RH}_{\max} \times \overline{\alpha}_V \times T^{RG}_{\max} \times \overline{\alpha}_T \\ &= V^{RH}_{\max} \times 0.77 \times T^{RG}_{\max} \times 0.89 \end{aligned}$$

E^{RH} (KJ/h) is the estimated exhaust heat per hour.

Table 6.3 The setting for working hours

Building category		Working hours(h)
Factory	the factory above 500cal/h	24
	Others	8
	Hospital, hotel	24
	Public Waste disposal facilities	24
Industry	Nursing homes for elders	24
	Shopping centers	12
	School	12
	Others	8

(3) The setting for working hours.

The working hours will be estimated according to the questionnaire among the factories. The investigation includes 170 factories and 213 industries, and the questionnaire recovery rate is 37.6% for factories and 20.3% for industries.

According to the result, the working hours can be set as in Table 6.3. The average working ratio can be set as 0.7.

(4) The estimation for yearly rejected heat energy.

The yearly rejected heat energy for every factory can be calculated as below:

$$E_{yearly}^{RH} = E_{\max}^{RH} \times 0.77 \times T_{\max}^{RG} \times 0.89 \times C_g \times 24h \times 356d \times 0.7$$

E_{yearly}^{RH} (KJ/h) is yearly gas discharge amount.

6.5.2.2 The Estimation Result of the Exhaust Heat

This survey covered 1552 facilities that have exhausted heat in Kitakyushu, among which 1412 were used for the estimation.

There are 7 districts in the city, and the 1412 factories are in the whole city. As the result of estimation, the yearly rejected heat energy is about 18,000TJ. Figures 6.20 and 6.21 show yearly rejected heat energy of every district.

Another important input to the energy system is the reutilization of the FEH. It is collected by PCM system and used in the community for heating, cooling, and hot water. The study selected out the four potential factory resources (within 20 km) [10]. Usually, the temperature for FEH is higher than 300 °C and daily exhaust heat is around 38.9 GJ. The tank type with the capacity of 1.4 MWh will be introduced into the system. As this research only discussed the environmental effect of the PCM system, thus it is supposed that there are enough tanks for collecting all the exhaust heat (the heat of the factory and the unused CHP recovery heat).

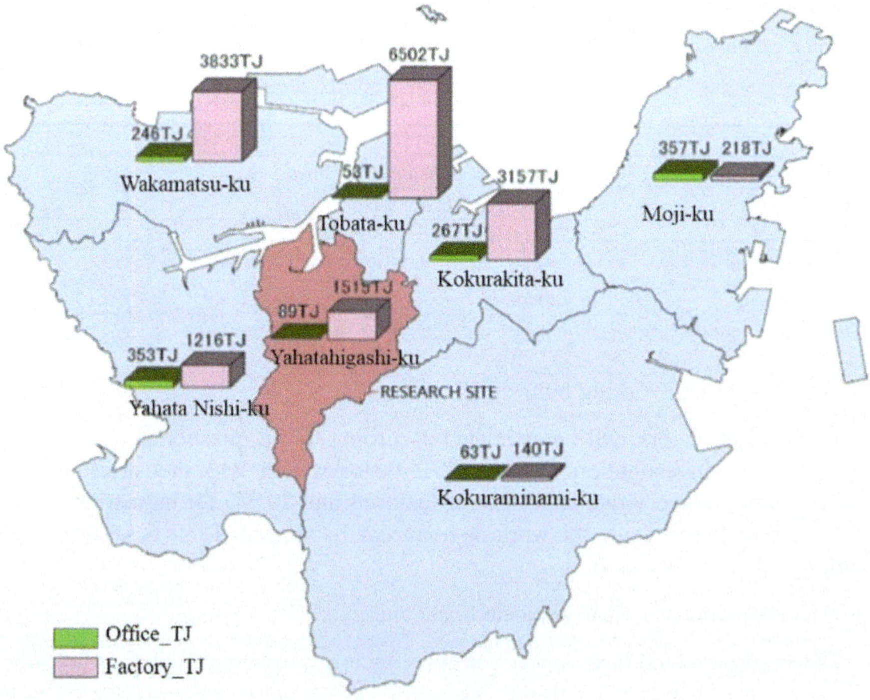

Fig. 6.20 Yearly rejected heat energy of every district

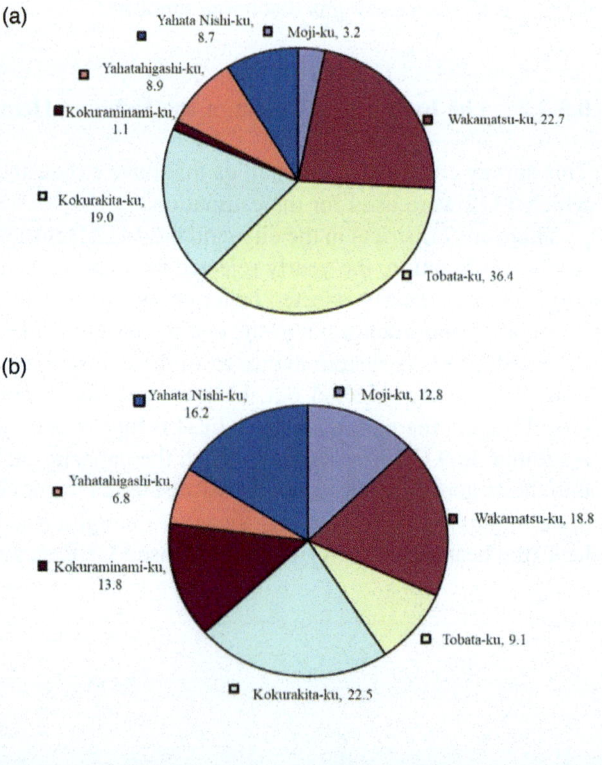

Fig. 6.21 The distribution of exhaust heat reserves and facilities by district (**a**) The exhaust heat reserve proportion (**b**) The exhaust heat facility proportion

Table 6.4 Technical parameters of system

Facility		Parameter	COP
Grid		η	0.35
CHP Plant			
Gas Engine (GE)	Electricity Generation	η_e	0.3
	Heat Recovery	η_{rec}	0.45
Fuel Cell (FC)	Electricity Generation	η_e	0.4
	Heat Recovery	η_{rec}	0.3
Hydrogen Fuel Cell (HFC)	Electricity Generation	η_e	0.48
	Heat Recovery	η_{rec}	0.42
Absorption Chiller		COP_{ac}	1.1
Heat Exchanger(H-EX)		COP_{he}	1
Boiler		η_{b1}	0.8
Multiple Air-conditioning System	Cooling	COP_1	4
	Heating	COP_2	3.9
Room Air Conditioner	Cooling	COP_1	3.22
	Heating	COP_2	2.83
PCM system		η_{rec2}	0.9

6.5.3 DEG Technologies and District Energy System

This district is the demonstration area that the latest technologies are expected to introduce into the area. The smart community has also undertaken the Kitakyushu Hydrogen Town project. The project is the world's first attempt to use a pipeline recycling the hydrogen generated in the iron manufacturing and operating the fuel cells as an energy supply to the district. The demonstration testing is processed jointly by Fukuoka prefectural and city gas utilities [13]. The pipeline is connected with the hydrogen station and hydrogen fuel cells that are installed in buildings in this district. These fuel cells generate electricity by combining hydrogen and oxygen. Table 6.4 shows DEG technologies assumed in this study and their properties, including gas engine (GE), fuel cell (FC), hydrogen fuel cell (HFC), and PV. All equipment is city gas fired.

6.5.4 Setting of Cases

In order to investigate the effect of technologies in the DRN, the following cases are assumed for analysis.

Base case: conventional system. Base case indicated conventional energy supply system. The electricity load is satisfied by grid. The buildings also used air conditioner for heating and cooling. Commercial buildings and office and public buildings use multiple air-conditioning systems, and residential buildings use room air conditioner. The hot water load is satisfied by gas boiler fired by city gas.

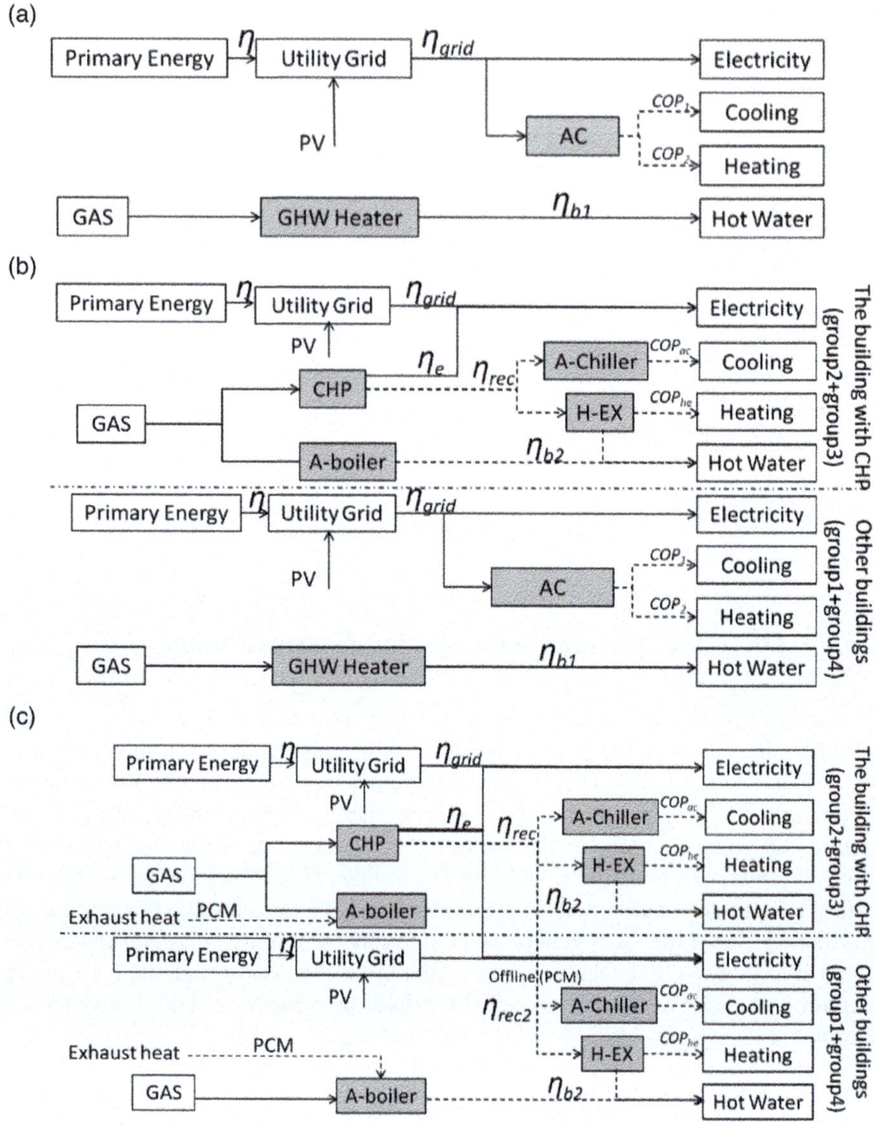

Fig. 6.22 Case setting

Case 1: The conventional system combined with PV systems. In this case, the community still keeps the conventional system, but facilitated with PV. The system is as in Fig. 6.22a. The electricity will be supplied by PV system, or by grid, or by both. The electricity from the PV system will be used by the buildings themselves, and remaining electricity will be sent back to the grid.

Case 2: Individually introduced DEG systems, displayed in Fig. 6.22b. In case 2, the CHP plants with GE are introduced in group 2 and group 3. The CHP plants and PV systems can satisfy the electricity load of these two groups. The thermal load can also be supplied by the recovery heat of the CHP plants and the deficiency supplemented by gas boiler. In this case, the electricity and recovery heat of the CHP plants cannot supply to other groups or return back to the grid. Therefore, the NCHP groups still get electricity from the PV and grid, kept as the conventional system.

Case 3: DRN system without using factory exhaust heat, described in Fig. 6.22c. In the DRN system, the community uses the same DEG technologies with case 2, but controlled and managed by CEMS. Under the CEMS, the electricity produced by the CHP plants cannot only be used for the CHP group but is also supplied to the NCHP group. The recovery heat of the CHP group will be used in the CHP group first and then recycled by the PCM system. The CEMS distributed the heat that is stored in the PCM system with thermal shortage and surplus profile of every group.

Case 4: DRN system with the utilization of the FEH, as in Fig. 6.22c. Besides the technologies and DRN system that are assumed in case 3, case 4 also makes use of the FEH by PCM system. The PCM system collects the exhaust heat from the factory resource that is set in part 3 and transports it to the CEMS in the community. Besides the surplus CHP recovery heat, this part of heat will also be distributed by CEMS.

Case 5 and case 6: The DRN system with the CHP plants of FC and HFC. Beyond the DRN systems built in case 4, case 5 introduced the CHP plant of FC and case 6 introduced HFC.

6.6 Results and Discussions

6.6.1 The Effect of Electricity Sharing in DRN System

Figure 6.23 is the electricity balance in the community with the individually introduced DEG systems (case 2) and the DRN system (case 3). Both cases use CHP plant with GE and PV. The comparison between the two cases can show the effect of the electricity sharing between them. It can suggest that PV system can provide 35% of the community electricity consumption and the individual CHP plant can produce 41% electricity. By electricity sharing, the CHP group can offer 2GWh electricity to the non-CHP group, which occupied 52% of their electricity consumption. As a whole, the community can produce 58% of the electricity by CHP, and only 7% from the grid, while the individual system needs 24%.

As we know, the electricity produced by DEG has less energy loss during the electricity delivery. Therefore, the system can save more energy as it gets less electricity from the grid. In the DRN system, the CEMS can operate the CHP plants and distribute the electricity to the whole community. Therefore, it will increase the output and working hours of the CHP plant and reduce the electricity from the grid.

The electricity sharing under the CEMS can balance the electricity consumption between the different groups, making the system more independent and reliable.

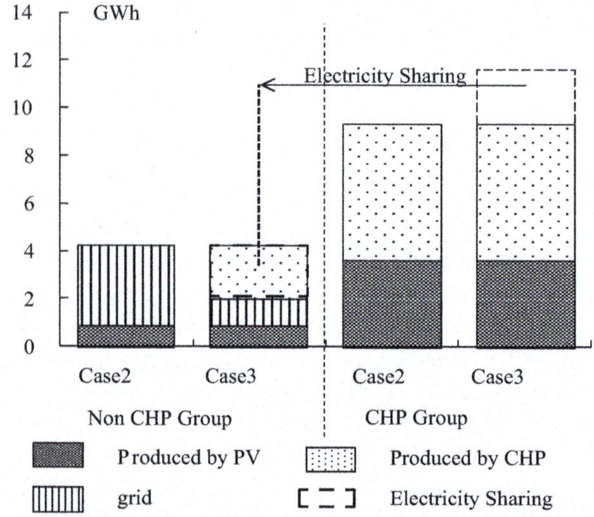

Fig. 6.23 Yearly electricity balances of case 2 and case 3

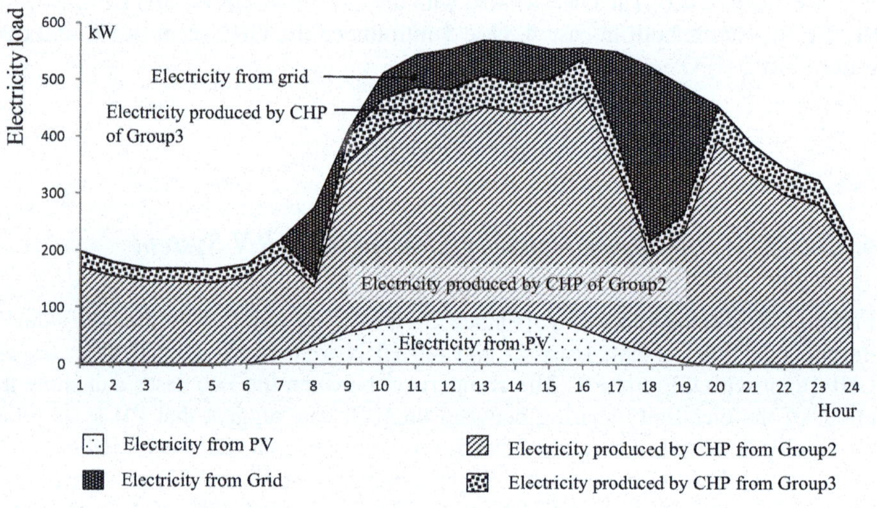

Fig. 6.24 Daily electricity balance of group 1 (summer)

Figure 6.24 displayed the daily electricity balance in group 1, taking summer as example. The buildings in group 1 have small roof areas that the PV can only offer a small part of the electricity. Compared with group 4 (the other NCHP group), group 1 has higher electricity shortage, thus preferentially getting electricity from the CHP group. CHP can satisfy most of its electricity demand. Especially, during the night, the electricity load is low in commercial buildings that the electricity demand in group 1 can be satisfied by CHP only.

The electricity sharing used in DRN system can shift the electricity demand from the peak. Just as Fig. 6.24 suggested, without CHP plant, the peak hour should come during the noontime, but now it shifts to 8 o'clock in the morning and 18 o'clock in the afternoon. Further, from the city level, the less relay on the grid will alleviate electricity shortage especially during the peak hours. That means with the DRN, the city can smooth out the electricity fluctuation.

6.6.2 The Effect of Heat Sharing in DRN System

The DEG with CHP plants not only reduce the energy loss but can make use of the recovery heat as well. In case 2, the individual CHP system can only use the recovery heat by the CHP group itself. However, under the CEMS, in case 3, the DRN system can distribute the recovery heat to other group with the PCM system. In that case, it improved the utilization rate of the recovery heat. As Fig. 6.25 illustrated, the individual CHP has 37.9GWh recovery heat every year, and 31.1 GWh is used for thermal consumption in CHP group. In DRN system, the yearly CHP recovery heat is 47.3 GWh, among which 6.4G Wh heat is offered to the NCHP group. This part of heat occupied 33.8% of heat consumption in NCHP. Under this condition, 85% of the CHP recovery heat can be reused which possessed 68.8% of the community heat demand.

Figure 6.26 illustrates the daily heat balance in the community, taking the wintertime as example. The plus value means the heat surplus of each group. Group 2 and group 3 are the CHP groups, and their heat surplus means the remaining

Fig. 6.25 Yearly heat balances of case 2 and case 3

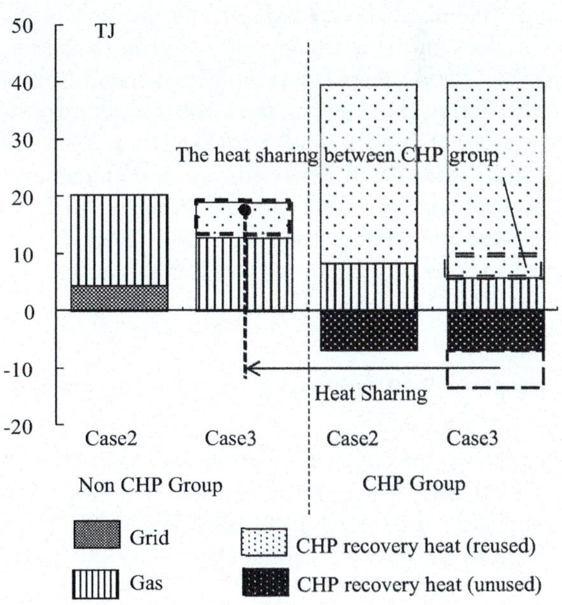

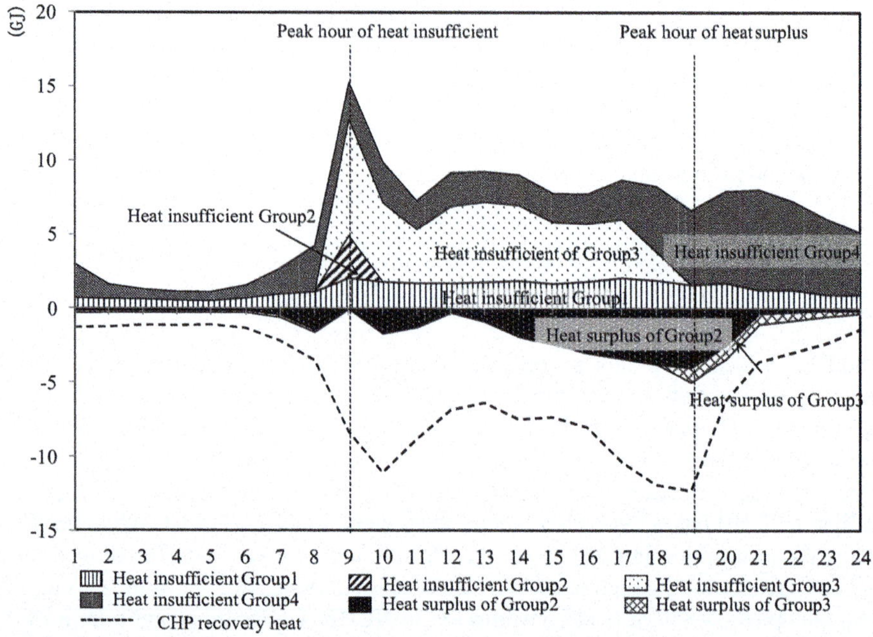

Fig. 6.26 Daily heat balances of the community

heat after their own utilization. PCM system can collect this part of heat which is used for heat supply in other groups. The minus part means the heat shortage. For group 2 and group 3, it means the heat deficiency after utilizing the CHP recovery heat. Figure 6.26 can suggest that the first peak of the heat shortage comes on 9 o'clock in the morning, and the peak of the heat surplus comes on 19 o'clock. Group 2 and group 3 have no heat demand from 19 o'clock to the next 9 o'clock; thus, during this time, all the CHP recovery heat will be supplied to NCHP group. From the 9 o'clock to 19 o'clock, group 3 has the largest heat shortage; thus, the stored heat in the PCM system will be preferentially supplied to group 3. That means the heat sharing is not only between the CHP group and NCHP group but also between the CHP groups. After the CEMS collected the heat and stored it in the PCM system, it only distributes the heat according to the heat insufficient volume.

6.6.3 *The Effect of Using Factory Exhausted Heat*

Until now, the city of Kitakyushu still has 1412 factories and industries, which have exhaust heat. The existing research put forward questionnaire to all the factories, estimating and setting up a database by GIS for the yearly exhaust heat. As a result, the yearly exhausted thermal energy is about 18,000TJ.

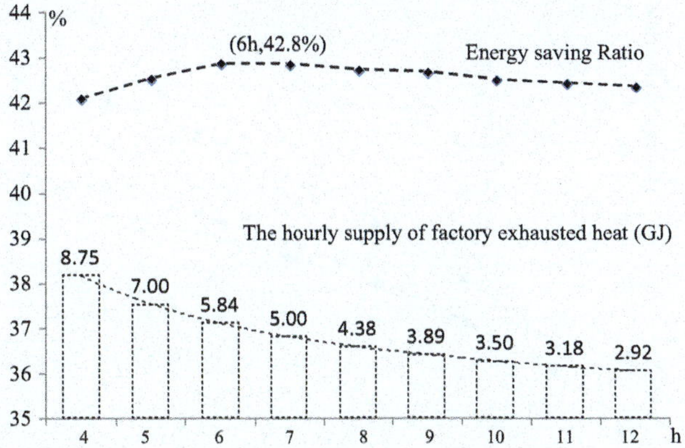

Fig. 6.27 Energy saving ratio and the supply span of factory exhaust heat

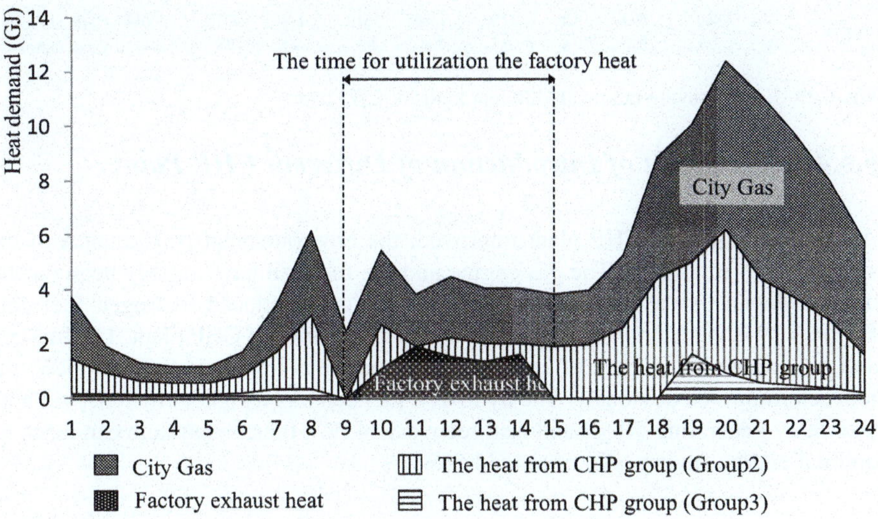

Fig. 6.28 Heat balance considering factory exhaust heat (group 4, winter)

For this community, four factories were set as the resources, and the total yearly heat amount that can offer to the community was 14.2 TJ (38.9GJ per day).

In this research, it is set that exhaust heat will be averagely supplied to the community from the first peak hour in the set time range. Figure 6.27 is the relationship between time range and the heat volume, as well as the energy saving result. It can suggest that in this case, 6 hours is the optimal time range and it can cut 41.4% of primary energy beyond the PV and CHP system.

Figure 6.28 is the daily heat balance with the utilization of the factory exhaust heat, taking group 4 in the wintertime as example. During the daytime, group 3 has higher heat load that the factory exhaust heat will firstly be supplied to group 3. However, the factory heat can still afford on part of the heat load of group 4. During the night, group 2 and group 3 have no thermal demand that the stored heat will firstly offer to group 1, but still another part can afford almost half of the heat demand in group 4.

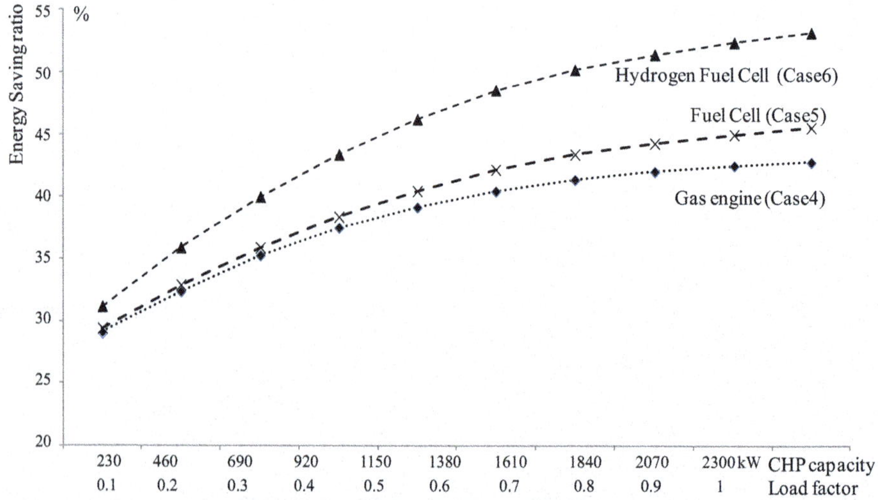

Fig. 6.29 Energy saving ratio with different kinds of CHP plant

6.6.4 The Effect of Introduction of Different CHP Plant

As the techniques of CHP plant improved, the environmental performance of the system changed as well. The gas engine and the fuel cell have already been widely used in Japan. As a trial project, the community introduced hydrogen fuel cell. Figure 6.29 is the energy saving ratio of these three kinds of CHP plant. The fuel cell and gas engine had similar effect when the capacity is low, but after 1000 kW, the fuel cell improved obviously. The hydrogen fuel cell had a higher efficiency on both electricity generating (48%) and heat recovery (42%); thus, the system can reach an optimal energy saving ratio of around 53%.

6.6.5 The Assessment from the Community Side

Figure 6.30 is the energy saving ratio for various cases. The PV system can cut off 22.6% primary energy consumption. The individual CHP plants and the PV system can totally cut 30.6% primary energy consumption. Based on this system, the execution of the DRN system can cut off 38.2%. By introduction of hydrogen fuel cell, the community can cut off 53.1% primary energy consumption as its target.

Figure 6.31 is the low carbon ratio for every technology. By introducing the PV system and the CHP plant (gas engine), it can cut off 29.4% of the carbon emission. The networking CHP system can reduce energy consumption and cut off another 7% carbon emission. Besides these, the reusing of factory rejected heat energy can cut off 41.1% CO_2 emission. With the introduction of fuel cell and hydrogen fuel cell, it is proved that the community can get 51.8% CO_2 emission reduction ratio.

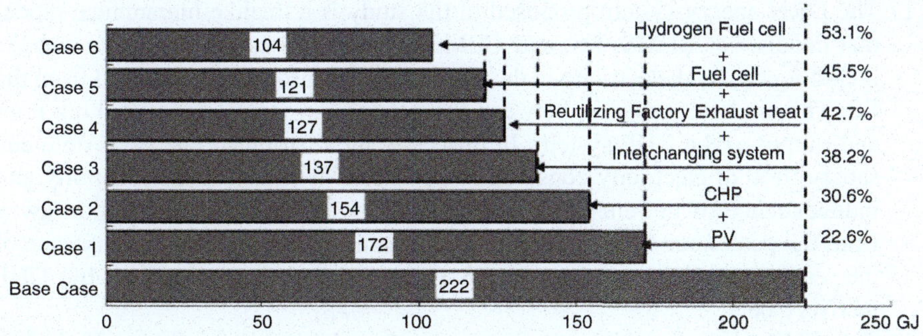

Fig. 6.30 The energy saving ratio for various cases

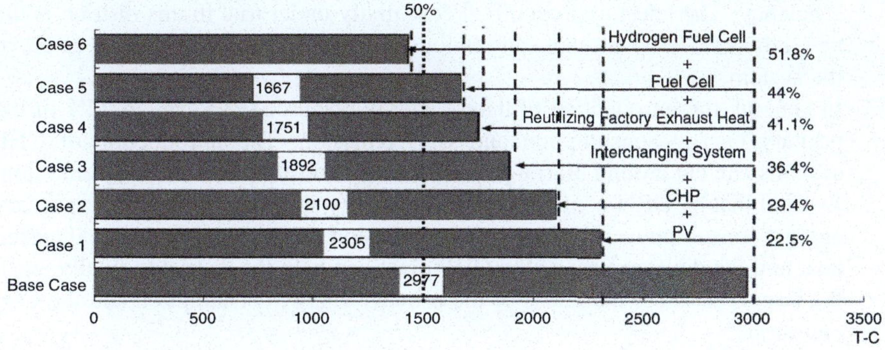

Fig. 6.31 Yearly CO_2 emissions for cases

6.6.6 Conclusions

This chapter proposed a DRN energy system model for smart community in Japan. One innovation is that the model not only has a smart grid but also has a smart heat energy supply chain by PCM system. The PCM system controlled by CEMS conducted the heat sharing between buildings. In that way, it can maximize onsite use of CHP recovery heat. Further, this model promoted a collaborative energy utilization mode between the industrial sector and the civil sector. The introduced PCM system will also collect the exhaust heat from the nearby factory. It not only made use of the untapped energy but also cuts off the CO_2 on the factory side (the exhaust heat) as well. In addition, the research chose the smart community in Kitakyushu as case study and executed the model. The simulation and the analysis of the model is embodied by temporal perspective of the low carbon techniques in Japan, including nature and untapped resource, CHP plants, and the PCM system. The result suggests not only the environmental effect of different technologies but also the potential of its overall performance.

(1) The DRN energy system proposed in the study is a treelike hierarchical model that consists of BEMS, ES, and CEMS. The CEMS can dispatch the energy, including heat and electricity in the district, by the information received from the ES. The electricity sharing between the groups can improve the working hour and output of the CHP system. In that case, the distributed energy system can satisfy 95% of electricity consumption by itself. It enhances the reliability and independence of the energy system and shifts the energy consumption away from the peak hour as well. Heat sharing can also enhance the independence of the energy system and satisfy 68.8% of the thermal demand by CHP recovery heat.
(2) The CHP plant is widely used and developed quickly in Japan. There are different kinds of CHP plants, as gas engine and fuel cells. They have different characteristics and different electricity generation efficiency and heat recovery efficiency. The latest hydrogen fuel cell, firstly under trial in this district, is the new kind of CHP plant that can obviously improve the environmental effect of the system.
(3) In general, the introduction of nature energy resource (PV) can cut 22% of the primary energy consumption and CO_2 emission. The introduction of CHP systems can cut around 30.6% primary energy consumption and CO_2 emission. Beyond that by DRN control, the district energy sharing can cut 38.2% primary energy consumption and 36.4% CO_2 emission. The use of factory exhausted heat and the development of the CHP plant can help the district to finally reach the target: cut more than 50% of the primary energy consumption and the CO_2 emission.

The area energy network will be widely used toward carbon neutral, with the district electricity and heat network. The offline heat transport system, which is not needed in the infrastructure investment, can be a useful way for heat sharing and waste heat utilization in low-density area.

References

1. Sollia C, Anantharaman R, Strømmana AH, Zhanga X, Hertwicha EG (2009) Evaluation of different CHP Op- tions for refinery integration in the context of a low carbon future. Int J Greenhouse Gas Contr 3(2):152–160
2. Evans R (1993) Environmental and economic implications of small-scale CHP. Energy Policy 21(1):79–91
3. Wissner M (2011) The smart grid—a Saucerful of secrets? Appl Energy 88(7):2509–2518
4. Urban MC (2015) Accelerating extinction risk from climate change[J]. Science 348(6234): 571–573
5. District heating and cooling, https://www.tte-net.com/solution/dhc.html, Japan, accessed on July,2022
6. Mathiesen BV, Lund H, Karlsson K (2011) 100% re- newable energy systems, climate mitigation and eco- nomic growth. Appl Energy 88(2):488–501

7. Lab O (2002) Consumption unit of electricity, heating, cool- ing and hot water. Waseda University, Tokyo
8. Ren H (2007) Effect of carbon tax and electricity buy-Back on the optimal economic adoption of PV system for residential buildings. J Environ Eng 622:49–55
9. Introduction of Offline Heat Transport System, SANKI. http://www.sanki.co.jp/product/thc/thc/point.html, Japan, accessed on July, 2013
10. Hiramitsu Kiyoto (2008) Study on the exhaust thermal energy utilization by using the PCM transportation system. Architectural Institute of Japan
11. Government's New Growth Strategy. http://www.kantei.go.jp/jp/sinseichousenryaku/sinseichou01.pdf, Japan. Accessed on July 2013
12. Fan L (2012) Potential analysis on the area-wide factory exhaust thermal energy utilization by PCM transportation system in a recycling-oriented community, AIJ Kyushu Chapter Architectural Research Meeting, ISSN 1343-8387, 341–344
13. Highlighting JAPAN July 2011, "Stepping Stones to 'Smart' ness," 2011. dl.gov-online.go.jp/public_html/gov/pdf/hlj/.../12-13.pdf , Japan. Accessed on July 2013

Chapter 7
Residential Multi-energy System Design for Energy Saving and Economic Optimization

Xueyuan Zhao and Xiaoyi Zhang

7.1 Characteristics and Composition of Residential Multi-energy System

Residential multi-energy system refers to a low-carbon residence system through the effective utilization of energy without affecting the living comfort. Buildings consume a large ratio of social energy consumption, for example, 75% in the USA. The building can be flexible in terms of energy use means having the ability to shift energy use in time and space, thus providing energy flexibility in aggregated form, known as grid-interactive efficient buildings [1]. Aggregated demand side participation as virtual grid asset is increasingly expected to play an important role in peak demand issues and offer a broader range of grid services for the incorporation of higher shares of variable renewable generations. Due to the decline of photovoltaic system cost and the implementation of feed-in-tariff (FiT), more families begin to install residential PV systems. With the rapid decline of the cost of fuel cell system, the number of houses using domestic fuel cell system also began to increase. Real estate developers have also begun to merge and sell residential photovoltaic systems, fuel cells, and household battery systems. It can be predicted that the smart house with home energy management system (HEMS) will be the development trend of residence in the future.

Energy management system (EMS) refers to the energy management system that uniformly manages the visualization of energy generation and consumption, power storage system, and control energy consumption equipment. The design and installation of the management system are divided according to the scale of the region,

X. Zhao (✉)
Department of Architecture, Hangzhou City University, Hangzhou, China

X. Zhang
Faculty Environmental Engineering, The University of Kitakyushu, Kitakyushu, Japan

W. Gao (ed.), *Distributed Energy Resources*, Green Energy and Technology,
https://doi.org/10.1007/978-3-031-21097-6_7

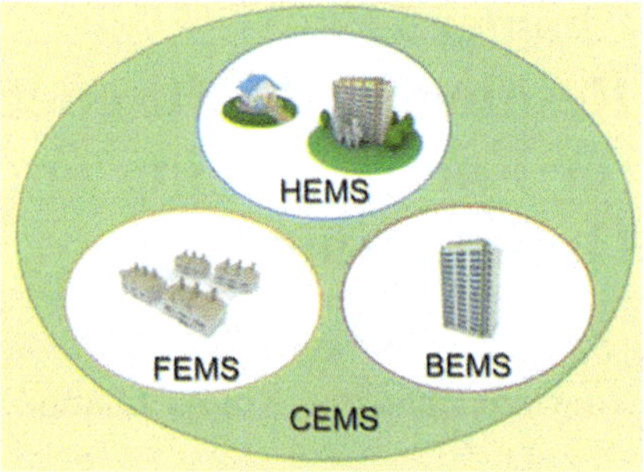

Fig. 7.1 The composition of EMS

including home energy management system (HEMS), building energy management system (BEMS), factory energy management system (FEMS), and community energy management system (CEMS). The house installed with HEMS is called smart house, the building installed with BEMS system is called smart building, and the area installed with CEMS system is called smart community (Fig. 7.1).

Specifically, the residential multi-energy system effectively manages the energy used, manufactured, and stored by household appliances, solar power generation system, residential energy supply system, and residential energy storage system through HEMS, so as to realize the rational use of energy and residential comfort.

7.1.1 Demand Side Management

Advancement and development in smart meter, Internet, and communication technologies enable the interconnection between electricity utility and distributed energy customer and provide chances for customers to participate more in local grid management via coordinated control of home appliances or local power resources in aggregated form. Contexts of DSM in community or microgrid generally focuses on the uptake of energy efficient appliances, coordinated control of local power technologies, and induced load pattern shift. Driven by cost-saving potential and relevant incentive policies, there is growing interest in the uptake of energy conservation technologies and real-time power consumption control in the demand side, generally called demand response, and district utilities pay more attention to potential benefits of demand response applications, such as cost-saving, load leveling, and carbon emission reduction. The aggregated uptakes of high energy efficiency appliances and on-site generators are expected to participate more in compensating increase in load and developing sustainable energy system. Current energy efficiency efforts and activities in building sector have been focused toward improving energy saving in appliances [2, 3], coordinated managements of grid-connected on-site generators [4–6], and demand response implementation driven by potential cost benefits brought to both plant and customer sides [7–9]. Figure 7.2 depicts the main activities for smart demand side management strategies, including variable renewable energy production dispatch, grid load leveling via valley fill and peak cut,

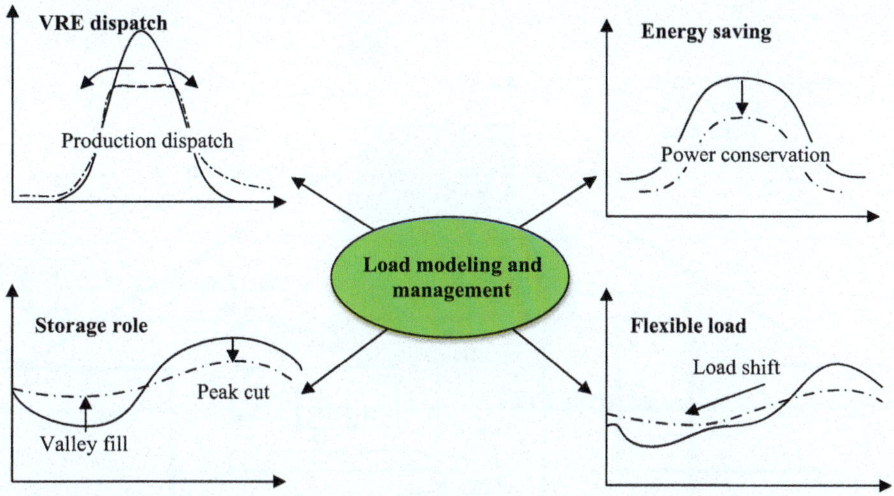

Fig. 7.2 Main activities for smart demand side management

energy conservation, and load shifting. The building sector is facing a trend toward decentralized, more efficient technologies to cover the electrical or heating loads, with increasing share of efficient power technologies installed in electrical distribution grid. Customers can participate in demand side management by offering flexible power and load profiles, through aggregation of owned distributed energy resources, storage system, and shift-able activity or comfort-based load.

Major changes in the energy sector in recent years mean that the need for smart, flexible energy is increasing and that energy production and storage facilities will have to become more coordinated. In the future smart grid, every facility and device will have its own IP address so that it can be monitored and controlled via the Internet. It is expected that markets related to "smart technology" will expand as it contributes to energy conservation and improves the convenience of everyday life. The Internet of Things (IoT) is an emerging technology in which smart devices are interconnected and communicate via the Internet. Devices could be incorporated into the IoT, from air conditioners and TVs to cars and solar panels. It can be used on different platforms to support a diversity of devices and the development of IoT applications. Technology optimists claim that IoT technology will be the vital missing link enabling us to meet the major challenges associated with climate change and energy efficiency. The new technology will also create new products, new services, and new applications. Small energy producers and urban districts with energy-plus houses produce more electricity than they consume, and motorists with electric cars that are part of a cooperative scheme can feed energy to the grid and act as energy companies. Smart management in demand side is embodied not only in integration of devices but also the information exchange between the utility and customers. Smart meter and wireless communication framework enable the real-time control of power technology consumptions and provide market cooperation potential

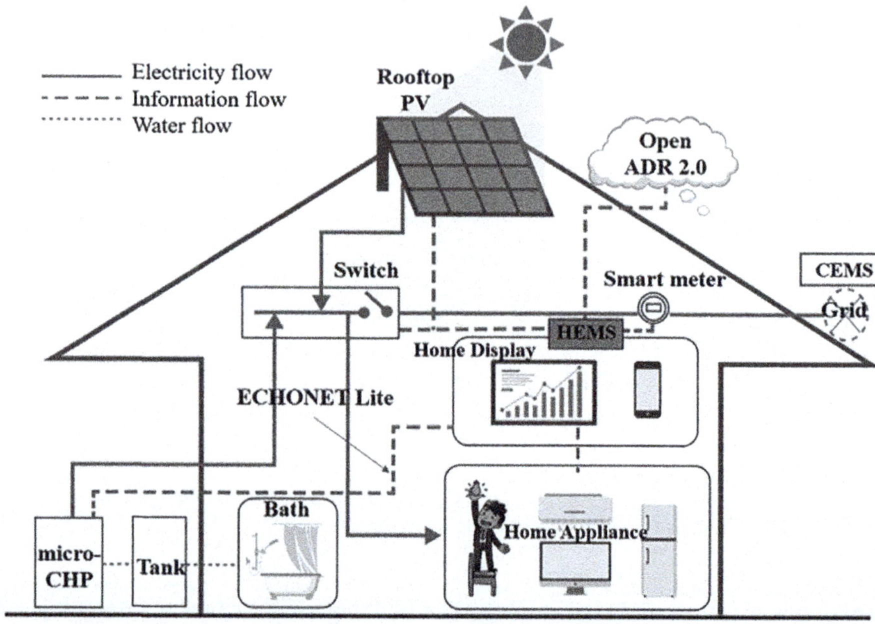

Fig. 7.3 Overview of home energy management system (HEMS) structure

between supply and demand sides based on demand side response strategies [10, 11]. As illustrated in Fig. 7.3, the installation of HEMS makes it possible to view the consumption of electricity and gas as visible information and control HEMS-compatible home appliances automatically [12]. Because it avoids wasting energy and reduces energy costs, HEMS is an essential part of smart house technology. Smart meters, air conditioners, water heaters, lighting devices, and other devices are connected to HEMS. Home energy management system (HEMS) enables smart houses, generally equipped with on-site generators, to gather the real-time energy production, consumption, and pricing that shift their energy consumption based on signal among communication network [13]. HEMS also enables customer to participate in the grid management from aggregated form, and applied HEMS algorithm receives the price information from the utility company in advance and controls the start and schedule of the power consumption of home appliances. Meanwhile, it brings cost-saving to customers via load shifting [14].

7.1.2 Composition of Residential Multi-energy System

The construction of residential multi-energy system is essentially a process of system integration. In this integration process, relevant technologies, equipment, and materials finally form a qualified smart system under the control of the integrator

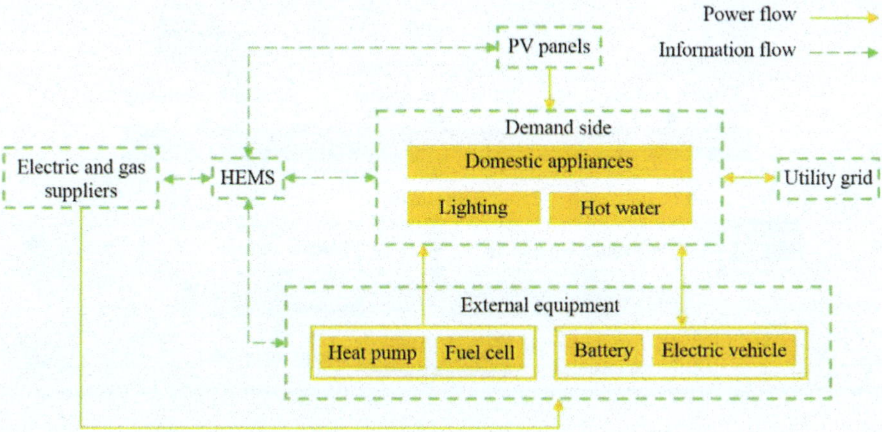

Fig. 7.4 Basic framework of residential multi-energy system

(Fig. 7.4). The information flow connects the electric and gas suppliers and the demand side through the HEMS, to realize a two-way communication of the energy data [15]. Users can validate the usage and power consumption of home appliances through the HEMS. Meanwhile, the HEMS sends the collected energy consumption data to the power company, which in turn formulates the electricity price and incentive scheme according to the power load characteristics of the users and feeds the data back to the household. Finally, the users can adjust the energy consumption behavior and select an economical electricity price plan through the information displayed by the HEMS, for optimizing the power cost. The energy flow comprises electric and gas suppliers, a utility grid, and the demand side. Power suppliers, public grids, and other energy supply equipment imported from residential buildings provide electricity to users. A PV battery uses solar energy to provide electricity for residence during the day, and users can sell the remaining power back to the grid. The external equipment imported by users can be classified into two types: energy supply only and energy supply and storage simultaneously. Energy-only devices include heat pumps and fuel cells, which provide electricity and heat to the users during the operation time. Energy storage devices, including batteries and electric vehicles, have the function of charging and discharging. They maximize the advantages of HEMS by storing energy when the cost of power is low and releasing it to users when the cost is high. The following will introduce the equipment and system operation mode in the energy system of smart house.

1. Home energy management system (HEMS)

The HEMS includes the most appropriate control system for the operation of household appliances. Lighting switch, solar power generation system, and fuel cell system are always visible equipment for power generation. The specific functions include the following:

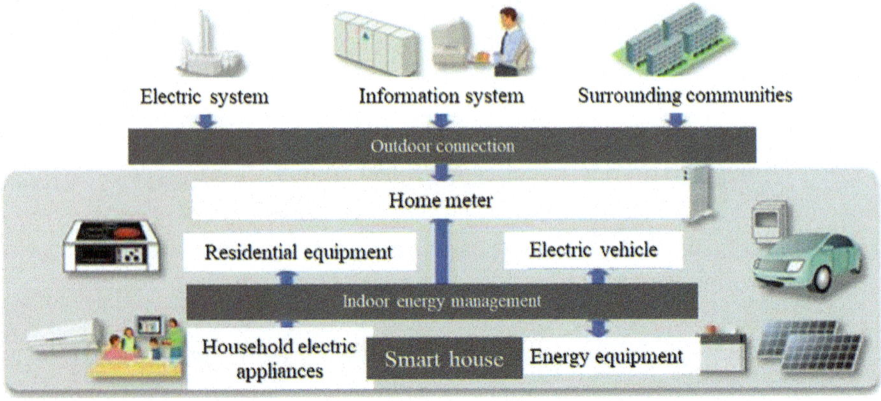

Fig. 7.5 Panorama of HEMS

1. HEMS can accurately grasp the power consumption of air conditioning and lighting and control the air conditioning and lighting through the comprehensive control and analysis of indoor temperature, humidity, and lighting.
2. The power and heat generated by the installed solar energy equipment, urban liquefied gas, and fuel cells shall be managed uniformly. Charge the battery with the remaining electric energy, and heat the water heater to meet the consumption during peak hours.
3. Based on the analysis of the actual situation of energy consumption in the past, HEMS comprehensively considers the changes of power generation caused by weather changes, so as to achieve the most reasonable control of household energy consumption.

Therefore, smart house is composed of HEMS, solar power generation system, battery system, energy-saving lighting system, solar energy utilization system, fuel cell system, heat pump system, electric vehicle charger, intelligent household appliances, etc. Smart housing industry is a new industry born under the premise of technological breakthrough and high technological integration and cross industry cooperation of enterprises (Fig. 7.5).

In addition to household appliances, HEMS can also relate to energy machines, residential equipment and instruments, electric vehicle, home gateway, and electricity meter to realize household energy management. In addition, the system can also connect outdoor power system, information system, and nearby community network.

The key technologies of HEMS include the following parts:

1. Data acquisition technology

   - Smart appliances: used for automatic monitoring of their own faults, automatic measurement, automatic control, automatic adjustment, and communication technology with remote control center. It has the characteristic of networking function, intelligence, openness, compatibility, energy saving, ease of use, and so on. The equipment has the function of intelligent mode operation, can automatically control the operation based on the environment, realizes the data collection of working conditions and energy consumption, and has the ability of fault self-diagnosis.

- Smart socket: it can directly reflect the operating power, current, voltage, and other information of electrical appliances on the socket. Be able to find electrical abnormalities in time to avoid abnormal power consumption. At the same time, it has a communication interface, which can transmit the monitoring data to the monitoring platform. As a transitional product, when smart appliances are not mature and unified, smart sockets will exist for a long time. The main function is to realize the electric energy measurement of electric equipment, mainly focusing on the measurement of voltage, current, power, and power factors.
- Electric energy meter: statistics the electric energy used by the equipment within the measurement range, so that the power supply department can charge according to the accumulated metering data and has the function of data remote transmission. Smart meter technology is mainly driven by the marketing business needs of State Grid Corporation of China. Its function is mainly measurement and cost control, which is updated according to the standards formulated by the State Grid Corporation of China.

2. Communications technology

- Home communication network: a home network access platform integrating home control network and multimedia information network to realize the interconnection and management of information equipment within the scope (Fig. 7.6).

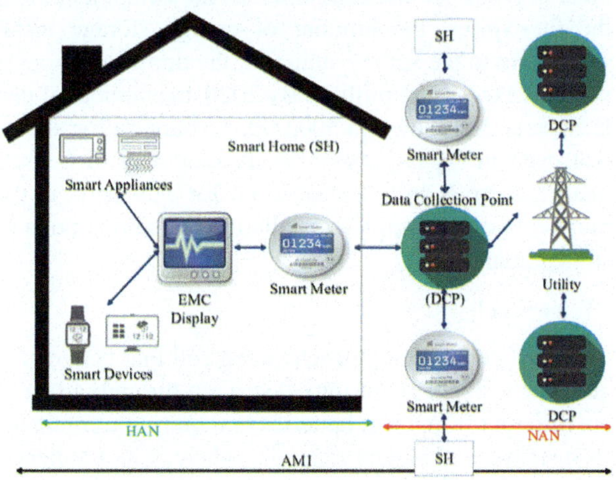

Fig. 7.6 The key technologies of HEMS

- Home energy gateway: it has the information collection of all metering equipment in the home and can realize the energy management technology of distributed energy access management and municipal power switching.

3. Visualization technology

- HEMS display terminal includes home energy display, mobile client software, and home interactive terminal. The display terminal can directly manage and manipulate all household equipment through wireless connection with HEMS and user interactive graphical interface.
- Display software: the open intelligent operating system control terminal design is adopted, and the application framework supporting component reuse and replacement is used.

4. Energy management technology

- Under the guidance and promotion of energy conservation and emission reduction policies, energy management technology will achieve rapid development, including energy management software development and design, energy consumption monitoring, and energy efficiency evaluation.

2. Photovoltaics (PV)

PV is the conversion of light into electricity using semiconducting materials that exhibit the PV effect, a phenomenon studied in physics, photochemistry, and electrochemistry. PV system employs solar modules, each comprising a number of solar cells, which generate electrical power. PV installations may be ground-mounted, rooftop-mounted, or wall-mounted. Japan has reduced the FIT price of household PV year by year and implemented subsidies for household energy storage installation in order to improve the spontaneous self-use rate and improve the power grid. This will also enable Japan to install about 130,000 household energy storage units in 2020. At the same time, as the equipment price decreases year by year and the fit expires, the number of energy storage installations will only increase. According to the survey statistics, the number of houses suitable for PV installation in Japan exceeds 25 million. By 2020, the number of installed PV and energy storage is close to 3 million and 500,000, accounting for about 10% and 2%. It can be seen that even in Japan, where the development of household PV has been relatively mature, there is still a lot of room for expansion in the future. In addition, a large number of distributed PV installations also bring potential for the future small energy storage market.

3. Fuel cell

A fuel cell is an electrochemical cell that converts the chemical energy of a fuel and an oxidizing agent into electricity through a pair of redox reactions. Fuel cells are different from most batteries in requiring a continuous source of fuel and oxygen to sustain the chemical reaction, whereas in a battery the chemical energy usually comes from metals and their ions or oxides that are commonly already present in the battery. A residential fuel cell is a scaled down version of industrial stationary fuel cell for primary or backup power generation. These fuel cells are usually based on combined heat and power-CHP or micro combined heat and power micro-CHP technology, generating both power and heated water or air. In May 2009, the world's first residential fuel cell sales began in Japan. In December 2018, the cumulative production of residential fuel cells exceeded 250,000 units.

Japan's interest in domestic fuel cells dates back to 1999, and its millennium project includes support for PEFC research. Japan's residential energy demand is large and has been growing. The Japanese government started a large-scale domestic fuel cell demonstration project in 2005. During the validation and pilot period from 2005 to 2009, nearly 5000 sets of distributed fuel cell systems were sold, reducing the system acquisition cost from 8 million yen in 2005 to 3.5 million yen in 2009, a decrease of 56.25%.

The promotion and popularization period of products is from 2010 to 2020. The Japanese government has provided subsidies of 1.4 million yen or half of the manufacturing cost to families installing fuel cell systems since 2010. In 2015, the price of the new generation of household fuel cell system launched by Panasonic, Toshiba, and other enterprises has been as low as about 1.5 million yen, down 81.25% and 57.14%, respectively, compared with 2005 and 2009. The efficiency of its thermoelectric system has increased from 70% to 95%. At the same time, the amount of government subsidies was also reduced to 500,000–600,000 yen. Since 2012, sales have almost doubled. In 2015, the subsidy amount of Ene-Farm reached 22.2 billion yen.

As Japan continues to encourage families to purchase fuel cells and manufacturers' mass production effect is expanded, the price of residential fuel cells is gradually falling. Forty-two thousand residential fuel cells were added in Japan from April 2016 to March 2017. The newly added quantity is about the same as that in 2015, and the cumulative number of installed units is 196,000, of which PEFC models with low price and short restart time account for a higher proportion.

4. Heat pump

A heat pump is a device that transfers heat energy from a source of heat to what is called a heat sink. Heat pumps move thermal energy in the opposite direction of spontaneous heat transfer, by absorbing heat from a cold space and releasing it to a warmer one. In Japan, heat pump is an electronic air conditioner, which is widely used in families and commercial spaces, so it constitutes a huge market. In recent years, heat pump products for domestic hot water supply have increased rapidly. As people pay more and more attention to environmental problems, heat pumps using natural refrigerants instead of fluorocarbon refrigerants have attracted the attention of consumers. The ozone depletion potential (ODP) of carbon dioxide heat pump is zero, and the global warming potential (GWP) is also very low.

The residential heat pump system was first introduced in Japan in April 2001. In September 2007, the cumulative shipments of the entire market exceeded one million units, and in October 2009, it exceeded two million units. In January 2014, the sales volume reached four million units (Fig. 7.7). Compared with other

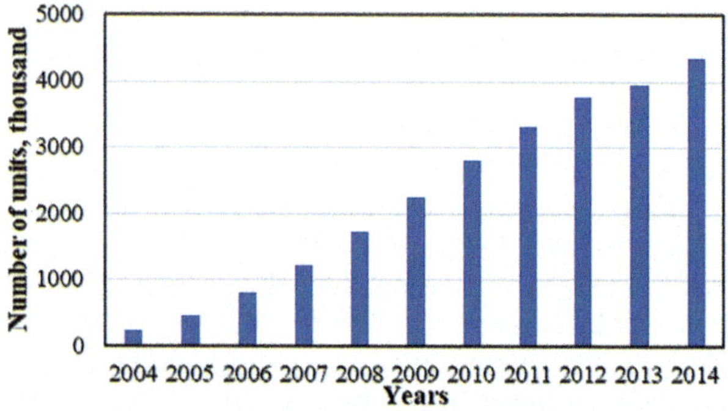

Fig. 7.7 Cumulative number of residential heat pump

countries in the world, domestic air energy heat pump water heater is the most popular in Japan. Small size, exquisite structure, leading technology, and excellent performance are the biggest characteristics of Japan's air energy heat pump water heater. "Eco-Cute" is the main model of air-water heat pump water heater using natural refrigerant (CO_2). The demand of Eco-Cute has increased steadily in recent years. CO_2 gas has the advantages of good safety, chemical stability, being harmless to the environment, large latent heat of evaporation, high refrigerating capacity per unit volume, good transportation and heat transfer properties, etc. And it still maintains high thermal efficiency in the heat exchange process with large temperature rise on the water side. The rated operating temperature can reach 600 °C, and the maximum outlet water temperature can reach 900 °C.

Eco-Cute has gradually become a new product of great concern in the Japanese heat pump market. Many new residential and commercial buildings take the heat pump water heater as the preferred home heating center. Since entering the market in 2001, the sales volume has continued to rise, and the household penetration rate reached 10% in 2007. Affected by the economic crisis, the sales volume of Eco-Cute products in 2008 still exceeded 510,000 units, although there was a certain gap with the expectation. According to the market ownership, about 1.8 million domestic carbon dioxide heat pump water heaters were sold in the Japanese market in 2008. In 2010, the sales volume was about 800,000 units, and the cumulative sales volume will reach 3.2 million units. In 2011, the output reached 1 million units, and the cumulative sales volume reached 5.2 million units.

As a rapidly growing market, heat pump water heater has attracted the attention of many manufacturers and promoted the development of relevant new products. It is certain that Eco-Cute will trigger significant changes in the central water heater market. In this market, gas water heater has occupied most of the share, and heat pump water heater has only accounted for about 6% for a long time. If the Eco-Cute market expands from now on and the total installed capacity can reach the expected 5.2 million units, the share of heat pump water heater in the central water heater market will rise to 30% or more.

5. Battery

Battery is an electric storage equipment that can be repeatedly used by charging. The battery is from a small size portable material to a large size of electric installation, and the size is approximately proportional to the storage capacity.

Household batteries are mainly divided into two types:

One is the "system disconnected type," which is the power storage type purchased from the power company. Use the socket for charging, and connect the appliances you want to use directly to the household battery. According to different models, household appliances can also be wired and connected in electrical engineering. If the line is connected in advance, it will switch automatically in case of power failure, which has certain safety.

The second type is "system connected." Through the residential distribution panel, it is connected with the electrical system in the family to supply power to household appliances and lighting and store the power produced by the solar power generation system. The stored electricity can also be used when using the electricity generated by sunlight. The system connection type is mainly used to store the power generated by solar power generation and users with large power storage capacity.

In recent years, the installation of household batteries in Japan, which can store the electricity generated by household solar panels, has increased sharply. According to FiT, during the 10-year contract period, the production company of household battery can buy the excess power of users' households at a high price. However, there are now families whose 10-year contracts have expired. However, compared with selling electricity, more families choose to use their own electricity. At present, household batteries have attracted much attention. Household batteries can generate and store electricity during the day and use it at night when electricity consumption is high. According to Sharp, the market of household batteries is expected to increase from 42,000 in 2017 to 150,000 in 2021. At the same time, due to frequent natural disasters in Japan, the demand for domestic batteries as a solution to power failure has also increased.

6. Electric vehicle

An electric car or battery electric car is an automobile that is propelled by one or more electric motors, using energy stored in batteries. Charging an electric car can be done at a variety of charging stations; these charging stations can be installed in both houses and public areas.

Out of all cars sold in 2020, 4.6% were plug-in electric, and by the end of that year, there were more than ten million plug-in electric cars on the world's roads, according to the International Energy Agency. Despite rapid growth, only about 1% of cars on the world's roads were fully electric and plug-in hybrid cars by the end of 2020. Many countries have established government incentives for plug-in electric vehicles, tax credits, subsidies, and other nonmonetary incentives, while several countries have legislated to phase out sales of fossil fuel cars to reduce air pollution and limit climate change.

Nowadays, electric vehicles can not only take the place of transportation but also supply power to the home. The vehicle to home (V2H) function with this feature has been applied in electric vehicles, and this function of plug-in hybrid vehicle (PHV) and fuel cell vehicle (FCV) is also under development. This is unmatched by previous gasoline vehicles. In the future, the energy conversion between cars and families is likely to add new value to cars.

At present, Japanese researchers have begun to explore the V2H system which is most suitable for popularizing electric vehicles to Volkswagen. Its advantage is that it can reduce the household electricity expenditure through "peak load shifting power consumption" and can be used as an emergency power supply in case of power failure.

At present, Japan has a variety of residential power generation equipment, such as solar power generation, household battery, household fuel cell, and natural gas power generation. Since the East Japan earthquake, the planned power outage policy implemented in Japan has increased the number of power outages caused by natural disasters such as typhoons and storms. Nissan Motor Company pointed out that "the actual demand of consumers for V2H is increasing."

After automobile manufacturers add V2H function to electric vehicles, some manufacturers aim to simplify the existing household power generation system, while others combine with the existing system to make the use more flexible and diverse. All manufacturers still hope to improve the added value of electric vehicles through the practical application of V2H, so as to promote the further popularization of electric vehicles.

The world's first V2H system produced by Nissan has sold about 2000 units so far. This device can convert up to 6 kW of power from the vehicle lithium battery for household use. The power conversion device (PCS) is used to convert DC power into AC power to supply power to the home. PCS configuration and automobile power output mode of different automobile manufacturers are different.

7.1.3 Location of Jono Zero Carbon Demonstration Projects

To analyze household using fuel cell, PV, and heat pump residential energy systems, we collected electricity and gas usage of users in the Jono zero carbon demonstration projects. The location of Jono region in Japan is shown in Fig. 7.8.

7.2 ZEH Low-Carbon Technology and Advanced Management

A ZEH is a house with an annual net energy consumption of around zero (or less) by saving as much energy as possible while maintaining comfortable living environment. This can be achieved through envelope thermal insulation, high-efficiency equipment, and creating energy with on-site distributed energy resources (DER).

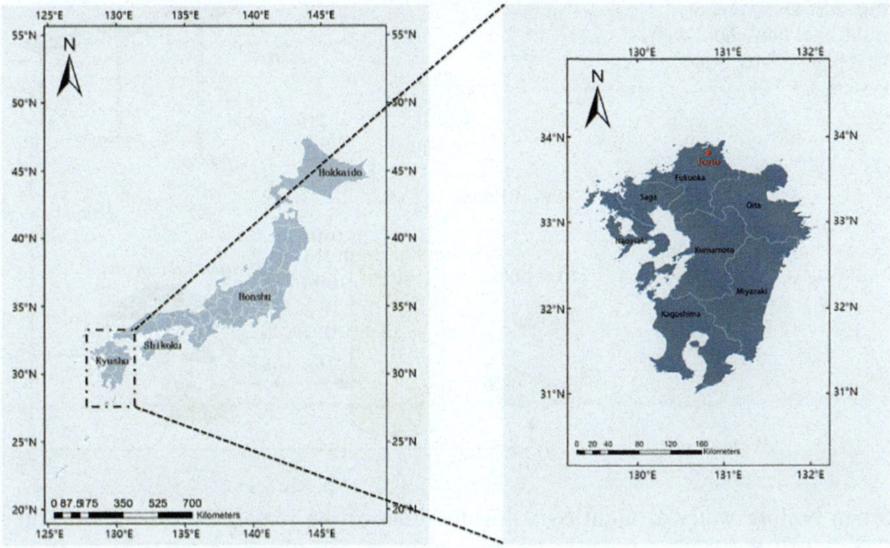

Fig. 7.8 Location of Jono region in Japan

The Japan Strategic Energy Plan sets a goal of achieving average zero emission in newly constructed houses by 2030. Buildings can be an important part of the solution in these future smart energy grids; aggregated buildings will be both energy user and producer [16]. This part aims to examine the performance of active demand side management for residential households in the social demonstration projects, which include solar PV production, thermal storage, and fuel cell system. Analysis selects objectives from different types of residential customers who have installed HEMS and cooperates with demand side management. The real history load and generation profiles were mainly collected from residential buildings, located in Jono zero carbon demonstration projects.

1. Eco-Cute water heater

Electrical power to heat and thermal energy storage are identified as effective measures to provide flexibility. Optimal control of heat pump and thermal tank enable building to bring flexibility toward the power grid. Figure 7.9 presents the function of Eco-Cute; it is suitable for residential hot water supply, especially for the house with all-electrification. Eco-Cute generally refers to a heat pump water heater, utilizing natural refrigerant (carbon dioxide) that is environmentally safe. Its capability has been improved based on customer needs, making them more multifunctional with features such as the ability to support floor heating, and providing more space-saving models. Because it makes effective use of heat extracted from air, the system can generate heat energy more than three times greater than the input electrical energy. Eco-Cute saves energy by always checking and keeping adequate bathwater temperature. Its warm-charge function saves energy

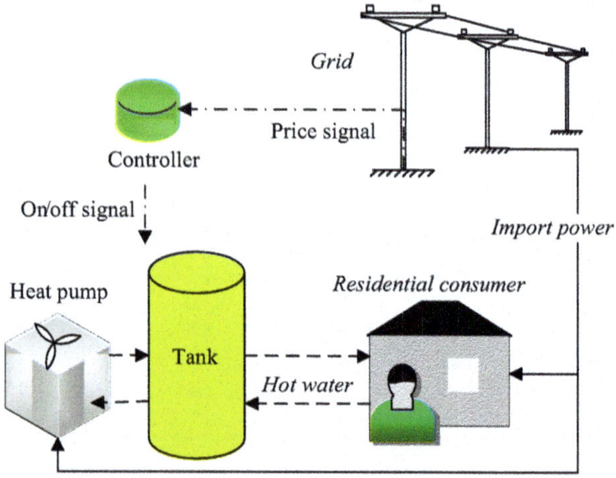

Fig. 7.9 Overview of residential household with heat pump water heater system

when boiling water at night by using the heat of the remaining warm bathwater. Temperature selection function enables users to select and set a mode from hot, standard, and warm temperature settings without adding water.

The popularity of heat pumps is generally supported by (designed specific electricity tariff scheme) energy market and policy implications. Heat pump water heater is generally considered as useful appliances for environmental protection and load shifting. Thermal storage applications are integrated to shift the daily energy consumption pattern and generally schedule the working time of heat pump water heater concentrates in lower pricing region (early morning and deep night) to provide potential economic benefits for customers. Figure 7.9 illustrates the scheme of the residential heat pump water heater system, the local controller can receive the real-time price single from the grid and switch on/off the heat pump, and the work period of heat pump is optimized according to the price information from the grid. The produced heating will be stored in thermal tank and then released for daily later use. Figure 7.10 presents the color scale distributions of power consumption of heat pump water heater in a typical residential household. Working period of heat pump mainly lasts from 0:00 to 7:00 am and locates in the valley period of demand load. Operating time becomes shorter with daily decreasing heating demand; heat pump water heater system shows higher power consumption density in winter, attributed to the higher heating demand and lower generating efficiency under low ambient temperature. Heat pump water heater tends to operate earlier in wintertime to meet the daily heating load, which may be highly dependent on activity-based load.

2. Rooftop PV and battery system

Rooftop solar PV system plays a significant role in raising local energy self-sufficiency. Generally, PV system has a connection with the public electricity grid via an inverter. The distributed PV battery is expected to reduce the customer's electricity bill and participate more in grid load optimization through optimal

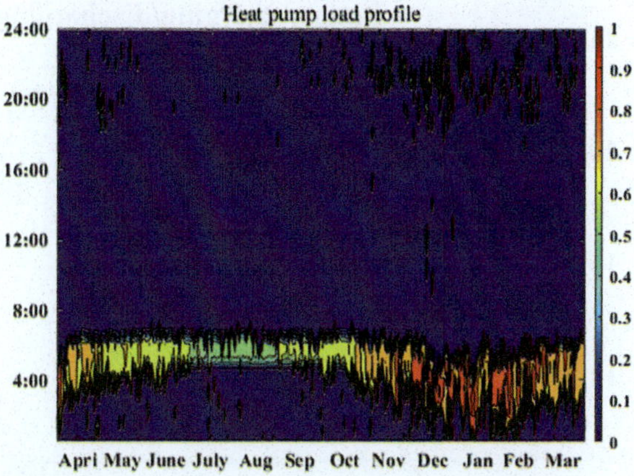

Fig. 7.10 Color scale distribution of power consumption of heat pump system in a typical residential house

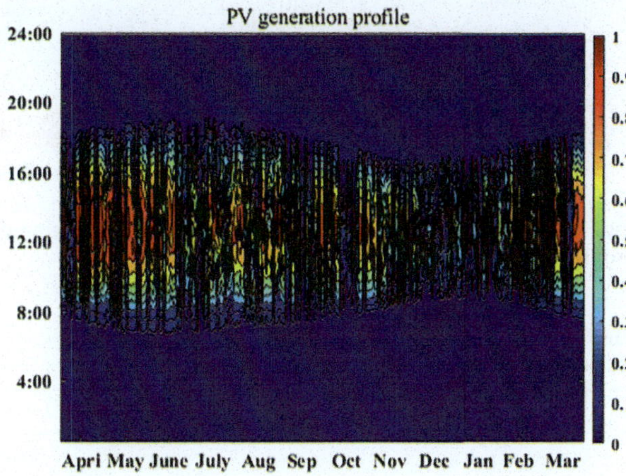

Fig. 7.11 Color scale distribution of PV power generation

management strategies in aggregated level. Figure 7.11 presents the variation of annual PV generation, high generation period concentrates in mid-season period.

The dynamic battery dispatch scenarios are shown in Fig. 7.12; the rule-based control forced the charging process of the battery into the off-peak price period and shifted the load upward during this period, and then its discharging process started during night peak price hours. The scheduling leads to peak power load concentrates over a shorter period at about 3 hours, discharging flow with the negative value corresponds with residential peak price period, and the cycling is expected to reduce power purchased from the utility with a high price under the time-of-use tariff scheme.

The Japan government aims to capture 50–70% of next-generation vehicles to total new car sales by 2030; electric vehicle plug-in hybrid vehicle is expected to account for 20–30% of diffusion target. Panasonic develops the smart house linked to cloud computing service; its function supports EV charging facilities and smart

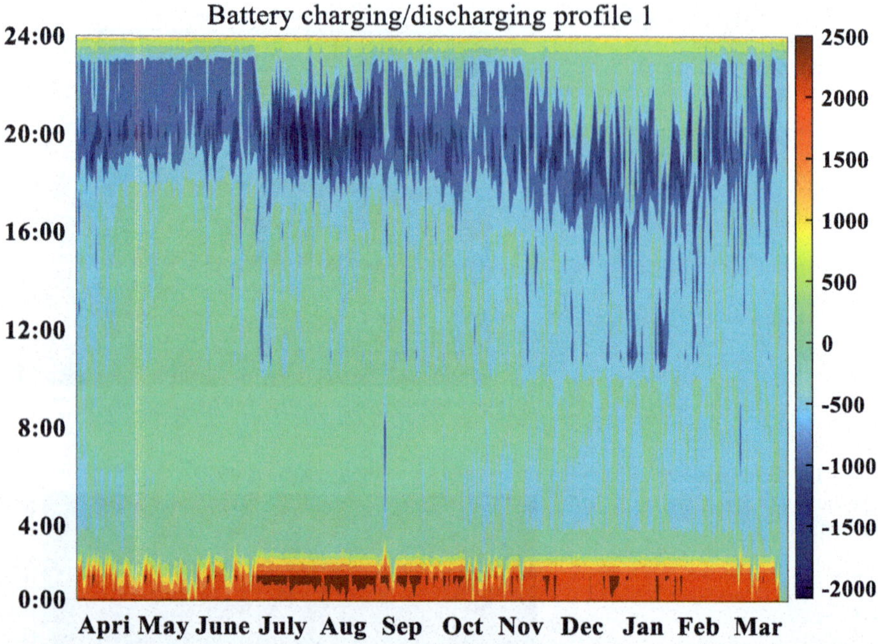

Fig. 7.12 Color scale distribution of battery storage charging-discharging dispatch power flow profile

Fig. 7.13 Scheme of residential household with smart V2G system

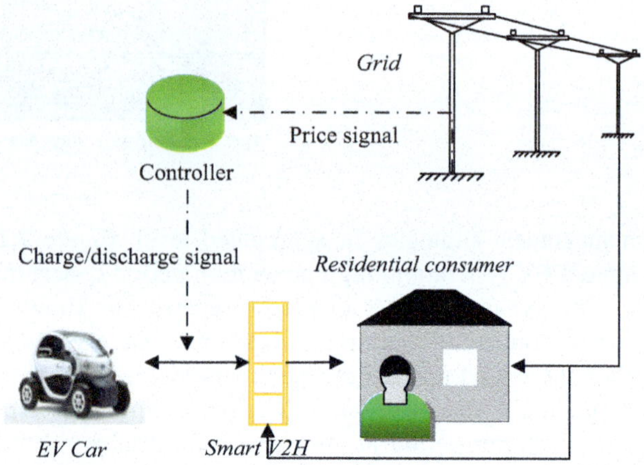

meters equipped with communication functions, making them useful as electricity supplies in communities. Figure 7.13 illustrates the scheme of EV application in residential sector. The plug-in electrical vehicle not only receives charge from the grid to power the vehicle but also provides backup power to the household. EV provides two-way flow from grid and to home, functioning electric power grid for peak shaving or enhancing local energy security.

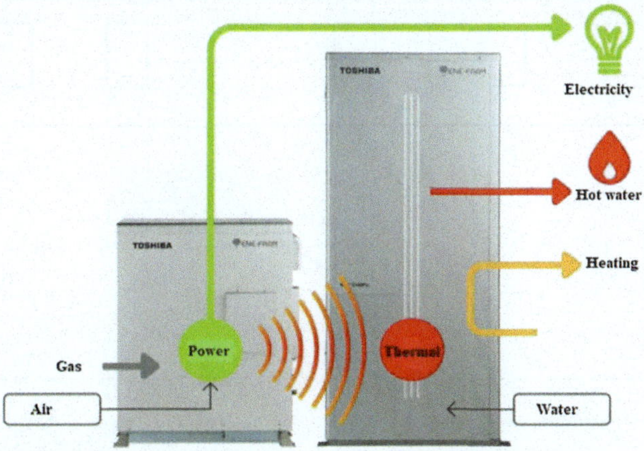

Fig. 7.14 The fuel cell cogeneration system

Potential opportunity to manage increasing electricity costs and demand spikes is the utilization of EVs to act as an aggregated energy store, providing peak shaving or demand shifting to both local buildings and to the power system when demand is high. Electrical vehicle for residential demand response could bring potential benefits to both of power supply and demand side, supporting peak reduction from aggregated form and reducing customer energy cost under time-of-use tariff scheme. This section will mainly analyze the techno-economic performance of residential EV system (V2H).

3. Micro-cogeneration system

Japan's Ene-Farm program is arguably the most successful fuel cell commercialization program in the world. As shown in Fig. 7.14, the fuel cell (Ene-Farm) devices, packaged in enclosures about the size of a refrigerator, convert natural gas into electricity and heat that can be used for hot water and space warming. Ene-Farm is the mainly applied micro-cogeneration technology in Japanese residential buildings, which couples the residential heat and electricity demand. The nominal power output capacity of the fuel cell is 700 W, thermal output is 998 W, the volume of combined water tank is 130 L, and stored water temperature is around 60 °C [17]. The cogeneration system runs in combined heating and power mode tracking thermal load. Panasonic, one of the main producers of the devices, claims 95% total energy efficiency.

As shown in Fig. 7.15, the power output of fuel cell is limited to the electricity demand and its nominal capacity (0.70 kW). The cogeneration system runs with thermal tracking strategy, and on/off operation cycle of the fuel cell is controlled by the energy consumption patterns and amount of hot water in thermal storage tank.

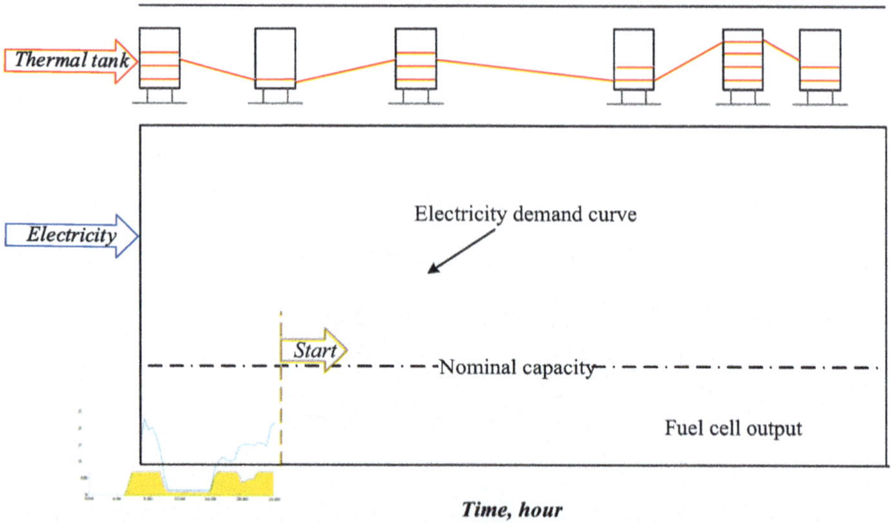

Fig. 7.15 Operational CHP mode of fuel cell in a typical day

Fig. 7.16 Fuel cell power output profiles

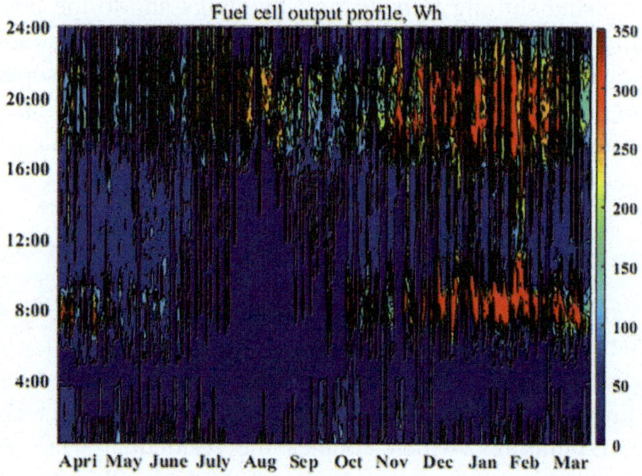

Annual output of fuel cell at 30-min is described in Fig. 7.16, the measured daily maximum output of the fuel cell concentrates in the time when the PV output is absent, its output is limited to the simultaneous heat energy demand, and large heat demand enables prolonged working period and increasing power generation in winter days.

7.3 Application and Evaluation of HEMS in Residential Multi-energy System

7.3.1 *IoT Smart Energy Management*

Internet of Things (IoT) is a technology consisting of sensing, network infra, and service interface that connects things to networks and shares information. Smart meter combined IoT enables the real-time information exchange between power supplier and customer. Private sector and local grid are expected to contribute to ensuring the efficient and sustainable use of natural resources and reducing carbon emissions. Home energy management systems (HEMS) also integrate IT networks with information on energy supply and use from appliances. Driven by the potential benefits from demand side management, HEMS is widely promoted to reduce household energy demand in the Japan national energy roadmap launched by METI in 2014. HEMS allows for automated measure and display of energy use information and thus stimulates energy conservation. The Japanese government has, through the introduction of subsidies, encouraged more construction companies and buildings to install HEMS devices. This has led to an increase in the number of smart houses in Japan as well as a target to have HEMS devices installed in every household by 2030.

Occupants' behavior contributes significantly to the energy conservation of the building energy system [14]. The function of HEMS involves the real-time control of on-site power generators and home appliances, monitoring and communicating building operating conditions and coordinating control of loads, and providing highly efficient energy supply resolution for customer, including space heating, cooling, and hot water supply. Currently the ZEH in Japan is a generally efficient energy system that features on-site renewable energy resource, cogeneration system, and smart energy management strategy under HEMS environment. Excess energy from local renewable generators, which cannot be directly self-consumed, is usually sold to the grid. Smart HEMS can also help the consumer to effectively control the cogeneration units, such as operating the micro combined heating and power unit via tracking the simultaneously electrical and thermal loads considering the balancing constraints; coordinated energy-saving control on home appliances, for example, maintaining the indoor temperature by 28 °C and after 30 min 20 °C air cooling conditioning operation in summer; and automatic lighting illumination adjustment for electricity saving. Customers can acquire the real-time energy consumption through APP visualization, turn on or off home electric appliances, and manage their operations in time, which can induce the customer to form an energy-saving habit [18].

A grid-interactive efficient building expands demand flexibility options beyond traditional demand response because of the smart technologies like advanced sensors and controls and visualization function that can actively manage DER and adjust a building's energy consumption. In order to facilitate controllability on the subset of smart houses and management of local energy system, the community energy

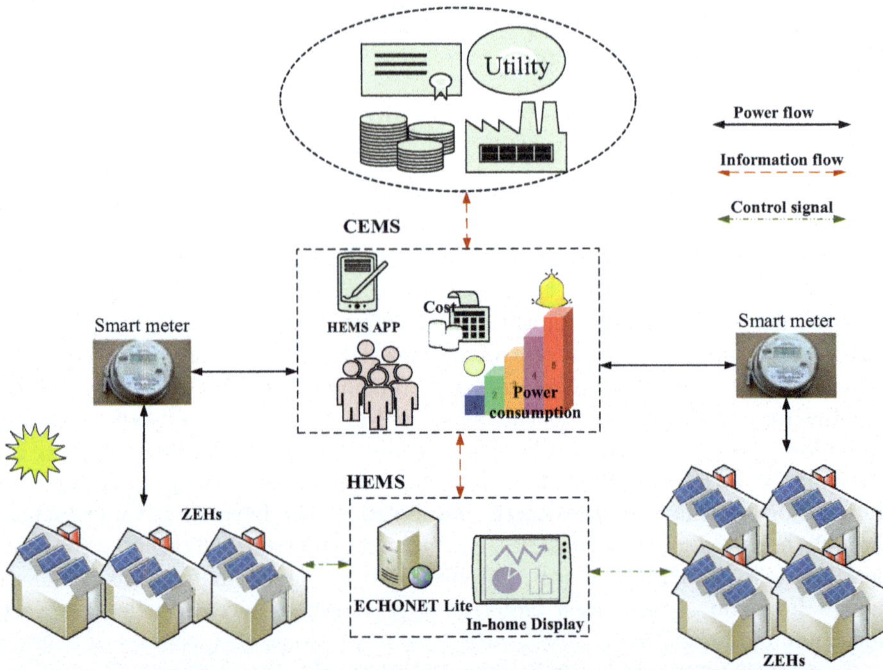

Fig. 7.17 Overview of system structure and integrated network between CEMS and HEMS

management system (CEMS) as an energy and information hub is introduced. As depicted in Fig. 7.17, the two-way information communication network between HEMS and CEMS could automatically record smart meter data and transmit it through IoT at both household and utility service level and enable consumers to easily access their own real-time electricity consumption behaviors and identify individual energy cost rank among community members through the online information.

Smartphone app using HEMS renders home energy use visible and intelligently controls equipment. As shown in Fig. 7.18, the user-friendly interface of smartphone supports awareness of energy issues, supports consumers to compare and manage their real-time power consumptions, and participates more responsibly in improving district energy system flexibility or energy consumption reduction [19].

7.3.2 Management Strategy and Suggestions

Major changes in the energy sector in recent years mean that the need for smart, flexible energy is increasing. Participating objectives in a flexible public grid would come from both supply and demand sides. It is expected that markets related to "smart technology" will expand as it contributes to energy conservation and

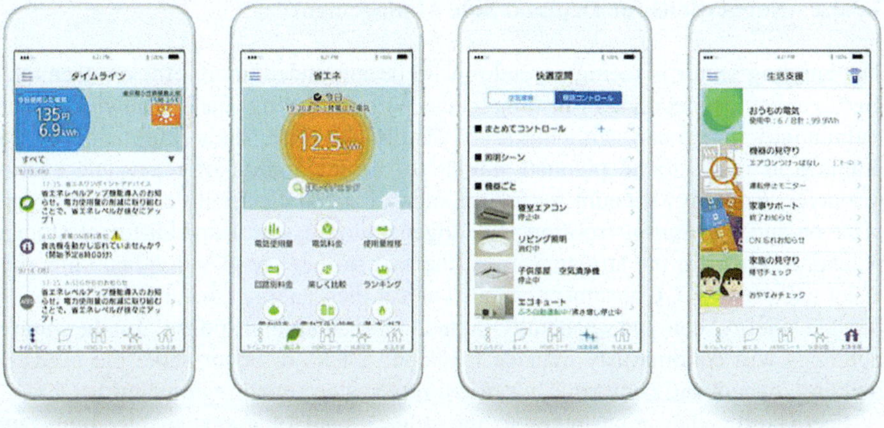

Fig. 7.18 Interface of HEMS APP

improves the convenience of everyday life. Low-carbon transformation in power sector and smart demand side management are expected to play increasing roles in developing renewable energy dominated public grid; energy production and energy consumption will become more coordinated.

7.3.2.1 Management from Supplier Perspective

The VRE intermittent output has a great influence on grid supply-demand balance, especially under high renewable energy penetration level. Performance investigation of VRE further integration can help utility understand future electricity market investment. Increasing renewable energy penetration level brings promising environmental benefit and energy self-sufficiency security at district level. However, the variable renewable energy differs from the conventional power supply technologies; variable renewable energy shows low peak capacity credit. Due to the grid flexibility constraint and low correlation between solar generation and load profiles, we can observe that the PV production can greatly shape the grid residual load, and PV integration mainly decreases the output from medium based plants.

Limited to flexible resources, actual market value of renewable energy would drop due to renewable power curtailment considering grid flexibility. The electricity spot trading price was reduced with increasing PV power share, especially at certain rising penetration ranges. To meet the above challenges, power to heat and power to hydrogen productions as potential solutions are expected to support the cost-effective integration of a large share of renewable energy production and comprehensively decarbonize the social energy system.

7.3.2.2 Suggestions for Demand Side Management

The building sector is facing a trend toward decentralized, more efficient technologies to cover the electrical or heating loads. With increasing share of efficient power technologies integrated with electrical distribution grid, their integration to the public grid needs to be planned similar to the integration of renewable energy resources. Buildings account for a large ratio of social electricity consumption and offer on-site generation (rooftop PV, cogeneration system) and different storage potentials, either in the structure itself (thermal storage) or in individual units (hot water tank, battery). Consumers can also adjust their energy consumption to have a flexible energy demand, generally based on incentive response. In the future, buildings will continuously manage loads and DERs to better serve the needs of building owners and occupants, electric utility systems, and regional grids.

Energy efficiency contributes to the achievement of a sustainable future and improves the social welfare through reducing carbon emission. With the widespread of behind smart meter and development of advanced information communication, high-efficiency appliances, integrated distributed power resources, and smart management are expected to be important flexible energy resources of future power supply system. Currently, high initial capital investment may be the main barrier to promote the energy efficiency products in demand side. Utilities and policymakers can implement financial incentive mechanisms via policies, rate designs, and programs for customers to shoulder part of capacity cost that may enhance the wider uptake of grid-supporting high-efficiency technologies. Direct subsidies to heat pump water heat system and cogeneration are essential for their wider development due to the high initial capital investment. Being government-led, customer participation, and being business-driven are expected to become the main features in sustainable development of smart community energy system.

7.4 Influence of Power Market Fluctuation on Residential Electricity Cost

To address the primary energy shortage problem, Japan has implemented a series of policies and measures for residential energy conservation and emission reduction. Among them, the home energy management system (HEMS) in a smart house as a hub connecting users and power companies to realize energy visualization has been widely studied.

The research object of this study is a two-story detached smart house integrated with HEMS in the "Jono smart house area" in Japan. A dynamic pricing model was developed to guide the users' electricity consumption behavior and adjust the grid load. The annual electricity charges of users under the three pricing schemes of multistep electricity pricing (MEP), time-of-use pricing (TOU), and real-time pricing (RTP) were calculated and compared.

Table 7.1 Unit price and basic charge incurred by the target household by selecting MEP and TOU

	Basic charge (yen/contract)	Unit price (yen/kWh)		
MEP	1782.0	0–120 kWh		17.46
		121–300 kWh		23.06
		300kWh-		26.06
TOU	1650.0	8:00–22:00	Spring/autumn	23.95
			Summer/winter	26.84
		22:00–8:00		13.21

7.4.1 *Comparison of Electricity Cost Among Different Pricing Models*

Power demand response refers to the market participation behavior of power users in response to the market price adjusted signals or according to the incentives of power companies to change their inherent usage patterns. It is a solution for demand side management. The price mechanism forms the core of the power market mechanism. In addition, reasonable electricity price can reflect the size of social benefits, realize the optimal allocation of power resources, allow users to select a reasonable power consumption time, and adapt to the intermittent characteristics of distributed power generation.

At present, the main pricing strategies used in the power market include MEP, TOU, and RTP. With the characteristics of stepped-type electricity price, MEP is applicable to most types of residential energy systems and is a common choice of users. Table 7.1 presents the unit price and basic charge incurred by the target household by selecting MEP and TOU [20]. The monthly electricity cost incurred by the customer is calculated by summing the basic charge and unit price.

In general, RTP has an economic advantage over TOU in terms of electricity cost, but due to the long operation time of the heat pump and large power consumption at night in winter, it is still higher than TOU. The cost of electricity in January was 22,200 yen for RTP and 24,724 yen for TOU. A comparison of the annual electricity cost of the three tariff models is illustrated in Fig. 7.19. In the absence of an evident gap between MTP and TOU, the application of RTP makes the system economy better than that in the other two modes. There is a huge difference in the electricity costs between MTP and RTP in winter; in contrast, the cost of TOU is evidently different from that of RTP in summer.

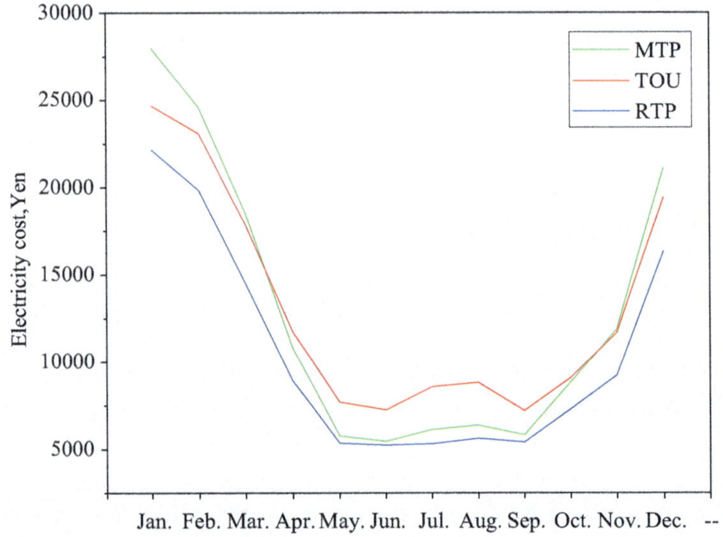

Fig. 7.19 Comparison of annual total electricity cost

7.4.2 Comparison Results and Power Market Suggestions

Faced with the problem of primary energy shortage, residential energy consumption has been a main objective of Japan's energy conservation and emission reduction strategy. HEMS as a hub in smart house which connects the supply and demand sides realizes the function of two-way communication between users and power companies. At present, most consumers select MTP and TOU as the modes to calculate electricity charges. According to the characteristics of real-time monitoring and feedback of energy consumption data in HEMS, Japan electric power company has not yet developed an applicable RTP scheme. Here, a smart house integrated with HEMS in the Jono area in Japan was selected as the case study, and the short-term load forecasting of user energy consumption was conducted based on the historical data. An RTP model was established based on the consumer's load forecast results of the next day, with an aim to reduce the annual electricity cost incurred by the user by adjusting their energy consumption behavior and transferring the peak load. An RTP model is established, and the load of the power grid can be adjusted to guide users' electricity consumption behavior through the model. This model is compared with MTP and TOU, and the annual electricity charges of the three pricing schemes are calculated according to the prediction results of the short-term load forecasting model. The result indicates that the annual electricity cost generated by RTP is less than that generated by MTP and TOU, and RTP's economic advantage becomes evident in case of high energy consumption.

References

1. Schwartz LC, Leventis G (2020) Grid-interactive efficient buildings: an introduction for state and local governments. Lawrence Berkeley National Lab (LBNL), Berkeley
2. Kim JJ (2016) Economic analysis on energy saving technologies for complex manufacturing building. Resour Conserv Recycl 123:249–254
3. Li Y, Gao W, Ruan Y (2017) Feasibility of virtual power plants (VPPs) and its efficiency assessment through benefiting both the supply and demand sides in Chongming country, China. Sustain Cities Soc 35:544–551
4. Houwing M, Negenborn RR, De Schutter B (2011) Demand response with micro-CHP systems. Proc IEEE 99:200–213
5. Sivaneasan B, Kandasamy NK, Lim ML, Goh KP (2018) A new demand response algorithm for solar PV intermittency management. Appl Energy 218:36–45
6. Carrilho da Graça G, Augusto A, Lerer MM (2012) Solar powered net zero energy houses for southern Europe: feasibility study. Sol Energy 86:634–646
7. Li Y, Gao W, Ruan Y, Ushifusa Y (2018) Demand response of customers in Kitakyushu smart community project to critical peak pricing of electricity. Energ Buildings 168:251–260
8. Asadinejad A, Tomsovic K (2017) Optimal use of incentive and price based demand response to reduce costs and price volatility. Electr Power Syst Res 144:215–223
9. Fotouhi Ghazvini MA, Soares J, Abrishambaf O, Castro R, Vale Z (2017) Demand response implementation in smart households. Energ Buildings 143:129–148
10. Iqbal A, Ullah F, Anwar H, Kwak KS, Imran M, Jamal W et al (2018) Interoperable internet-of-things platform for smart home system using web-of-objects and cloud. Sustain Cities Soc 38: 636–646
11. Song Q, Sheng W, Kou L, Zhao D, Wu Z, Fang H et al (2016) Smart substation integration technology and its application in distribution power grid. CSEE J Power Energy Syst 2:31–36
12. Kawashima A, Yamaguchi T, Sasaki R, Inagaki S, Suzuki T, Ito A (2015) Apartment building energy management system in group optimization with electricity interchange using in-vehicle batteries. SICE J Control Meas Syst Integr 8:52–60
13. Ahmed M, Kang Y, Kim Y-C (2015) Communication network architectures for smart-house with renewable energy resources. Energies 8:8716–8735
14. Shakeri M, Shayestegan M, Abunima H, Reza SMS, Akhtaruzzaman M, Alamoud ARM et al (2017) An intelligent system architecture in home energy management systems (HEMS) for efficient demand response in smart grid. Energ Buildings 138:154–164
15. Zamanloo S, Askarian Abyaneh H, Nafisi H, Azizi M (2021) Optimal two-level active and reactive energy management of residential appliances in smart homes. Sustain Cities Soc 71: 102972. https://doi.org/10.1016/j.scs.2021.102972
16. Li Y, Zhang X, Gao W, Xu W, Wang Z (2022) Operational performance and grid-support assessment of distributed flexibility practices among residential prosumers under high PV penetration. Energy 238:121824
17. Panasonic (2020) Residential fuel cell cogeneration system.
18. Li Y, Gao W, Zhang X, Ruan Y, Ushifusa Y, Hiroatsu F (2020) Techno-economic performance analysis of zero energy house applications with home energy management system in Japan. Energ Buildings 214:109862
19. Panasonic (2019) AiSEG2 HOME IoT, home energy management System
20. (13 December 2019). *Kyushu island electricity price plan overview and unit price list*. Available: http://www.kyuden.co.jp/user_menu_plan_index.html

Chapter 8
Maintenance and Reliability in Distributed Energy Resource System

Jinming Jiang, Zhonghui Liu, and Weijun Gao

8.1 Maintenance and Reliability of a Complex System

8.1.1 Distributed Energy Resource System, a Complex System

The DER system is a complex energy conversion system. As introduced above, the DER system is a kind of energy that can be close to the user side, contain various energy types, and produce electricity, heat, cold, hot water, and other forms of energy.

J. Jiang (✉)
Innovation Institute for Sustainable Maritime Architecture Research and Technology, Qingdao University of Technology, Qingdao, China

Faculty of Environmental Engineering, The University of Kitakyushu, Kitakyushu, Japan

Z. Liu
Faculty of Environmental Engineering, The University of Kitakyushu, Kitakyushu, Japan

School of Municipal and Environmental Engineering, Jilin Jianzhu University, Changchun, China

W. Gao
The Engineering Academy of Japan, Chiyoda City, Tokyo, Japan

The University of Kitakyushu, Kitakyushu, Japan

Qingdao University of Technology, Qingdao, China

International Research Center for Green Buildings and Low-Carbon Cities, Zhejiang University, Hangzhou, China

Xi'an Jiaotong University, Xi'an, China

W. Gao (ed.), *Distributed Energy Resources*, Green Energy and Technology,
https://doi.org/10.1007/978-3-031-21097-6_8

8.1.1.1 The Complexity of Energy Types

The DER systems can use a variety of energy sources and convert them into a variety of energy forms. The primary energy can use renewable energy sources, such as solar, wind, and hydrogen, and non-renewable energy sources such as carbon, natural gas, oil, etc. For an efficient and low-carbon DER system, the primary energy used is often a combination of various energy sources. At the same time, distributed energy can transform primary energy into electricity, space heating or cooling, domestic hot water, and other energy forms to meet users' needs. For example, the DER system of Kitakyushu Science and Research Park in Japan is a hybrid DER system, which combines natural gas and solar energy as the primary energy and, to meet the electricity demand, space heating in winter, space heating in summer, and domestic hot water [1]. Therefore, one of the complexities of the DER systems is the complexity of its primary energy types and energy utilization forms.

8.1.1.2 The Complexity of Technologies

Because of the complexity of primary energy types and energy utilization forms, the technical complexity of the DER systems is determined. Generally, the DER system includes electric energy conversion units, heat energy conversion units, energy storage units, etc., including various energy generation technologies, energy conversion technologies, and energy storage technologies. Akorede et al. [2] gave us a review of the distributed generation technologies and energy storage technologies. The different energy conversion technologies correspond to other energy types and provide different energy types. Any distributed energy system embraces multiple technologies; that is to say, the DER system is complex technology integration. Therefore, one of the complexities of distributed energy systems is technological complexity.

8.1.1.3 The Complexity of Maintenance

Maintenance is a complex activity and process. The complexity of maintenance is mainly reflected in the following three aspects: (1) the complexity of the maintenance subjects; (2) the complexity of the maintenance types; and (3) the complexity of the maintenance levels.

The Complexity of Maintenance Subjects

Maintenance subject refers to the party who undertakes or partially undertakes the maintenance responsibility of the equipment during the maintenance of the system. For example, the manufacturer or supplier of a device may be the main maintenance

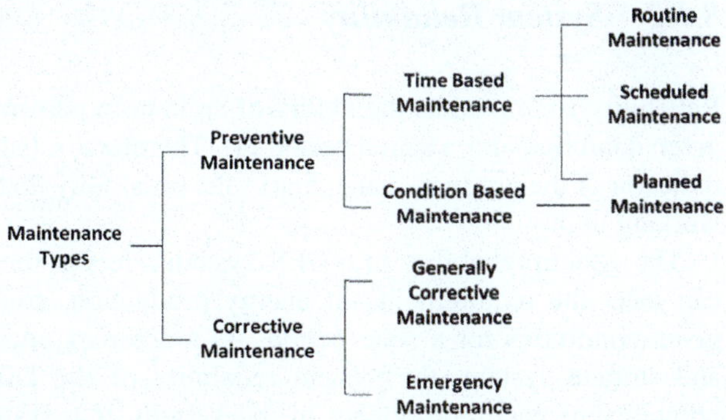

Fig. 8.1 The classification of maintenance types

subject of the device, and the user and maintainer of the device are also the main maintenance subject of the device. They all take responsibility for the health of the equipment. There are many maintenance subjects for a DER system, and they often take different responsibilities.

The Complexity of the Maintenance Types

A maintenance definition which has been proposed by Geraerds (1985) is widely recognized as "all maintenance activities aimed at keeping an item in, or restoring it to, the physical state considered necessary for the fulfillment of its production function" [3]. To ensure the system running in health, there are multiple maintenance types based on the requirements of different devices and components. For the maintenance strategy of a system, a single maintenance type cannot meet the requirements of different devices and maintenance layers. Therefore, the maintenance process of a system is often a combination of multiple maintenance types. Therefore, the maintenance of a DER system is complex. Figure 8.1 shows some maintenance types for a system; more details are presented in Sect. 8.1.

The Complexity of the Maintenance Levels

The maintenance level is the level at which a device or system is restored to its original state through maintenance actions here. The different maintenance behaviors cause different maintenance levels. Reliability levels are commonly used to describe system status. For example, overhaul maintenance maybe can make the reliability level recovery to the initial or better. The corrective maintenance can restore the reliability level to some extent but lower than the initial state when a maintenance action or process occurs. The level of maintenance performed can result in different maintenance results, which can make the reliability level of the device higher or lower. Therefore, the maintenance level in the maintenance process is complex and often difficult to calculate or detect.

8.1.2 System Reliability

Reliability is defined as the ability of an item to perform a required function under given conditions for a stated period [4]. Therefore, reliability is an important quality indicator of the product, which marks the possibility that the product will not lose its working ability.

The system reliability of a DER system refers to the extent to which the system can meet the requirements of energy production, conversion, and storage under given conditions for a state period. As a complex energy production, conversion, and storage system, the system reliability of the DER system depends on the reliability of each component in the system. If a system component fails, it can lead to the failure of a unit, a sub-system, or the whole system.

If the failure of a component will lead to the failure or shutdown of the entire system, then this component is a critical component for the whole of the system and needs more attention in maintenance. If a component fault has little or no impact on the entire system, the component is not a critical component. You only need to periodically check the component during maintenance to meet the maintenance requirements. In system reliability analysis, whether a component is a critical component depends on its impact on system reliability. The connection relationship between components determines the reliability of the entire system. Components in the system are connected to other components in series or parallel. More information about reliability calculation will be introduced in Sect. 8.3.

8.1.3 Operation and Maintenance

Maintenance, replacement, and reliability are the key factors of an equipment or a system to make the equipment or system complete the function within a function or production period. For a DER system, the operation and maintenance (O&M) are the main activities during the life cycle of a system.

Maintenance is an orderly and systematic administrative, financial, and technical framework for assessing, planning, organizing, monitoring, and evaluating maintenance and operation activities and their costs continually. The maintenance plan also depends on the maintenance experience, and the purpose of the maintenance plan is to make sure the equipment can be operated on the system requirement functions. Therefore, an analysis of system maintenance strategy, system operation status, failure, maintenance cost, and effect of the function should be performed in the maintenance process.

Maintenance is seen as an opportunity to save energy [5]. Better maintenance can reduce the operating and maintenance costs of the system and thus reduce the system's life cycle costs. In addition, good maintenance can reduce the system's failure rate and improve its reliability.

From the point of view of system reliability, it is indispensable to improve the system's reliability and improve the utilization efficiency of products. Maintenance can only enhance the system's reliability in operation and maintenance stage. However, an excessive focus on improving reliability can lead to excessive maintenance and thus increase the total life cycle costs; inadequate maintenance may reduce system reliability. Therefore, there needs to be a balance between maintenance costs and system reliability requirements.

The maintenance of any system should consider several points:

1. When to carry out maintenance and the frequency of maintenance.
2. How to use maintenance.
3. How much is the repair cost?
4. Downtime during maintenance (downtime often comes with costs).
5. Demand for types and numbers of maintenance personnel.
6. Required maintenance preparation (tools, spare parts, etc.)

The main objectives of operation and maintenance include the following:

1. To adopt reasonable maintenance strategies to improve system reliability and reduce life cycle costs.
2. To avoid the waste of energy caused by system failure and performance degradation and manage for energy saving.
3. To provide a basis for system design optimization, update, and transformation.

8.1.4 Significance and Challenges of Maintenance and Reliability Research of the DER System

8.1.4.1 Significance of Maintenance and Reliability Research

The DER system has been built up and applied for decades; many application programs on the DER system are closed to the design life of the DER system. As indicated in the above presentation, there are many devices or components in a DER system. Thus, a DER system is a complex system that can meet multiple functions. When a device or component has failed, the whole system or a sub-system will be failed. The energy supply or a part of the function will be interrupted. Therefore, one of the primary purposes of DER system utilization is to ensure the reliability and availability of components in the DER system and keep components in a good (or working) state. However, most previous studies are focused on the optimization and analysis of the DER system to reduce the total life cycle cost; few articles discuss the reliability and availability of DER systems. Thus, the research on maintenance, reliability, and renewal of distributed energy systems is an important research direction to improve the reliability of distributed energy systems and further reduce the operation and maintenance costs and optimize the design of distributed energy systems.

8.1.4.2 Challenges

Reliability and maintenance management is a complex multidisciplinary research system. Usually, its research content can increase availability and security, constituting the study of reliability, availability, maintenance, and security (RAMO); the four are almost inseparable and complement each other in the research.

The research on reliability and maintenance management of distributed energy systems should focus on the whole life cycle cost of the system. The optimization of a maintenance strategy is the key to reducing the operation and maintenance cost of the DER system. Secondly, for complex repairable systems, clarifying the maintenance priority of internal sub-systems or components of the system is an effective means to avoid waste of maintenance resources and save costs. Third, system reliability analysis and maintenance should provide methods for system design optimization. Fourth, in changing from a traditional maintenance strategy to an intelligent maintenance strategy driven by big data, the maintenance of distributed energy systems should be improved intelligently.

8.2 Maintenance Optimization of a DER System

8.2.1 Maintenance Strategy

A maintenance strategy is a specified maintenance process and project for the deterioration of a system or device. At the initial stage of system maintenance, maintenance strategies can be used to maintain the device according to the suggestion of the device manufacturer. Maintenance strategy optimization can be performed based on system conditions to save costs and improve reliability.

8.2.1.1 Corrective Maintenance

Corrective maintenance (CM) refers to the maintenance task of identifying, isolating, and repairing faults so that the faulty equipment, machine, or system can be restored to the normal operating state within the allowable error range. The CM is also called break-down maintenance, and the maintenance is based on whether the fault is intact or available. Maintenance is based on whether the fault is intact or usable. CM adopts the fault maintenance strategy of maintaining when a fault occurs, restoring the original state only after the device's partial or complete failure. It can be divided in detail into immediate corrective maintenance and deferred.

CM maintenance is classified into the following categories according to the recovery of the operating status of the maintained devices.

1. Perfect Repair or Maintenance. Maintenance that restores a system or operating state to a state as new.
2. Minimal Repair or Maintenance. An activity of maintenance that restores the system to the failure rate at which a fault occurred. After minimal repair or maintenance, the operating state of a system is often referred to as the obsolete operating state.
3. Imperfect Repair or Maintenance. A maintenance activity in which the system is restored to a non-new state, but the system's operating state is improved.
4. Worse Repair or Maintenance. Maintenance that restores the system to a non-new state but does make the system better.
5. Worst Repair or Maintenance. Maintenance activities that unintentionally cause a system or equipment to fail or collapse.

The previous maintenance strategies have different maintenance results (levels), reflecting the maintenance complexity.

8.2.1.2 Preventive Maintenance

Preventive maintenance is to keep the equipment or system in a satisfactory running state, according to the production plan and maintenance experience before the occurrence of serious failure, according to the specified time or interval to stop testing, inspection, closure, replacement of parts, in order to prevent damage, secondary damage, and production loss.

Preventive maintenance is widely used in power systems and distributed energy systems. The DER system is complex, and its preventive maintenance process is as follows:

1. Identify the system's fundamental fault modes according to the system's structure and functional characteristics.
2. Select the appropriate residual life prediction method according to the obtained degradation or failure data samples.
3. The cost function of preventive maintenance and timely repair for each failure mode of the steel member is distributed by the predicted remaining life.
4. Use an optimization algorithm to obtain the best preventive maintenance scheme to minimize the cost of maintenance decisions.

8.2.1.3 Condition-Based Maintenance

Condition-based maintenance (CbM) is a preventive maintenance strategy based on state detection technology. Conditional maintenance is premised on the fact that many failures do not occur instantaneously. If this failure process is developing, measures can be taken to prevent the failure or avoid serious consequences. Therefore, condition-based maintenance technology includes data acquisition, feature extraction, fault detection, fault diagnosis, and fault prediction.

Data acquisition may involve various data galaxies, such as temperature, pressure, velocity, etc. Feature extraction includes Fourier transform, data filtering/smoothing, temperature/pressure ratio, etc. Fault detection algorithms alert users to potential risks. Fault diagnosis algorithms isolate and identify faults of specific components or sub-systems. The fault prediction algorithm estimates the device's usable life RUL or failure probability based on the history and current operating state. Fault identification and diagnosis technologies play an important role in condition-based maintenance.

8.2.1.4 Predictive Maintenance

Predictive maintenance (PdM) is preventive maintenance based on CbM. When the equipment or system is running, it periodically or continuously detects and diagnoses the status and faults of major components or parts. Determine the status of the device or system and predict the future development trend of the device status. In addition, a predictive maintenance plan is made based on the device status trend and possible failure modes. Thus, determine the time, content, method, and necessary technical services and material support for the maintenance of the machine.

The technologies of PdM include condition detection technology, fault diagnosis technology, condition prediction technology, and maintenance decision technology. Condition monitoring technology and fault diagnosis technology are the foundation, condition prediction technology is the key, and maintenance decision technology gives the final suggestion. Fault identification and diagnosis technologies are the key technology of CbM.

8.2.1.5 Intelligent Predictive Maintenance

The Internet of Services (IoS), big data, artificial intelligence (AI), Internet of Things (IoT), cloud computing, and other new information technology have promoted the rapid development of intelligent manufacturing technology. Cyber-physical systems (CPS) can monitor the objects and processes, create a virtual copy of the physical world, and decentralize control and decision-making. Compared with PdM, intelligent predictive maintenance mainly uses CPS, AI, IoT, IoS, and other technologies to monitor and analyze the status, making the whole system more intelligent. The research and application of intelligent predictive maintenance have been carried out, representing the development trend of maintenance strategy in the future.

8.2.1.6 The Selection Method of Maintenance Strategy

A suggestion of the selection method was shown in Fig. 8.2; the selection of maintenance is related to the failure frequency (failure rate) and the effect of failure. A complete system should not have a high failure rate and failure effect, so the design needs improvement. Components with a high failure rate and low effect

Fig. 8.2 The selection method of maintenance strategies

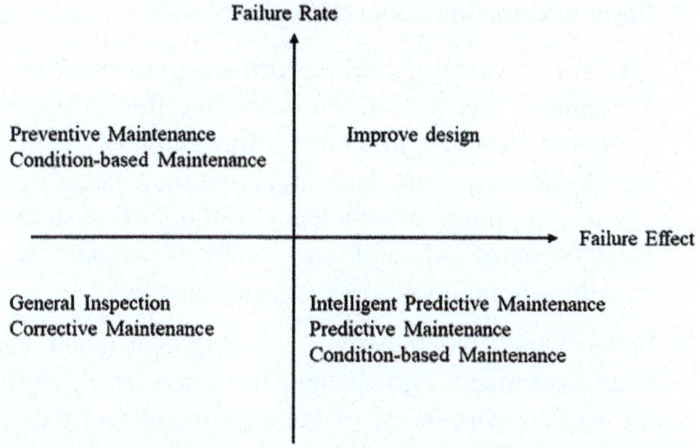

should be monitored and replaced. Components with a low failure effect and low failure rate do not need to monitor and collected data. They only need to perform the general inspection, replacement, or preventive replacement of components in a period. Components with a high failure effect when a failure occurs, maybe with the extended downtime and higher cost for maintenance, should be detailed of failure mode and effect analysis to decide the data collection objects and maintenance strategies.

8.2.2 Failure Diagnosis and Predictive Techniques

8.2.2.1 Fault Identification and Diagnosis

When a system failure occurs, it needs to be detected, isolated, and identified. Failure detection is to detect a failure in the system; failure isolation is to locate the faulty components, sub-systems, and systems; failure identification is to identify the failure mode that has occurred.

Whether it is a degraded failure or a sudden failure, the general steps and methods for fault diagnosis are as follows:

1. Collect relevant data and materials.

1. The types of collected data.
   (a) Numerical data: the data collected is a single value, for example, temperature, pressure, humidity, etc.
   (b) Waveform data: the data collected in a specific time domain is a time series, usually called a time waveform, such as vibration and acoustic data.
   (c) Multi-dimensional data: the collected data is multi-dimensional, and the most common multi-dimensional data is image data, such as infrared hot soil, X-ray images, etc.

2. Basic information about the system.
   (a) System structure and performance information, including the working principle of the system, the status and role in the production process, the basic parameters, the structural composition of the system, etc.
   (b) System operation data, including the change of load, the situation of starting and stopping, the efficiency of the DER system, etc.
   (c) Environmental conditions: ambient temperature, humidity, relationship with surrounding facilities and equipment, etc.
3. Failure and repair records, including past maintenance records, overhaul time, what adjustments and changes have been made during the maintenance process, the weak environment of the equipment and the type and location of failures expected to occur, the failure records of other equipment of the same model and working conditions, etc.

2. System testing.

Detect the system's state after the failure, additional testing, etc.

3. Failure analysis and diagnosis.

According to the collected data, analyze the signal and data changes, and analyze the waveform and stability of the signal in a specific time domain, frequency domain, etc. On this basis, failures are diagnosed through diagnostic methods, such as pick-up models, signal processing, expert knowledge, etc.

For traditional diagnostic methods, the model-based analysis method is better if a more accurate mathematical model of the monitored object can be obtained, such as state estimation, parameter estimation, consistency check, etc. The disadvantage of this method is that the quality and capacity of the sample are limited and the presence of noise and the system's complexity result in poor accuracy.

Signal-based processing methods, such as frequency domain analysis, wavelet analysis, adaptive time-frequency analysis, etc., can be used for data preprocessing but cannot be used as a separate diagnostic method.

When it is difficult to establish a mathematical model of the diagnostic object, knowledge-based diagnostic methods can be used, including expert system, neural network, fuzzy algorithm, genetic algorithm, rough set, artificial immune algorithm, fault tree, support vector machine, and other fault diagnosis methods. With the development of new-generation information technologies such as the Internet, big data, and artificial intelligence, new-generation data-driven failure diagnosis methods based on deep learning have received more attention.

8.2.2.2 Failure Prediction

Failure prediction is used to determine whether a failure is imminent and estimate the time and possibility of failure. Unlike fault diagnosis, failure prediction is made before the failure occurs, while failure diagnosis is made after the failure occurs.

At present, there are two mainstream failure prediction categories. One is to predict the remaining service life of equipment, that is, to predict how much time is left before the equipment fails based on past operating data and current equipment status. The other is to predict the probability of failure at a particular time in the future, the failure rate.

Failure prediction is the core method to realize the prediction of system performance degradation state and remaining life.

8.2.3 Maintenance Strategy Optimization Based on Risk Analysis

8.2.3.1 Failure Modes and Effects Analysis, an Approach Based on Risk Assessment

Failure modes and effects analysis (FMEA) was one of the first highly structured and systematic techniques for maintenance analysis. It is a systematic method for analyzing and ranking the wind associated with various products, processes, and failure patterns. Prioritize for remedial action, take action on the highest ranked items, reevaluate those items, and return to the priority steps continuously until marginal returns begin.

The mentioned method was designed to model the system's operation to determine its reliability characteristics. Each method has its advantages and disadvantages and has been widely analyzed in the scientific literature. The advantages and disadvantages have been introduced, and the compression of FTA, FMEA, and FMECA has been presented. FTA method may not find all possible initiating failures and cannot analyze the more complex system. The FMECA method can give us more details about the system's component reliability. However, it needs to investigate many trivial cases, but for a complex DER system, it may not have a lot of actual case data to be analyzed.

This method does not allow the evaluation of reliability functions of complex systems but allows for the identification and analysis of all system failures, assesses their importance in system reliability, and then focuses on maintenance practices and their impact on system reliability. In addition, FMEA allows dealing with uncertainties, including the complexity of systems and the ambiguity of human judgment [6]. The FMEA method uses actual data from equipment during operation to analyze the relevant faults of components, so there is no theoretical causality. The effectiveness of the FMEA method comes from the practice-based approach, which allows the selection of cost-effective actions for the correct maintenance plan.

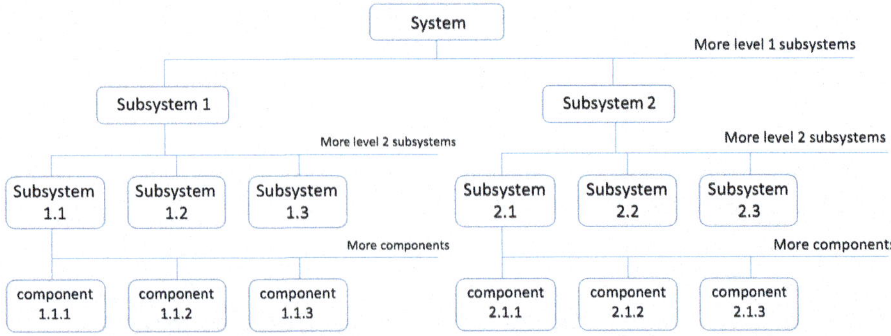

Fig. 8.3 Component block diagram for a complex system

8.2.3.2 Analysis Method and Process

The reliability of a complex system depends on the reliability of its components and the way those components are connected throughout the system. The FMEA method adopts the risk priority number (RPN) to evaluate and optimize the failure of components by analyzing the failure modes, causes, and effects. The steps of the FMEA method can be written as follows:

Step 1: Review the system's components details, such as the requirement, maintenance data, system function, component's function, failure events, etc.

Step 2: Break down the structure of the equipment or system, and draw the reliability equipment block diagram. The equipment block diagram for a complex system is shown in Fig. 8.3.

Step 3: Determine the components' failure mode, failure cause, and failure effect. The effect of failure can relate to a part, a sub-system, and a whole system.

Step 4: Calculate the risk priority number (RPN), severity (S), occurrence (O), and detection (D) for each failure mode.

Step 5: Perform the FMEA process to assess the reliability and maintenance.

The FMEA process for the DER system in KSRP is shown in Fig. 8.4. The FMEA process is applied to assess the importance level of the component. After performing the FMEA assessment, a team review will be applied to the maintenance strategy. The team members can include the engineers, operators, manufacturers, designers, managers, maintainers, etc. Professional maintenance knowledge, equipment knowledge, maintenance experience, and management experience put forward improvement strategy for equipment maintenance strategy and evaluate new RPN.

Three factors are used to calculate the RPN, including severity, occurrence, and detection. Severity (S), occurrence (O), and detection (D) factors are rated separately using numerical scales, usually ranging from 1 to 10 [7].

The details of the three factors can be described as follows:

- (S): Result generated from failure
- (O): Opportunity or probability of a failure
- (D): Opportunity for an unidentified failure because of the difficulty in detection.

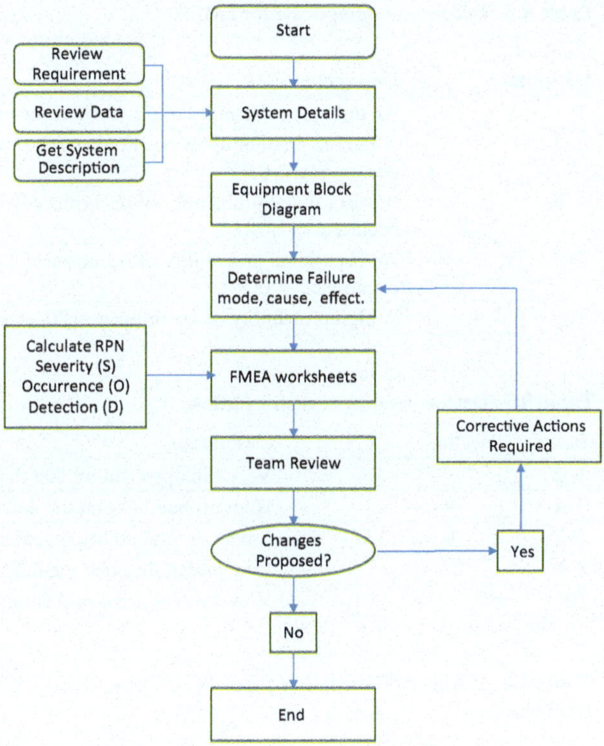

Fig. 8.4 Component block diagram for a complex system

Table 8.1 Severity rating scale for FMEA

Rank of severity	Description
1–2	Failure is of such minor nature that the operator will probably not detect the failure
3–5	Failure will result in slight deterioration of part or system performance
6–7	Failure will result in operator dissatisfaction or deterioration of part or system performance
8–9	Failure will result in a high degree of operator dissatisfaction and cause the non-functionality of the system
10	Failure will result in significant operator dissatisfaction or significant damage

The RPN can be expressed as:

$$\mathrm{RPN} = S \times O \times D \tag{8.1}$$

The evaluation rank is based on a scale of 1–10, with the corresponding description from Mauro Villarini et al. [6], Kapil Dev Sharma et al. [8], and Jichuan Kang [9] and the investigation in KSRP. The rank of severity (S), occurrence (O), and detection (D) is shown in Table 8.1, 8.2, and 8.3, respectively. As a consequence of

Table 8.2 Occurrence rating scale for FMEA

Rank of occurrence	Description
1	An unlikely probability of occurrence: the probability of occurrence < 0.001
2–3	A remote probability of occurrence: 0.001 < probability of occurrence < 0.01
4–6	An occasional probability of occurrence: 0.10 < probability of occurrence < 0.10
7–9	An occasional probability of occurrence: 0.10 < probability of occurrence < 0.20
10	A high probability of occurrence: 0.20 < probability of occurrence

Table 8.3 Detection rating scale for FMEA

Rank of detection	Description
1–2	Very high probability that the defect will be detected
3–4	High probability that the defect will be detected
5–7	Moderate probability that the defect will be detected
8–9	Low probability that the defect will be detected
10	Very low (or zero) probability that the defect will be detected

Table 8.4 Effect of different RPN levels

Level	RPN	Effect
I	1~10	No effect
II	10~100	Duty is unfulfilled
III	100~250	It is failing a critical mission
IV	250~1000	Abandonment of duties

the scale indicators, the RPN values are ranked between 1 and 1000. The effect of different RPN levels is divided into four levels, including no effect (RPN 1–10), duty unfulfilled (11–100), failing an important mission (101–250), and abandonment of duties (251–1000), as shown in Table 8.4.

Through FMEA analysis of the DER system, component priority can be obtained according to the level of RPN, thus helping the maintainer to determine which component should be paid more attention to in the maintenance process. For components with a high RPN level, improve maintenance policies based on device performance, expert advice, and the experience of the maintainer. For example, add a condition monitoring system, increase inspection frequency, replace parts, etc.

8.3 Reliability and Maintenance Prioritization Analysis of the DER System

8.3.1 *Concept of Maintenance Prioritization Analysis*

Maintenance is the leading way to keep the components or systems healthy and safe. Maintenance prioritization is defined as determining the maintenance sequence based on the effect of component failure during system maintenance. Of course, maintenance and inspection are generally carried out in actual operation in a period. The maintenance priority can provide a reference for us to recommend the maintenance interval and the content of each maintenance. Generally, maintenance priorities within a system are determined based on manufacturer advice or expert experience. This chapter introduces maintenance priority calculation methods based on reliability analysis and cost requirements.

8.3.2 *Reliability Calculation of Complex Systems*

8.3.2.1 Reliability Concepts

Failure rate, repair rate, mean time to failure (MTTF), mean time to repair (MTTR), and mean time between failure (MTBF) are the general reliability indices for components and system reliability engineering. The failure is defined as a component or system loss of the function during the operating period that the production cannot be carried out. The failure rate is the frequency in which a component or a system fails during a period, expressed in failures per time, the Greek letter λ (lambda) [4]. The repair rate is the frequency that the failed component or system gets repaired; the repair rate unit is the same as the failure rate. It is often denoted by the Greek letter μ (mu).

There are three common basic categories of failure rates: mean time between failures (MTBF), mean time to failure (MTTF), and mean time to repair (MTTR). The relationship between MTBF, MTTF, and MTTR is shown in Fig. 8.5. Figure 8.5 shows that MTBF is the sum of MTTR and MTTF. Although MTBF was designed for repairable items, it is commonly used for repairable and non-repairable items. For non-repairable items, MTBF is the time until the first failure after t_i.

The following formula can explain the calculation relationship among failure rate, repair rate, MTTF, MTTR, and MTBF:

$$\lambda = \frac{1}{\text{MTTF}} \tag{8.2}$$

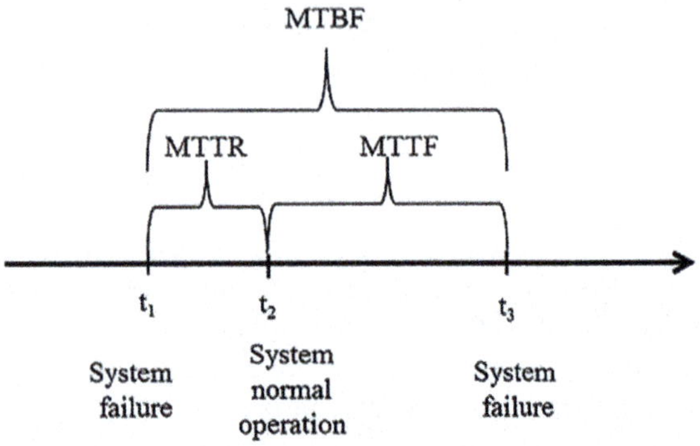

Fig. 8.5 Differentiating the MTBF, MTTF, and MTTR

$$\mu = \frac{1}{\text{MTTR}} \tag{8.3}$$

$$\text{MTBF} = \frac{\sum (t_n - t_{n-1})}{n} \tag{8.4}$$

$$\text{MTBF} = \text{MTTF} + \text{MTTR} = \frac{1}{\lambda} + \frac{1}{\mu} \tag{8.5}$$

For a component or system which has a much higher repair rate than failure rate, the MTBF can be approximated by the MTTF.

$$\text{MTBF} \approx \text{MTTF} = \frac{1}{\lambda} \tag{8.6}$$

8.3.2.2 Reliability and Availability

Reliability analysis also is known as survival analysis. When the study concerns are focused on the biological event with the object of humans or animals, it is usually called survival analysis [10]. The survival analyses focus on a non-parametric estimation approach. The reliability analyses focus on a parametric approach.

The reliability function ($R(t)$) is the probability that the individual survives after time t. It is defined that T is the entire life cycle; the $R(t)$ is probable when the T is more than t ($T > t$). The function can be estimated by the non-parametric Kaplan-Meier curve or one of the parametric distribution functions. The system reliability decreases exponentially as time increases [11]. The following equation can express it:

$$R(t) = e^{-\lambda t} \tag{8.7}$$

Availability is defined as the ability of a project (in the integrated aspects of its reliability, maintainability, and maintenance support) to perform its required functions at a specified time point or within a specified period.

$$A = \frac{\text{MTTF}}{\text{MTTF} + \text{MTTR}} = \frac{\mu}{\lambda + \mu} \tag{8.8}$$

Here, the MTTR sometimes can be substituted by the mean downtime (MDT) to make it clean to show the downtime. In addition to reliability and availability, other concepts can describe the reliability of components or systems, such as maintainability, security, security, reliability, and quality [4].

8.3.2.3 Reliability Analysis Method

The Markov process, based on the state-space method (SSM), is suitable for analyzing a DES system's reliability. The stochastic process reliability analysis method can be used to analyze the reliability of repairable systems, such as the homogeneous Poisson process, renewal process, and Markov process. The Markov process can model the system with multiple states and transitions between states, so it is widely used [4]. Reliability assessment methods of multistate systems are based on two different approaches: the stochastic analytical process and the Monte Carlo simulation method. The Markov process is the leading analytical stochastic process since it can be used to perform the reliability analysis of a system that has changed continuously or discretely with the passage of time and space. The state-space method (SSM) is applicable for assessing large and complex systems' reliability, availability, and maintainability. It is considered an irreplaceable method for evaluating repairable and complex systems.

The state of a system depends on its components; each component has two states: functioning (1) and failed (0). Since each component has two states (functioning or failed), when a system has n quantity of components, the system still will have at most 2^n possible states. The state of a system is transferred randomly with time in those states. A Markov model based on a state-space method (SSM) is performed for the reliability analysis of a system with two components. The Markov model and possible states of the system are shown in Fig. 8.6 and Table 8.5, respectively. The failure rate (λ) is represented by the transition rate of one component from a functioning state to a failed state. Similarly, the repair rate (μ) is represented by the transition rate of one component from a failed state to a functioning state. Thus, the failure rate and repair rate of a component are used to describe the transition rate between two system states. The reliability and availability analysis model using the Markov process and SSM can be decomposed into the following steps:

1. List and classify all system states; the same state should be merged, and the non-related state should be removed.
2. Construct the state space diagram of the system; confirm the transition rate between states.

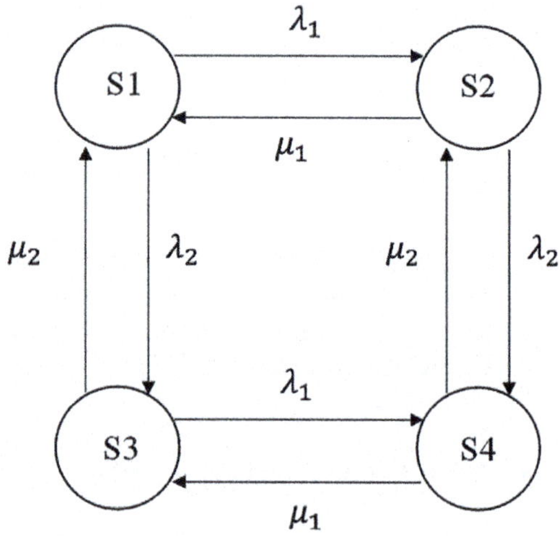

Fig. 8.6 Markov model of a system with two components

Table 8.5 Possible states of a system with two components

Items	Component 1	Component 2
Failure rate	λ_1	λ_2
Repair rate	μ_1	μ_2
State 1	1	1
State 2	0	1
State 3	1	0
State 4	0	0

3. Calculate the probabilities of the states during a lifetime.
4. Calculate the reliability and availability indices, such as the failure rate (generally represented by mean time to failure), repair rate (generally represented by mean time to repair), and availabilities of components.

A steady-state distribution system is used to limit the Markov processes. Generally, a set of linear, order differential equations is established to determine the probability distribution of the system. The probability distribution equation is shown:

$$P(t) = [P_1(t), P_2(t), \ldots, P_n(t)] \tag{8.9}$$

where the $P_i(t)$ is the probability of the system in state i at time t and $P(t)$ is the state probability matrix at time t.

A density matrix, Q, is defined as the following:

$$Q = \begin{bmatrix} q_{11} & q_{12} & \cdots & q_{1n} \\ q_{21} & q_{22} & \cdots & \\ \vdots & \vdots & \vdots & \\ q_{n1} & q_{n2} & \cdots & q_{nn} \end{bmatrix} \tag{8.10}$$

where $q_{ij} = \lambda_{ij}(i \neq j)$ and $q_{ii} = -\sum_{i \neq j} \lambda_{ij}$.

The following state equations are presented for the steady-state probability of the system:

$$\begin{cases} P \cdot Q = 0 \\ \sum P_i = 1 \end{cases} \tag{8.11}$$

Thus, for a system with two components, as shown in Fig. 8.6 and Table 8.5, the state transition density matrix is presented as:

$$Q = \begin{bmatrix} -(\lambda_1+\lambda_2) & \lambda_1 & \lambda_2 & 0 \\ \mu_1 & -(\mu_1+\lambda_2) & 0 & \lambda_2 \\ \mu_2 & 0 & -(\mu_2+\lambda_1) & \lambda_1 \\ 0 & \mu_2 & \mu_1 & -(\mu_1+\mu_2) \end{bmatrix} \tag{8.12}$$

Based on Eq. (8.11), the following equations can be acquired:

$$\begin{cases} -(\lambda_1+\lambda_2)P_1+\mu_1 P_2+\mu_2 P_3=0 \\ \lambda_1 P_1-(\mu_1+\lambda_2)P_2+\mu_2 P_4=0 \\ \lambda_1 P_1-(\mu_1+\lambda_2)P_2+\mu_1 P_4=0 \\ \mu_2 P_2+\mu_1 P_3-(\mu_2+\mu_1)P_4=0 \\ P_1+P_2+P_3+P_4=1 \end{cases} \tag{8.13}$$

The state probabilities of the system are obtained through solving the equations in (8.13), with the following results:

$$P_1 = \frac{\mu_1\mu_2}{(\lambda_1+\mu_1)(\lambda_2+\mu_2)} \tag{8.14}$$

$$P_2 = \frac{\lambda_1\mu_2}{(\lambda_1+\mu_1)(\lambda_2+\mu_2)} \tag{8.15}$$

$$P_3 = \frac{\lambda_2\mu_1}{(\lambda_1+\mu_1)(\lambda_2+\mu_2)} \tag{8.16}$$

$$P_4 = \frac{\lambda_1\lambda_2}{(\lambda_1+\mu_1)(\lambda_2+\mu_2)} \tag{8.17}$$

The probabilities of the system availability can be calculated. Therefore, the availability of component 1 is written as follows:

$$A_{\text{component 1}} = P_1 + P_3 \tag{8.18}$$

where the availability of component 2 is written as follows:

$$A_{\text{component 2}} = P_1 + P_2 \tag{8.19}$$

where the availability of the whole system (both component 1 and component 2 are functioning) is written as follows:

$$A_{\text{whole system}} = P_1 \tag{8.20}$$

where the availability of the system (whole system or part of the system is functioning) is written as follows:

$$A_{\text{system}} = P_1 + P_2 + P_3 \tag{8.21}$$

Generally, a complex system consists of multiple series and parallel sub-systems. Thus, the reliability calculation methods for a series sub-system and parallel sub-system are different [12]. The reliability of a series system with n components is presented as follows:

$$R_{\text{series}} = R_1 R_2 R_3 \cdots R_n \tag{8.22}$$

The reliability of a parallel system with n components is presented as follows:

$$R_{\text{parallel}} = 1 - [(1 - R_1)(1 - R_2) \cdots (1 - R_n)] \tag{8.23}$$

8.3.3 Component Reliability Importance Indices

Two reliability importance indices based on cost considerations are proposed and described to determine maintenance priorities in complex systems [13]. One is the reliability importance index of failure cost, and the other is the potential failure cost importance index.

The component reliability importance index based on failure cost is defined as follows:

$$I_k^F = \frac{\partial C_{\text{TF}}}{\partial \lambda_k} \tag{8.24}$$

where I_k^F is the reliability importance index that is considered the failure cost. The unit I_k^F is failure cost per failure (failure cost/f).

Reliability importance index of failure cost (I_k^F) is affected by the repair cost of the component and the repair rate but is not related to the failure rate. The second reliability importance index is related to failure rate and proposes a maintenance prioritization with comprehensive consideration of failure rate, repair rate, and required repair cost. The potential failure cost importance index defines the expected cost of failure before the failure occurs. It is presented as the following:

$$I_k^P = I_k^F \lambda_k \tag{8.25}$$

The failure cost of a component of a CCHP system is defined as the system's total cost during the failure time, including the component repair cost and the added cost for an unserved load (electricity or space cooling and heating or hot water).

The total failure cost is defined as the following:

$$C_{\mathrm{TF}}(k) = \sum_n [C_{\mathrm{R},n}(k) + C_{\mathrm{A},n}(k)] \tag{8.26}$$

where C_{TF} is total failure cost of component k's failure, C_{R} is the repair cost for the k component, and the C_{A} is the added cost of the outage (electricity or space cooling and heating or hot water).

The added cost during an outage is defined as the cost that should be paid in order to meet the insufficiency of the energy load when the failure occurs, such as when the power generator experiences a failure leading to an outage state, the electrical grid will meet the insufficient electricity, and the insufficient heat from the waste heat will be met by the gas boiler or gas-fired absorption chiller. Thus, the total cost of electricity and natural gas for meeting the insufficiency is the unserved load's added cost.

The added cost of the outage (electricity or space cooling and heating or hot water) is calculated as the following:

$$C_{\mathrm{A}} = u \times L_u \times \mathrm{MTTR} \tag{8.27}$$

where u is the unit price of electricity or city gas and L_u presents the amount of electricity or city gas is purchased, and MTTR is the meantime to repair. Generally, MTTR is defined as the total amount of time spent performing all corrective or preventative maintenance repairs divided by the total number of those repairs [14].

Failure cost importance and potential failure cost importance indices are developed to provide accurate cost indicators for managers to optimize the maintenance strategy to reduce the total maintenance cost and improve the system reliability.

8.4 Availability and Cost Analysis of DER System with Redundant Design

8.4.1 Reliability and Availability of System with Redundant Design

Redundant design is an important method to improve system reliability and availability. The reliability and availability of the system with redundant design have changed and can be calculated using k-out-of-n: G (good) or k-out-of-n: F (failure) methods. A case of the redundant design was presented in Ref. [15].

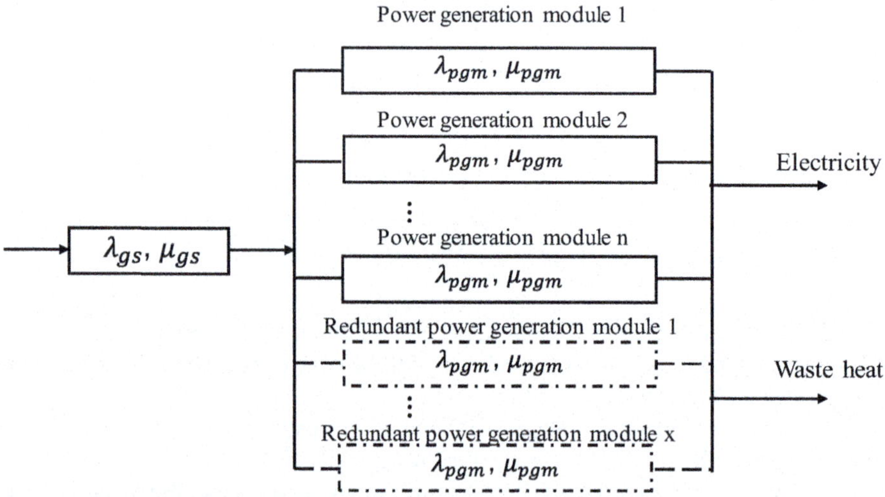

Fig. 8.7 The structures of power generation unit with *n* PGM and *x* PGM

A redundant design of the power generation unit in the DER system was presented in this part. The power generation unit is assumed to be a parallel system composed of multiple power generation modules (PGM). The power generation modules are independent of the power generation unit. The structure of the power generation unit with *n* PGM and *x* PGM is presented in Fig. 8.7.

For a parallel and independent power generation unit with $n + x$ power generation modules, the system is satisfying the *k*-out-of-*n*: G model. This model is defined as for an n-component system that good components (or is "works"), if and only if at least k of the n components good (or are work) is called a *k*-out-of-*n*: G model [16]. Therefore, the reliability of the power generation unit can be presented as the following:

$$R_{n+x} = \sum_{j=n}^{n+x} \binom{n+x}{n} R^i (1-R)^{n+x-j} \tag{8.28}$$

For a parallel and independent system, the repair rate is equal to the repair rate of the power generation module; the failure rate of the $n + x$ parallel power generation unit can be expressed as:

$$\lambda_{n+x} = \frac{1}{\mathrm{MTBF}_{n+x}} = \frac{1}{\frac{1}{\lambda}\sum_{j=n}^{n+x}\frac{1}{j}} \tag{8.29}$$

The availability of the $n + x$ parallel power generation unit can be expressed as:

$$A_{n+x} = \sum_{j=n}^{n+x} \binom{n+x}{n} A^j (1-A)^{n+x-j} \tag{8.30}$$

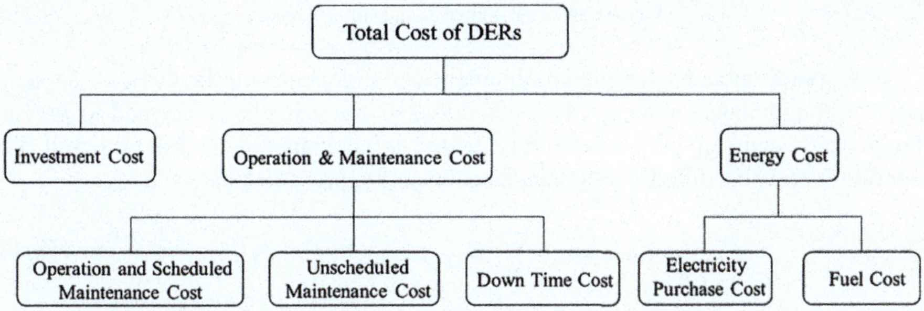

Fig. 8.8 Diagram of the total cost of the DER system

8.4.2 *Cost Analysis of DER System with Redundant Design*

8.4.2.1 Total Life Cycle Cost

Minimum total life cycle cost is the economic objective of optimizing a DER system. The diagram of the total cost of a DER system is presented in Fig. 8.8. The total cost of the DER system includes the investment cost, operation and maintenance cost, and energy cost.

The total cost of the DER system can be expressed as follows:

$$C_{\text{Total}} = C_{\text{Inv}} + C_{\text{O\&M}} + C_{\text{Energy}} \tag{8.31}$$

8.4.2.2 Investment Cost of DER System with Redundant Design

The investment cost of the system includes the cost of all the devices. Therefore, the investment cost of the devices can be presented as:

$$C_{\text{Inv}} = \sum\nolimits_i C_{\text{unitInv},i} \times Q_i \times \left(1 + \frac{x_i}{n_i}\right) \tag{8.32}$$

where C_{Inv} is the total investment cost; i is the device type, for instance, power generator, absorption chiller, and so on; and $C_{\text{unit},\ i}$ is the unit investment cost of device i (unit, \$/kW). Q_i is the installed capacity of device i (kW). n_i is the number of the device i, and x_i is the number of the redundant device i.

8.4.2.3 Operation and Maintenance Cost

The O&M cost of a DER system includes the operation and scheduled maintenance cost ($C_{\text{O \& schM}}$), unscheduled maintenance cost (C_{unschM}), and downtime cost (C_{Down}). The O&M cost is presented as follows:

$$C_{O\&M} = C_{O\&schM} + C_{unschM} + C_{Down} \tag{8.33}$$

The operation and scheduled maintenance cost depends on the devices' capacity and the maintenance strategy. The scheduled maintenance is performed to prevent faults from occurring [17]. It is usually treated as a constant in the design period. The operation and scheduled maintenance costs can be presented as:

$$C_{O\&schM} = \sum\nolimits_i C_{unitO\&schM,i} \times Q_i \times L_i \tag{8.34}$$

where $C_{unitO\ \&\ schM,\ i}$ is the unit value of operation and scheduled maintenance cost of device i (unit, \$/kW) and L_i is the design life cycle.

The unscheduled maintenance cost also can be called the corrective maintenance cost; the unscheduled maintenance cost occurred after the failure [18]. Thus, the unscheduled maintenance cost is the cost of repairing and recovering the device from the failed state to the operation state. So, the unscheduled maintenance cost depends on the device or system's failure rate and repair rate.

$$C_{unschM} = \sum\nolimits_i C_{repair,i} \times \lambda_i \times L_i \tag{8.35}$$

where $C_{repair,\ i}$ is the mean repair cost per failure for the device i (unit, \$/failure).

The downtime cost of a power outage for a power supply is difficult to estimate, generally deviled into direct and indirect costs [19]. The indirect cost for the DER system is challenging to define. However, the direct downtime cost of the energy supply system is composed of two parts: one is the fixed loss cost of the investment of devices, and another is increased energy cost. Increased energy cost is defined as the cost which should be paid to meet the insufficiency of energy load when a failure occurs [13]. The increased energy cost can be expressed as follows:

$$C_{Inc,i} = P_u \times Q_{outage,i} \times \mathrm{MTTR}_i \times \lambda_i \times L_i \tag{8.36}$$

where $C_{Inc,\ i}$ is the increased energy cost of device i (\$); P_u is the price of energy (grid electricity or fuel); and $Q_{outage,\ i}$ is the failed outage capacity of device i.

The fixed loss cost (C_{Loss}) is defined as the average investment cost of device i during the design life cycle. The fixed loss cost can be presented as:

$$C_{Loss,i} = C_{Inv,i} \times \mathrm{MTTR}_i \times \lambda_i \tag{8.37}$$

where $C_{Inv,\ i}$ is the investment cost of device i.

Therefore, the downtime cost of the system can be presented as:

$$C_{Down} = \sum_i \left(P_u \times Q_{outage,i} \times L_i + C_{Inv,i}\right) \times \mathrm{MTTR}_i \times \lambda_i \tag{8.38}$$

References

1. Hongbo R et al (2005) Integration of distributed energy resource and distribution system, part 25: analysis the feasibility of adopting biomass energy in Kitakyushu Science and Research Park. Rep Kitakyushu Branch Archit Inst Jpn 47:277–280
2. Akorede MF, Hizam H, Pouresmaeil E (2010) Distributed energy resources and benefits to the environment. Renew Sust Energ Rev 14(2):724–734
3. Jardine AKS, Tsang AHC (2013) Maintenance, replacement, and reliability: theory and applications, 2nd edn. CRC Press, Boca Raton
4. Rausand M, Hsyland A (2004) System reliability theory models and statistical methods, 2nd edn. Wiley, Hoboken
5. Darabnia B, Demichela M (2013) Maintenance an opportunity for energy saving. Chem Eng Trans 32:259–264
6. Villarini M et al (2017) Optimization of photovoltaic maintenance plan by means of a FMEA approach based on real data. Energy Convers Manag 152:1–12
7. Arabian-Hoseynabadi H, Oraee H, Tavner PJ (2010) Failure modes and effects analysis (FMEA) for wind turbines. Int J Electr Power Energy Syst 32(7):817–824
8. Sharma KD, Srivastava S (2016) Failure mode and effect analysis (FMEA) for enhancing reliability of water tube boiler in thermal power plant. SAMRIDDHI J Phys Sci Eng Technol 8(2):79–86
9. Kang J et al (2017) Risk assessment of floating offshore wind turbine based on correlation-FMEA. Ocean Eng 129:382–388
10. Software NS (2019) Distribution (Weibull) fitting. https://www.ncss.com/wp-content/themes/ncss/pdf/Procedures/NCSS/Distribution-Weibull-Fitting.pdf
11. Wang J-J et al (2013) Reliability and availability analysis of redundant BCHP (building cooling, heating and power) system. Energy 61:531–540
12. Sondalini M (2009) Plant and equipment wellness. Engineers Media, Seattle
13. Jiang J et al (2018) Reliability and maintenance prioritization analysis of combined cooling. Heat Power Syst Energies 11:1519
14. Garg A, Deshmukh SG (2006) Maintenance management: literature review and directions. J Qual Maint Eng 12(3):205–238
15. Jiang J et al (2019) Reliability and cost analysis of the redundant design of a combined cooling, heating and power (CCHP) system. Energy Convers Manag 199:111988
16. Kuo W, Zuo MJ (2003) Optimal reliability modeling: principles and applications. Wiley, Hoboken
17. Lin B et al (2019) optimization of high-level preventive maintenance scheduling for high-speed trains. Reliab Eng Syst Saf 183:261–275
18. Spinelli R, Eliasson L, Magagnotti N (2019) Determining the repair and maintenance cost of wood chippers. Biomass Bioenergy 122:202–210
19. Praktiknjo AJ, Hähnel A, Erdmann G (2011) Assessing energy supply security: outage costs in private households. Energy Policy 39(12):7825–7833

Chapter 9
Multi-criteria Evaluation of a Distributed Energy Resource System Focusing on Grid Stabilization and Carbon Emission Reduction

Liting Zhang, Yongwen Yang, Weijun Gao, and Fanyue Qian

9.1 The Evaluation of the Distributed Energy Resource

9.1.1 Promotion Difficulties of the Distributed Energy Resource

Nowadays, the rapid depletion of fossil fuels and environmental deterioration are two global challenges.

Since the dawn of the industrial revolution, fossil fuels have been the driving force behind the industrialized world and its economic growth. According to the Statistical Review of World Energy, the primary direct energy consumption of the fossil fuels was from insignificant levels in 1800 to an output of nearly 140,000 TWh in 2019 [1]. Global fossil fuel consumption is on the rise, but new reserves are

L. Zhang (✉) · Y. Yang
Energy and Environment Engineering Institute, Shanghai University of Electric Power, Shanghai, China
e-mail: yangyongwen@vip.163.com

W. Gao
The Engineering Academy of Japan, Chiyoda City, Tokyo, Japan

The University of Kitakyushu, Kitakyushu, Japan

Qingdao University of Technology, Qingdao, China

International Research Center for Green Buildings and Low-Carbon Cities, Zhejiang University, Hangzhou, China

Xi'an Jiaotong University, Xi'an, China

F. Qian
School of Mechanical and Energy Engineering, Tongji University, Shanghai, China

W. Gao (ed.), *Distributed Energy Resources*, Green Energy and Technology,
https://doi.org/10.1007/978-3-031-21097-6_9

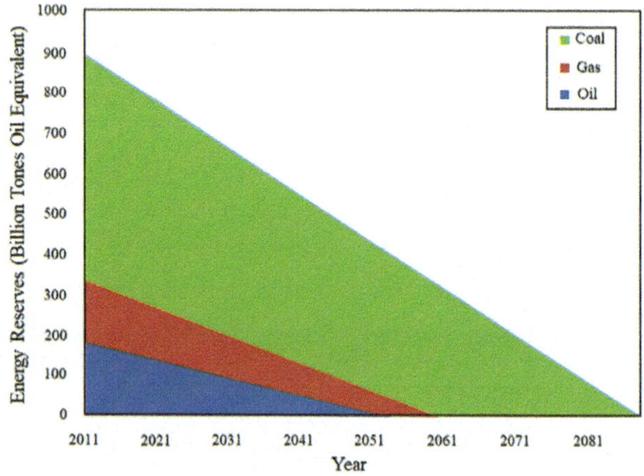

Fig. 9.1 Future reserves for coal, gas and oil. (Source: CIA World Factbook and Statista)

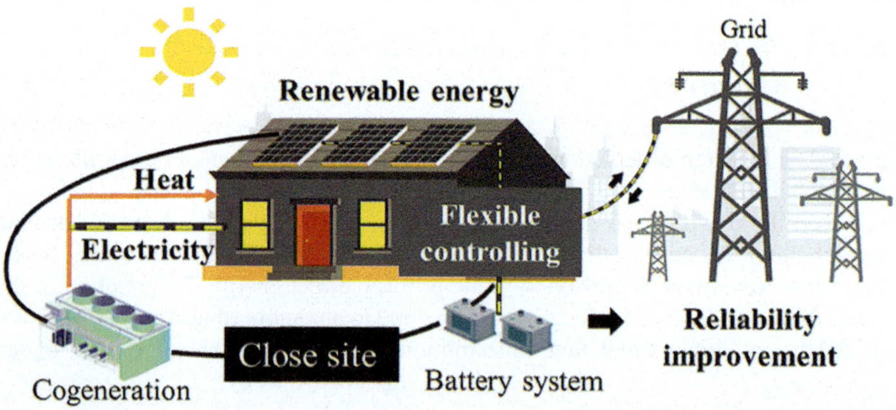

Fig. 9.2 Diagram of proposed distributed energy resource system

becoming harder to find. Figure 9.1 shows the future energy reserves for coal, gas, and oil. Those that are discovered are significantly smaller than the ones that have been found in the past. Global reliance on fossil energy is causing serious environmental deterioration. Figure 9.2 shows the CO_2 emission trends from 1800 to 2018 by fuel type [2]. In 2018, nearly 35 billion tons of CO_2 were emitted from fossil fuel consumption, and this has 3.5 times since 1950. In fact, energy production using fossil fuel is the dominating source of CO_2 [3]. Energy demand will double by 2050. If the current proportion of fossil fuels remains the same, carbon emissions will certainly exceed the upper limit allowed to keep the global average temperature rise below 2. Such high emissions will have a disastrous impact on the global climate. Moreover, with the sharply increasing demand load, large-scale and centralized power grid system relying on fossil fuels is suffering a security concern. The network needed for its transmission and distribution is relatively complex. Most of the users are concentrated in a specific area, so the flexibility of load change and the safety of energy supply are poor. Once a small failure will happen in the supply chain, all users in the area will be suffering electricity loss. Therefore, countries around the world are looking forward to an alternative approach of more clean, efficient, and reliable energy consumption.

PV system is inexhaustible and non-polluting; it has been playing proactive roles in sustainable energy development. Increasing the proportion of PV in the energy structure is inevitable for all countries in the world. To address the intermittent output of PV, battery system is a key technology to keep the grid stability. It also can save the overproduction of the PV and shave the peak load. In the future, battery system will be cheaper and popularly used. Moreover, in order to improve the energy self-sufficient of the users, co-generation which can provide electricity and thermal load is the best choice due to its high efficiency. It can reduce the impact of demand load fluctuation on power grid with its flexible controlling. Therefore, a distributed energy resource system (Fig. 9.2) combined with three technology is proposed. It is a clean and low emission system which stays close to the consumer side, can effectively solve the energy and environmental problem, as well as improves grid reliability.

Due to the advantages of high efficiency, energy conservation, and environmental protection, distributed energy resource system has been vigorously developed by the government. However, the practices of DER systems have shown that the actual operation performance is not as good as expected in many cases. Nearly half of more than 40 DER projects in China have been out of service due to economic problems [4]. There are some main barriers:

1. Economic aspect.

Even though lower fuel and operating costs may make the DER cost competitive on a life cycle basis, higher initial capital costs can mean that the DER system provides less installed capacity per initial dollar invested than conventional energy system. Thus, investments of the DERs generally require higher amounts of financing for the same capacity. Depending on the circumstances, capital markets may demand a premium in lending rates for financing the DER projects because more capital is being risked up front than in conventional energy projects [5].

2. Technology aspect.

The unreasonable capacity of the system equipment is the most important issue. Distributed energy supply system has a wide range of optional system forms, main and auxiliary equipment, and capacity. There is no universally applicable technical scheme. Its configuration is closely related to the climate characteristics, load demand, energy price, and supply of the user's area, which puts forward high requirements for the system configuration determination. For the optimal configuration of regional distributed energy supply system, the main task is to determine the system structure and form reasonably; optimize the type, capacity, and number of main equipment; and obtain the comprehensive performance of economic, environmental, and other aspects of the whole year, so as to provide decision-making reference for owners, provide selection basis for design, and provide guidance for operation strategy formulation. Improper configuration of distributed energy supply system will lead to waste of equipment investment, failure to give full play to economic benefits, low system operation efficiency, and other problems and even lead to system failure in extreme events.

3. Evaluation aspect.

The economic performance of the DER can be directly reflected through quantitative indicators such as annualized cost or payback period, but the social benefits brought by the advantages of energy-saving, environmental protection, and improving the reliability of the power grid cannot be directly compared with the economic benefits. As a result, the current evaluation method of the DERs usually uses energy-saving or economic benefits only, which is relatively simple and one-sided [6, 7]. The single criterion cannot reasonably and accurately reflect the comprehensive benefits of the DERs. Economic sustainability, energy security, and environmental protection are the most important aspects of the distributed energy resource system. However, they often mutually influence each other. For example, energy costs and carbon emissions are evaluation indicators from two different perspectives and often conflict with each other. How to reach a reasonable compromise is critical [4]. Trade-offs between different performances can be addressed by the multi-criteria evaluation analysis.

To deal with the energy depletion and environmental problems as well as reducing the grid weakness, this chapter proposed a DER system composed of PV, battery, and ICE. Its grid stabilization and carbon reduction potentials were analyzed. Then, focusing on these advantages, a multi-criteria evaluation method is established to optimize the system. Finally, different utilization case studies of the DER were demonstrated. It is hoped to improve the core competitiveness of the DER and promote its development.

9.1.2 Model Establishment of Distributed Energy Resource System

To study the application potentials and promotion of DER, it is necessary to establish the model of the devices and systems. The distributed electricity generation system and combined cooling, heating, and power (CCHP) system used in the follow-up study were proposed.

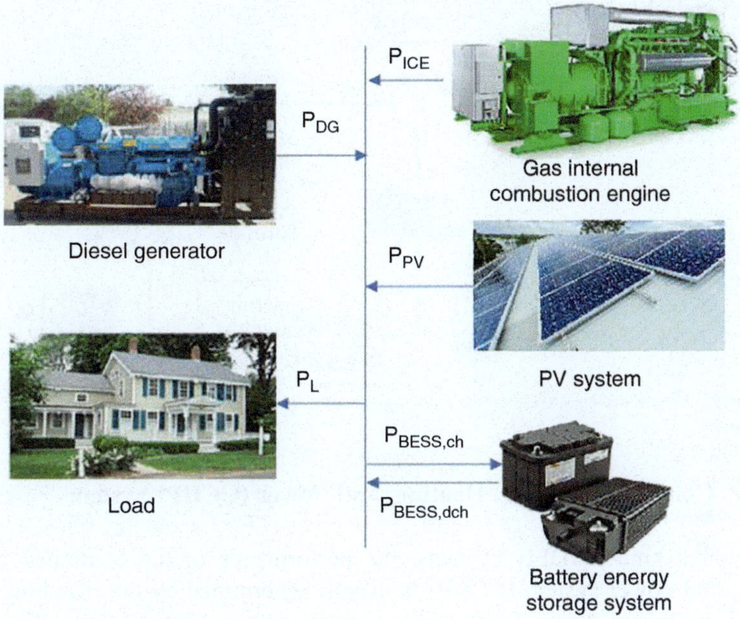

Fig. 9.3 Distributed electricity generation system

9.1.2.1 Distributed Electricity Generation System

The structure of the proposed regional distributed electricity generation system is depicted in Fig. 9.3. The distributed generators including diesel generator (DG), photovoltaic (PV) system, battery energy storage system (BESS), and internal-combustion engine (ICE) are designed to meet the load requirements.

The energy balance of the proposed power system is presented in Eq. 9.1 as:

$$\sum_{z=1}^{N_{\text{load}}} E_L^t = \sum_{i=1}^{N_{\text{DG}}} E_{\text{DG}}^t + \sum_{j=1}^{N_{\text{PV}}} E_{\text{PV}}^t + \sum_{k=1}^{N_{\text{ICE}}} E_{\text{ICE}}^t - \sum_{m=1}^{N_{\text{BESS}}} E_{\text{BESS,ch}}^t + \sum_{n=1}^{N_{\text{BESS}}} E_{\text{BESS,dch}}^t + E_{\text{grid}}^t \quad (9.1)$$

where E_L^t is the demand load at t-time, kW. E_{DG}^t is the electricity generated by the DG at t-time, kW. E_{PV}^t is the electricity generated by the PV system at t-time, kW. E_{ICE}^t is the electricity generated by the ICE at t-time, kW. $E_{\text{BESS,dch}}^t$ and $E_{\text{BESS,ch}}^t$ are the electricity discharged and charged by the BESS at t-time, kW. E_{grid}^t is the electricity imported from the utility grid. In addition to this, N_{DG}, N_{PV}, N_{ICE}, and N_{BESS} depict the number of DG, PV, ICE, and BESS units utilized in the DER, while N_{load} denotes the number of load points.

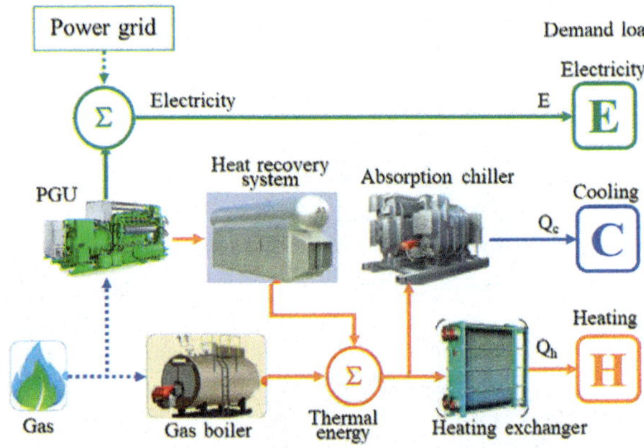

Fig. 9.4 Combined cooling, heating, and power (CCHP) system

9.1.2.2 Combined Cooling, Heating, and Power (CCHP) System

As a typical kind of DER systems, the performance of the combined cooling, heating, and power system (CCHP) is usually determined by the matching degree between the energy demand side and the supply side, as shown in Fig. 9.4.

The energy demands of users are mainly divided into three parts: (1) electric demands, E; (2) cooling demands for space cooling, Qc; and (3) heating demands for space heating and domestic hot water, Qh. The CCHP system consists of a power generation unit (PGU), a waste recovery system, a back-up gas boiler, a cooling system, and a heating system. In this study, the PGU is the internal combustion engine (ICE). The PGU consumes the natural gas and produces the electricity to the demand side. The system relates to the grid, so the sufficient electricity can be imported from the grid and the excess electricity can be sent back to the grid. Then, the recovered heat from the ICE is used to provide heating and drive the absorption chiller. In addition, the auxiliary boiler and the electric chiller are used as back-up devices to provide the additional heating and cooling load of the demand side, respectively. The energy balance in CCHP system is as follows:

$$E_{\text{load}}^{t} = E_{\text{ICE}}^{t} - E_{\text{ec}}^{t} - E_{\text{p}}^{t} + E_{\text{grid}}^{t} \tag{9.2}$$

$$Q_{\text{c}}^{t} = Q_{\text{ac}}^{t} + Q_{\text{ec}}^{t} \tag{9.3}$$

$$Q_{\text{h}}^{t} = Q_{\text{ah}}^{t} + Q_{\text{bh}}^{t} \tag{9.4}$$

$$F_{\text{m}}^{t} = F_{\text{ICE}}^{t} + F_{\text{b}}^{t} \tag{9.5}$$

where E_{load}^{t} is the electricity load of the demand side, kW. E_{ICE}^{t} is the electricity generated by the ICE, kW. E_{p}^{t} is the parasitic electric energy consumption of CCHP system. E_{grid}^{t} is the electricity imported or sold back to the grid, kW. Q_{c}^{t} is the cooling load of the demand side, kW. Q_{ac}^{t} is the cooling produced by the absorption chiller, kW. Q_{ec}^{t} is the cooling produced by the back-up electric chiller, kW. Q_{h}^{t} is the heating load of the demand side, kW. Q_{ah}^{t} is the heating generated by the absorption chiller, kW. Q_{bh}^{t} is the supplementary heat from the back-up gas boiler. F_{m}^{t} is the total natural gas consumption at *t*-time, kWh. F_{ICE}^{t} is the natural gas consumption of the ICE at *t*-time, kWh. F_{b}^{t} is the natural gas consumption of the back-up gas boiler at *t*-time, kWh.

9.1.3 Evaluation Criteria

9.1.3.1 Economic Criteria

The economic evaluation method is divided into static evaluation analysis and dynamic evaluation analysis. Static evaluation analysis method is generally used for the evaluation and analysis of the initial period of the system. Dynamic evaluation analysis method is to convert the inflow and outflow of current funds in different periods into the value of funds at the same time, such as the annualized total cost. Annual basic cost (ABC) includes the annualized investment cost (AIC), the annualized maintenance cost (AMC), and the annualized operation cost (AOC) of each equipment in the system, expressed as Eq. 9.6. Among them, the annualized investment cost and annualized maintenance cost refer to that the total equipment investment cost and maintenance cost are evenly amortized throughout the lifetime of the system, calculated as Eqs. 9.7, 9.8, and 9.9.

$$\mathrm{ABC} = \mathrm{AIC} + \mathrm{AMC} + \mathrm{AOC} \tag{9.6}$$

$$\mathrm{AIC} = \mathrm{CRF} \cdot \sum_{n=1}^{N} \mathrm{NC}_n \cdot C_n \tag{9.7}$$

$$\mathrm{AMC} = \beta \cdot \sum_{n=1}^{N} \mathrm{NC}_n \cdot C_n \tag{9.8}$$

$$\mathrm{CRF} = \frac{r(1+r)^y}{(1+r)^y - 1} \tag{9.9}$$

where NC_n is the nominal capacity of the nth equipment in the system, kW. C_n is the initial capital investment cost of the nth equipment, \$. β is the proportion of annual maintenance cost to the initial investment cost of each equipment in the system. CRF is the capital recovery factor. r is the interest rate, %. y is the lifetime of each equipment in the system, year.

The annualized operating cost of the system refers to the cost of fuel consumed by equipment of the system, such as natural gas consumed by the ICE and the purchasing electricity from the external grid, which is calculated as follows:

Table 9.1 Emission conversion factors of natural gas and electricity [8–10]

Countries	Natural gas (t/m^3)	Electricity (g/kWh)
China	2.20	980
Japan	2.19	462

$$\mathrm{AOC} = \sum_{t=1}^{8760} \left(E_{\mathrm{grid}}^{t} \mathrm{EC}_{\mathrm{e}}^{t} + F_{\mathrm{m}}^{t} \mathrm{EC}_{\mathrm{f}}^{t} \right) \tag{9.10}$$

where $\mathrm{EC}_{\mathrm{c}}^{t}$, $\mathrm{EC}_{\mathrm{h}}^{t}$, $\mathrm{EC}_{\mathrm{e}}^{t}$, and $\mathrm{EC}_{\mathrm{f}}^{t}$ are the energy price of cooling, heating, electricity, and natural gas at *t*-hour, respectively, \$/kWh.

9.1.3.2 Environmental Criteria

At present, the world is vigorously advocating a low-carbon economy to achieve sustainable social development. The environmental performance evaluation of the system refers to the amount of pollutant emitted by the burning of fossil energy during the operation of the system. The pollutants produced by burning fossil energy will have a variety of effects on the environment, such as soil eutrophication, greenhouse effects, and ozone layer depletion. The system burns fossil energy to produce a lot of pollutants, mainly including CO_2, NOX, CO, and particulate matter (PM). Among them, the proportion of CO_2 is as high as 99.5%, while other pollutants account for a small proportion, which can be ignored. Therefore, this research takes carbon dioxide emissions (CDE) as an environmental evaluation index. It can be calculated as Eq. 9.11:

$$\mathrm{CDE} = \sum_{i=1}^{I} \sum_{t}^{8760} E_{\mathrm{fuel},i}^{t} \cdot \mu_{\mathrm{fuel},i} \tag{9.11}$$

where i is the ith kind of fuel used in the DER system. $E_{\mathrm{fuel},i}^{t}$ is the energy consumption of the fuel at *t*-time, kWh. $\mu_{\mathrm{fuel},\,i}$ is the emission conversion factor of the fuel. Usually, the energy fuel of the DER system is natural gas and electricity. Their emission conversion factors in China and Japan are listed in Table 9.1 [8–10].

Carbon dioxide emission cost (CDEC) is the charge that the user pays for the CDE by carbon taxes, which is an effective way in contributing to reduce the CO_2 emissions [11]. Carbon taxes represent a cost-effective way to steer the economy toward a greener future [12]. Through carbon taxes, the environmental performance can be transformed into economic performance to evaluate the comprehensive performance of the DER system. The CDEC is calculated as follows:

$$\mathrm{CDEC} = \mathrm{CDE} * \mathrm{Tax}_{\mathrm{CO_2}} \tag{9.12}$$

where CDEC is carbon dioxide emission cost and $\mathrm{Tax}_{\mathrm{CO_2}}$ is the carbon tax (\$/kg).

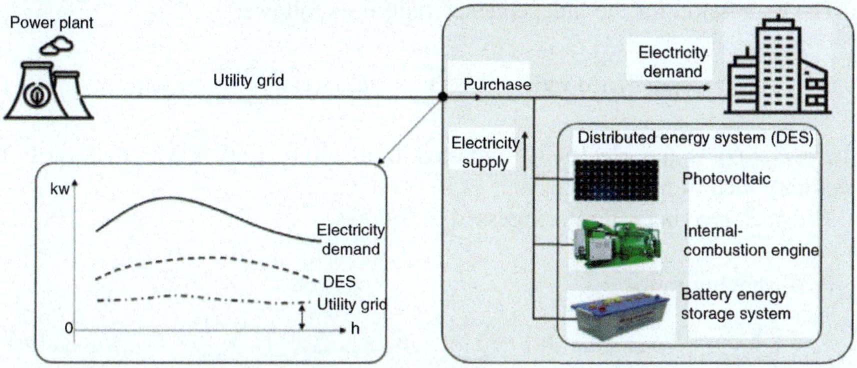

Fig. 9.5 Application of a DER system for grid stabilization

9.1.3.3 Reliability Criteria

Power system reliability can be defined as the ability to provide high-quality and continuous power to the demand side. Improving the grid stabilization is one of the most important contributions of DERs. On the one hand, DERs can reduce the peak load demand to stabilize the utility grid during grid connected. On the other hand, they can be emergency power systems to provide power for critical facilities during times of large-scale power disruptions and outages.

The Independence and Peak Shaving Performance During Grid Connected

Distributed energy resource system is considered as one of the effective control means to adjust peak power consumption and reduce grid load. Power can be stored during peak hours through storage or other technologies, released during peak hours, or replenished to balance the grid with power generation equipment.

As presented in Fig. 9.5, the power produced by localized power generation systems can be self-consumed directly by the demand side in the DER system, which can reduce the electricity demand of the utility grid. The dependence on the grid can be significantly fell down for the user. Peak shaving by operating the power generation systems can further help to reduce the peak load pressure of the power grid and stabilize power grid fluctuation. These two performances can achieve the grid stabilization, which contributes to the reliability of the power supply in the service region. Therefore, we propose two indices of "independence ratio" and "peak shaving ratio" for the grid stabilization of the DER system, which are represented the power self-supply ability of users and the effect of shaving peak load by the DER system. The impact of the demand load fluctuation on the utility grid is less with the improvement of the independence ratio of the DER system. And the peak load pressure of the grid reduces with the growth of the peak shaving ratio.

The expression for the independence ratio is as follows:

$$\text{Independence ratio} = 1 - \sum_{t=1}^{8760} E_{\text{grid}}(t) / \sum_{t=1}^{8760} E_{\text{Load}}(t) \tag{9.13}$$

where $E_{\text{grid}}(t)$ is the electricity imported from utility grid, kWh. $E_{\text{Load}}(t)$ is the electricity load of users, kWh.

The peak shaving ratio is expressed as follows:

$$\text{Peak shaving ratio} = \left(\sum_{m=1}^{12} E_{\text{original, max}}(m) - \sum_{m=1}^{12} E_{\text{grid, max}}(m)\right) / \sum_{m=1}^{12} E_{\text{original, max}}(m) \tag{9.14}$$

where $E_{\text{original, max}}(m)$ is the monthly maximum electricity imported from utility grid before employing the DER system, which is equal to the monthly maximum demand load, kWh. $E_{\text{grid, max}}(m)$ is the monthly maximum electricity imported from utility grid after application of the DER system.

Peak load cost (PLC) refers to the electricity charge paid monthly by users because it occupies the electricity capacity. It is calculated based on the maximum capacity of the demand side in 1 month, which is equal to the monthly peak load. The peak load price is the unit capacity cost of maximum power. The higher the peak load price, the more obvious is the benefit of the peak shaving effect. The peak load can be reduced by the operation of a reasonably power generation system in the DER system, which can reflect the effect of the DER system on peak shaving.

$$\text{PLC} = \sum_{m=1}^{12} \max\left(E_{\text{m,grid}}\right) \cdot \text{Price}_{\text{ps}} \tag{9.15}$$

where $P_{\text{grid, max}}$ is the maximum capacity of the demand side in 1 month (kW), Price_{pl} is peak load price (yen/kW).

The Reliability Assessment Index as Emergency Power Systems During Power Outage

Power outage loss refers to the total economic loss that the society bears when the power supply is not completely reliable or expected to be not completely reliable. To assess the reliability of a power supply system, it is necessary to estimate the economic loss caused by power outage and power supply interruption to customers. The economic losses caused by interruption of power load in a region can be divided into direct economic losses and indirect economic losses. The direct outage loss is usually determined by the short-term effect of the unexpected outage, while the indirect outage loss is caused by the longer-term consideration of the expected power outage. The cost of energy not supplied (CENS) in a power system represents the average cost during the power outage. It can be estimated by modelling the power outage loss as a function of the unsupplied energy regardless of the power outage duration and frequency [13]. The CENS consists of all financial damage that consumers experience during the power outage or load shedding. In order to estimate the CENS in the distribution system, the following details must be supplied: (i) direct economic loss; (ii) indirect economic loss; and (iii) information on the non-supplied energy during the time period of the analysis [14]. The expression is as follows:

$$\text{CENS} = \frac{\sum_{i=0}^{n} F_i D_i + \sum_{j=0}^{n} I_j}{\sum_{t=0}^{T} \text{EENS}} = \frac{\left(F_{\text{res}} D_{\text{res}} + F_{\text{ind}} D_{\text{ind}} + F_{\text{com}} D_{\text{com}} + F_{\text{g\&i}} D_{\text{g\&i}} + F_{\text{other}} D_{\text{other}}\right) + \left(\sum_j I_{jx} - \sum_j I_{jy}\right)}{\sum_{t=0}^{T} \text{EENS}} \tag{9.16}$$

where CENS is the value or cost of energy not supplied for the case study, \$/kWh. F_i is vulnerability factor of consumer at the load point, %. i is type of consumers such as residential (res), industrial (ind), commercial (com), government and institution (g&i), and other (others). D_i is the direct economic losses of consumers, \$. I_j is indirect economic losses, \$. I_{jx} is total indirect cost brought about during the time of investigation, \$. I_{jy} is the indirect costs that had no association with the investigation, \$. n is number of customers in the system. EENS is the non-supplied energy during the period of analysis, kWh.

The expected energy not supplied (EENS) referred to the power outage loss load is the expected unsupplied energy in a year owing to generation unavailability or inadequacy or lack of primary energy, which is calculated as

$$\text{EENS} = \sum_{k=1}^{n} P_k \times E_k \tag{9.17}$$

where P_k is the power outage with a probability (usually using the failure rate). E_k is the average load of the load point, kWh.

9.2 Economic and Environmental Analysis of the Distributed Energy Resource System Focusing on Grid Stabilization

Through application of the DER system, the power produced by localized power generators can be self-consumed directly by the demand side, which significantly reduces the power import from the grid and lessens the dependence of the users on the utility grid. At the same time, the DER system can stable the fluctuation of the grid load by flexibly operating the distributed generators to shave the peak load.

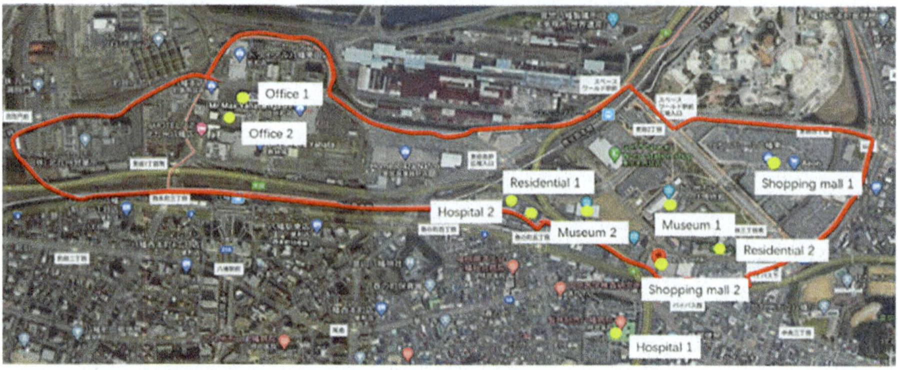

Fig. 9.6 The Higashida District, Yahata, Kitakyushu City

Therefore, the DER system can minimize the shock of demand load surge on the grid and effectively achieve grid stabilization. However, the cost of DER system to realize the different requirements of grid stabilization is different. This section analyzed the economic and environmental performance of the DER system focusing on grid stabilization.

9.2.1 Research Object

Taking a smart community in Higashida, Japan, as research area, five different types (including two offices, two residential buildings, two hospitals, two museums, and two shopping malls) of buildings were selected, as shown in Fig. 9.6.

The electrical load of these ten buildings was investigated from April 2013 to March 2014 at hourly intervals, collected by smart meters and cluster energy management system of Higashida District. Figure 9.7 shows the variations in daily or seasonal power demand based on the history of grid load. The grid demand load has the characteristics of "peak during daytime, valley at night," especially in summer.

The proposed DER system model consists of PV system, battery system, and ICE. The energy balance of the model was as Eq. 9.18. In order to reflect the grid stabilization effect of the DER system, two aspects should be evaluated. Therefore, we proposed two novel indices. The "independence ratio" is represented by the power self-supply ability, and the "peak shaving ratio" is reflected by the effect of peak load reduction, calculated as Eqs. 9.13 and 9.14.

$$E_{\mathrm{L}}^{t} = E_{\mathrm{PV}}^{t} + E_{\mathrm{ICE}}^{t} + E_{\mathrm{BESS,dch}}^{t} - E_{\mathrm{BESS,ch}}^{t} + E_{\mathrm{grid}}^{t} \tag{9.18}$$

where E_{L}^{t} is the demand load at t-time, kW. E_{PV}^{t} is the electricity generated by the PV system at t-time, kW. E_{ICE}^{t} is the electricity generated by the ICE at t-time, kW. $E_{\mathrm{BESS,dch}}^{t}$ and $E_{\mathrm{BESS,ch}}^{t}$ are the electricity discharged and charged by the BESS at t-time, kW. E_{grid}^{t} is the electricity imported from the utility grid.

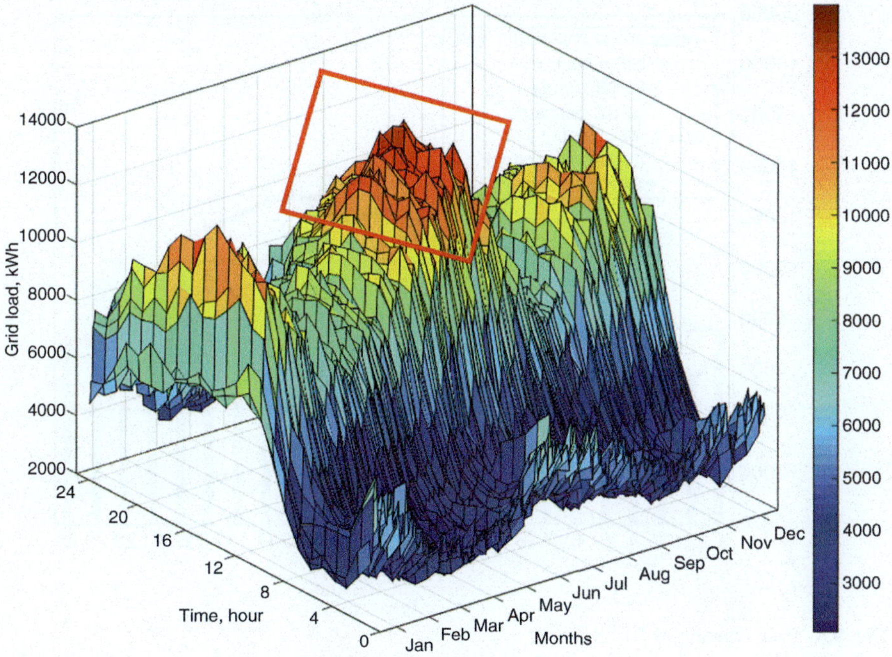

Fig. 9.7 Proposed distributed energy resource system

Considering the economic operation strategy, the battery system can only migrate the grid load without increasing the DER system output, depicted in Fig. 9.8. Therefore, the independence ratio is only determined by the power generation of PV and ICE. So, we first analyzed the performance of PV and ICE under different independence and peak shaving ratios. Then, the impact of the battery system on the peak shaving was explored. Finally, the optimal combinations under different grid stabilization effect can be obtained (Fig. 9.9).

9.2.2 *Economic and Environmental Analysis Focusing on Grid Stabilization*

9.2.2.1 Distributed Energy Resource System with PV and ICE

After simulation, peak shaving ratio and investment cost changes are shown in Fig. 9.10. We can see that the output of PV and ICE system can improve the independence ratio at the same time increasing peak shaving ratio. Because the ICE only operates in the daytime and cannot cover the night load, the independence and peak shaving ratio have maximum values. The investment cost of the DER system increases with the independence ratio, and the growth rate accelerated quickly when the independence ratio is high.

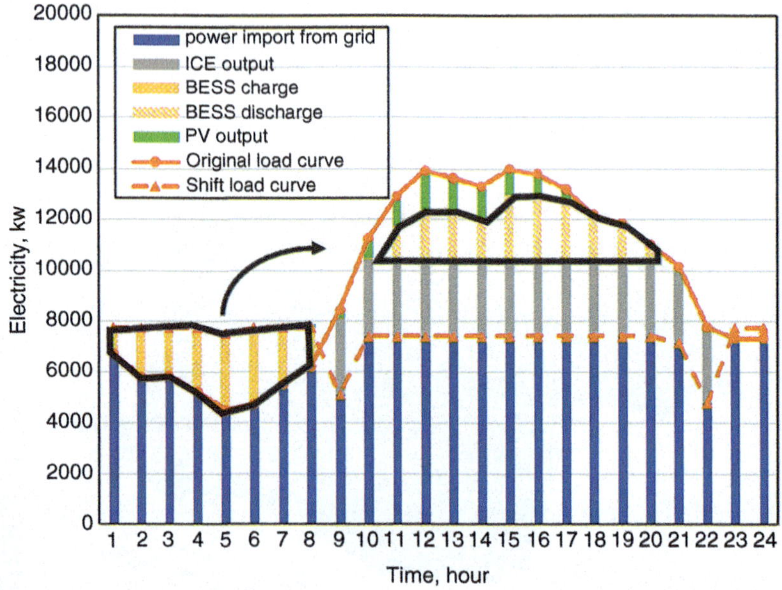

Fig. 9.8 Peak shaving of BESS

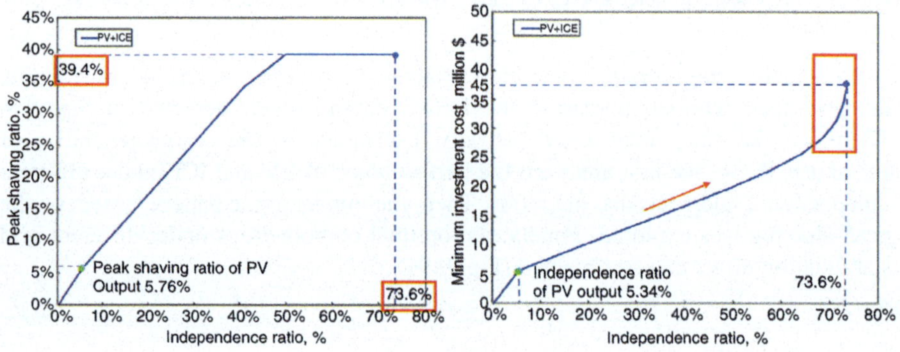

Fig. 9.9 Peak shaving of BESS

The annual basic cost (ABC) and carbon emission reduction changes are shown in Fig. 9.11. The rising of independence ratio will increase investment cost but can greatly reduce energy bills. Therefore, the annualized total cost first decreased. When the independence ratio reaches 64%, the annualized total cost is the least. To continue increasing the independence ratio, the economic performance of the DER system will become less. Moreover, the carbon emission reduction rate will be increased linearly with independence ratio rising.

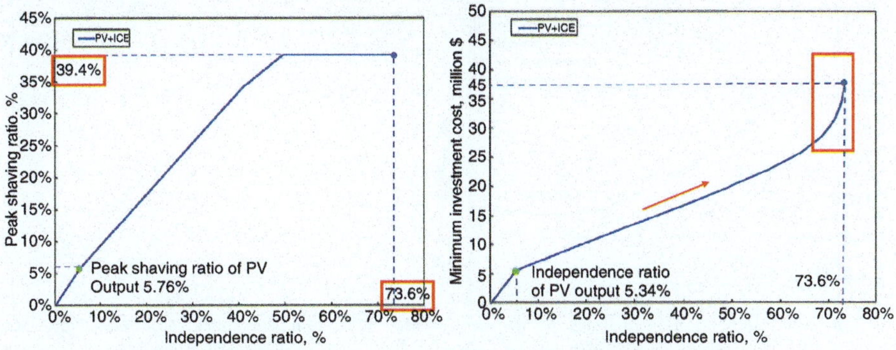

Fig. 9.10 (**a**) Peak shaving ratio changes; (**b**) investment cost changes

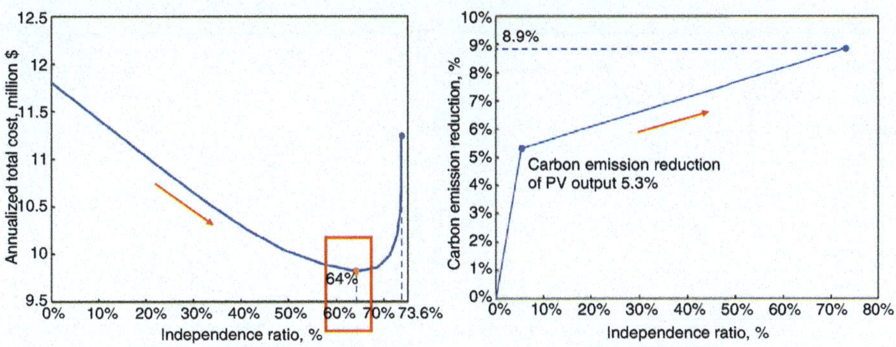

Fig. 9.11 (**a**) ABC changes; (**b**) carbon emission reduction changes

9.2.2.2 Distributed Energy Resource System with PV, ICE, and BESS

Figure 9.12a shows the ABC changes with the increase of BESS capacity. Figure 9.12b shows the ICE and BESS optimal capacity changes with the increase of independence ratio. Due to the peak shaving and valley-filling operation of the battery, the more expensive power in the daytime can be replaced by the cheaper power at night, which can save the energy bills. Therefore, the annualized total cost after introducing the battery system is reduced. And the optimal capacity of battery is different with the independence ratio changing. As this figure shows, with the increase of independence ratio, the installed capacity of the BESS gradually decreases. This is because the peak shaving and valley filling operation of the BESS will reduce the daytime load, which leads to the reduction of the DES output. It will affect the independence ratio of the system. Therefore, the installed capacity of the BESS is limited by the independence ratio.

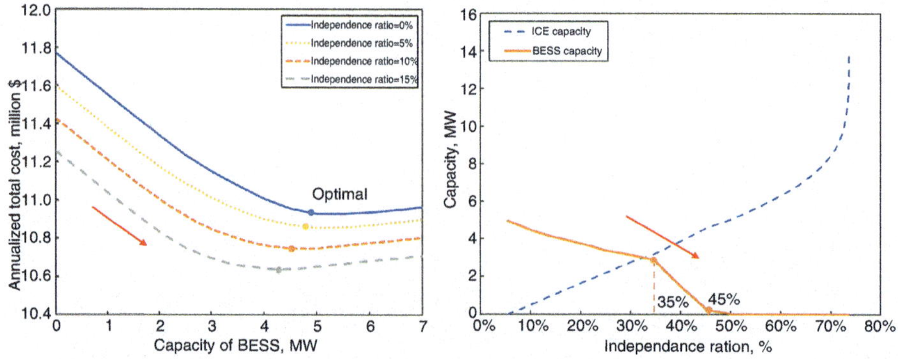

Fig. 9.12 (**a**) ABC changes with the increase of BESS capacity; (**b**) ICE and BESS optimal capacity changes with the increase of independence ratio

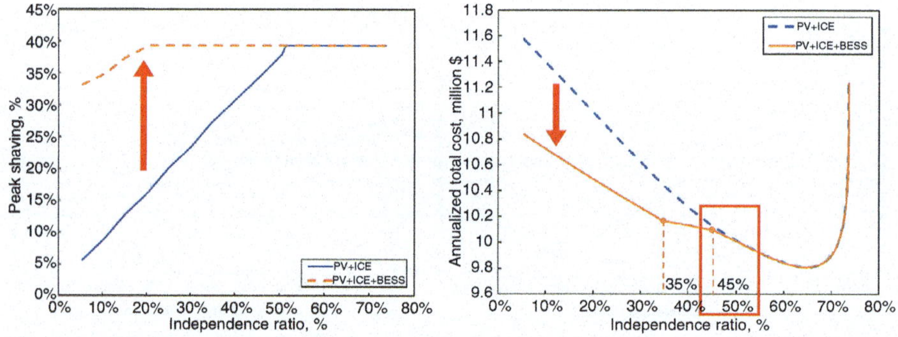

Fig. 9.13 (**a**) Peak shaving ratio changes before and after introducing the BESS; (**b**) ABC changes before and after introducing BESS

After introducing the battery system, the peak shaving ratio is greatly improved. However, when it comes to the maximum peak shaving ratio, increasing the capacity of the battery is not useful. Therefore, the annualized total cost can be reduced at the low share of independence ratio by introducing the battery system, but it is not economic when the independence ratio is above 45% (Fig. 9.13).

9.2.3 Summary

DER has the application potentials for grid stabilization and environmental improvement. Two novel indices called "independence ratio" and "peak shaving ratio" were proposed to analyze the grid stabilization effect of the DER. The results showed that increasing the installed capacity of the ICE could improve the independence ratio and peak shaving ratio. The battery system has no contribution to the independence ratio and only plays a role in peak shaving due to its economic operation strategy. As for the economic benefits, the effect of the DER system investment cost increase is becoming smaller with the growth of the independence ratio. Therefore, the introduction rate of the DER on the demand side has an optimal proportion. If continuing increasing the independence ratio to enhance the grid stability and environment effects, the gains come at the expense of economic benefits. Therefore, it is essential to balance all of them.

9.3 Multi-criteria Assessment for Optimizing Distributed Energy Resource System

Cost saving, grid stabilization, and CO_2 emission reduction are causing the increasing attention of the DER. As analyzed in the previous chapter, the improvement of grid stabilization and environmental effects will weaken the economic performance of the DER system. To achieve comprehensive assessment, we proposed a multi-criteria evaluation method combined with these three different perspectives and transferred the advantages into economic benefits using peak load price and carbon tax, as depicted in Fig. 9.14.

9.3.1 Different Criteria

1. **Objective Function Setting**

To compare the impact of grid stabilization and carbon reduction performance on optimizing the structure of DER system, three different objective functions were proposed and compared.

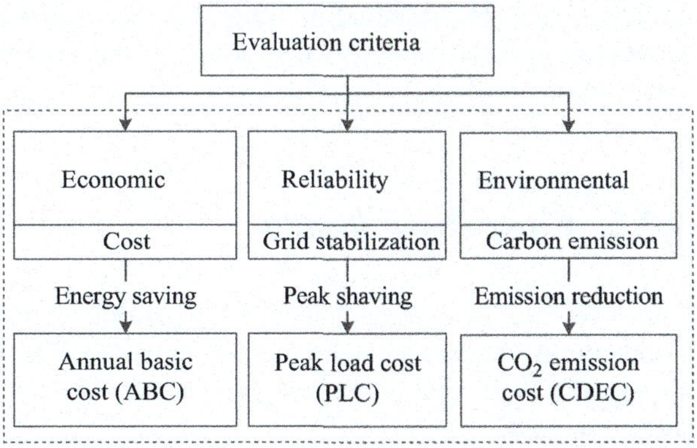

Fig. 9.14 Multi-criteria evaluation method

The cost consists of annual basic cost only:

$$F_1 = \min(\mathrm{ABC}) \tag{9.19}$$

The costs consist of annual basic cost and peak load cost:

$$F_2 = \min(\mathrm{ABC} + \mathrm{PLC}) \tag{9.20}$$

The total costs consist of annual basic cost, peak load cost, as well as CO_2 emission cost:

$$F_3 = \min(\mathrm{ABC} + \mathrm{PLC} + \mathrm{CDEC}) \tag{9.21}$$

where ABC is the annual basic cost which is the amount of money that users need to spend for their demand each year, calculated as Eq. 9.19. PLC is the peak load cost referred to the fee paid by users according to the monthly maximum grid load, calculated as Eq. 9.24. CDEC is annual cost of carbon emission, calculated as Eq. 9.11.

The first Function (F1) only includes the system basic cost, which is a single index including economic performance. The second Function (F2) adds the peak load cost, which means that the evaluation considers both economic and grid stabilization performance. And the third Function (F3) considers the annual basic cost, peak load cost, and carbon emission cost at the same time, which is a multi-criteria evaluation method.

2. **Basic Information.**

The output of PV system will significantly influence the performance of the DER. High share of PV penetration may cause overproduction and aggravate the grid fluctuation, as demonstrated in Fig. 9.15. Therefore, we set different application scenarios with 0 ~ 50% PV penetration in Higashida smart community for research.

The technical specifications and cost data details for each unit of the proposed DER systems are presented in Table 9.2. According to the Kyushu Electric Power Company, the electricity price is adopting the time of use (TOU) event, which is presented in Table 9.3 [15]. Other prices are listed in Table 9.3 as well.

9.3.2 Comparison Results

In this section, we assume that the PV penetration changes from 0% to 50% and it is divided into 11 scenarios with increasing by 5%. The surplus PV production cannot be sold out to the grid in this case. The results of the different scenarios are obtained and compared.

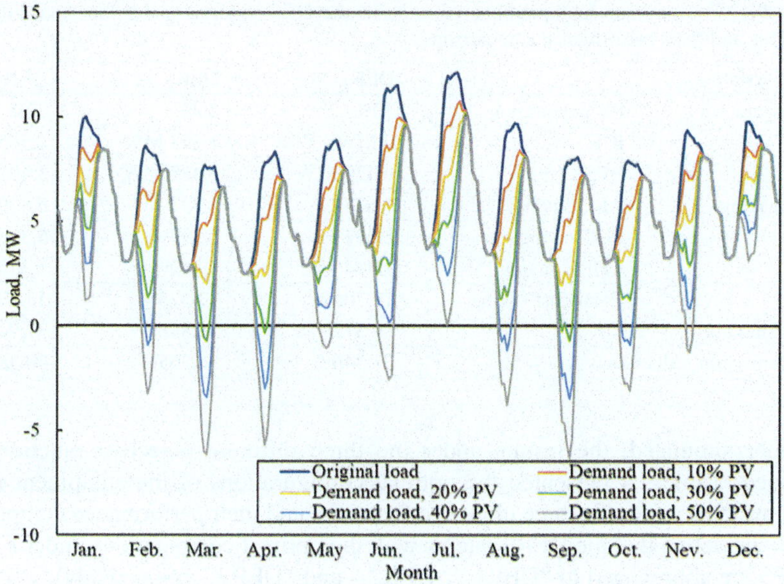

Fig. 9.15 Daily average curves of load in each month with 0 ~ 50% PV generation

Table 9.2 Technical parameters of the units in the distributed energy resource systems [16–19]

System	Capital cost[a] ($/kW)	Lifetime (years)	Other operation parameters
PV	1904.8	15	Conversion efficiency = 16%
			Angle of incidence = 30°
ICE	2857.1	15	Electricity efficiency = 0.4
			Lower heating value = 45 MJ/Nm^3
BESS	1428.6	9	$d = 0.8$
			Max SOC = 95%
			Max SOC = 15%
			Charge/discharging efficiency = 100%

[a]The US dollar exchange rate against RMB is 1:7; the US dollar exchange rate against JPY is 1:105

Table 9.3 Energy prices [15, 20]

Item	Time			Prices ($/kWh)
Electricity	Summer	Peak time	13:00–16:00	0.248
		Daytime	8:00–13:00, 16:00–22:00	0.213
	Other seasons	Daytime	8:00–22:00	0.203
	Night		22:00–8:00	0.0863
	Peak load price		Monthly	12.571
Natural gas	Always			0.575 $/m^3

Table 9.4 The optimal configurations of units in the DER systems under the three objective functions and their performance comparison

DER systems		DER_1	DER_2	DER_3
Configuration	PV penetration	30%	30%	30%
	ICE	0 MW	1.573 MW	1.565 MW
	BESS	9.479 MW	3.488 MW	3.491 MW
Cost saving	ABC saving	7.677%	6.837%	6.839%
	PLC saving	2.634%	5.392%	5.390%
	CDEC saving	1.394%	1.349%	1.350%
	Total cost saving	11.706%	13.578%	13.579%
Peak shaving		19.372%	37.848%	37.945%
CO_2 emission reduction		31.346%	30.365%	30.380%

After simulated, the results under the three different objective functions are obtained. Table 9.4 presented the optimal configurations of the equipment in the DER systems under the three objective functions and their performance comparison (here, we define that the DER systems with the optimal combinations under F1, F2, and F3 are abbreviated to "DER1," "DER2," and "DER3," respectively).

The total cost saving of the DER3 is the most, which improves 1.87% compared with the DER1. The peak shaving performance of DER3 is the most significant; it reduces 37.945% of the peak load. But the CO_2 emission reduction of DER1 is most; it declines 31.346% of the carbon emission. The results demonstrated that when the peak shaving capacity and emission reduction effect are converted into economic benefits, the peak load price and carbon tax will have a greater contribution. Therefore, the comprehensive evaluation criteria should be considered when determining the configuration of DER. The utilization of multi-criteria evaluation when optimizing DERs can fairly balance the different performance and improve its competitiveness.

9.3.3 Summary

This section proposed a multi-criteria evaluation method from different points of view and converted them into economic benefit by introducing peak load price and carbon tax. After analyzing the characteristics of grid load for five multi-types of buildings in a smart community in Japan, the configurations of each unit in the DER system were optimized and compared under three different evaluation methods. The results presented that the multi-criteria method can trade off the different performance of the DER. Taking grid stabilization effect as evaluation index can improve the economic advantage of ICE, and the carbon tax is helpful to promote the PV penetration. Optimizing the configuration with multi-criteria evaluation can maximize the application potentials of the DER and improve its market competitiveness.

9.4 Utilization of Multi-criteria Assessment Method for Two Cases of Distributed Energy Resource Systems

9.4.1 Promotion and Utilization of the Distributed Energy Resource System: A Case Study of Combined Cooling, Heating, and Power System

As a typical DER system, the CCHP system can meet the energy demand by generating electricity and simultaneously provide the cooling and heating load by recovering waste heat. It is identified as a sustainable energy development with its high efficiency. However, poor economic performance has limited its diffusion. In order to improve the economic performance of the CCHP system, this section evaluated the comprehensive performance of the CCHP system through a multi-criteria method. And the impacts of different factors on its promotion were discussed. The research object takes a typical CCHP system in an amusement park resort in Shanghai, China. First, we established the energy flow model of CCHP system. The detailed model was presented in Sect. 9.1.2.2.

9.4.1.1 Evaluation Criteria

Dynamic payback period and carbon emission were calculated to evaluate its economic and environmental performance.

1. **Economic Criteria.**

The payback period, an index of economic performance, is the time required for the project to recover the initial investment cost. By calculating payback period, the economic performance of the projects can be compared. A short payback period means that the economic benefits of the system are high. The dynamic payback period is calculated when the cumulative net present value (NPV) is zero, as shown by Eq. 9.22:

$$\mathrm{NPV}(n) = 0 \rightarrow \mathrm{PB} = n \tag{9.22}$$

The payback period cannot be longer than the lifetime of the system, which is 25 years in Japan.

The net present value (NPV) is the difference between the present value of cash inflows and the present value of cash outflows during a period, which mainly represents the balance between the present value of total profit and the initial investment. It can be expressed as [21]:

$$\mathrm{NPV} = \sum_{n=1}^{\mathrm{PB}} \frac{\mathrm{ATP}_n}{(1+i)^n} - \mathrm{IN} \tag{9.23}$$

where ATP_n is the annual total profit in \$. i is the discount rate percentage. IN is the total investment cost in \$.

2. **Environmental Criteria.**

The amount of carbon dioxide emissions (CDEs) from the CCHP system can be determined using the emission conversion factors as follows [8, 22]:

$$CDE = \sum_{t}^{8760} E_{\text{grid}}^{t} \cdot \mu_{CO_2,e} + \sum_{t}^{8760} F_{\text{m}}^{t} \cdot \mu_{CO_2,\text{gas}} \tag{9.24}$$

where $\mu_{CO_2,e}$ and $\mu_{CO_2,\text{gas}}$ are the emission conversion factors for electricity from the grid and natural gas, respectively, in g/kWh.

A carbon tax is one of the effective means to reduce carbon emissions. At present, at least 20 countries in the world have imposed carbon taxes [23]. The carbon tax can be calculated into the total annual profit, as follows:

$$ATP' = ATP - \Delta CDE * Tax_{co_2} \tag{9.25}$$

where ATP' is the total annual profit considering the carbon tax. ΔCDE is the difference in carbon dioxide emissions before and after utilization of the energy supply system. Tax_{co_2} is the carbon tax.

9.4.1.2 Case Study

1. **System Setting.**

To promote the CCHP, we proposed three systems with different penetrations for comparison. Generally, centralized energy supply systems with electric chillers and gas boilers are commonly used in amusement parks to provide cooling and heating [24]. Nowadays, a few major theme park companies are embracing a more energy-saving and environmentally friendly way to solve energy problems. The CCHP system is an alternative to conventional systems. The research case of this chapter is a hybrid CCHP system with penetration of 50% (the cooling and heating load provided by waste heat from PGUs account for 50% of the total demand). To study the promotion of CCHP systems, this chapter proposed three CCHP systems with different penetration, using the current system for comparison:

- **System 1:** conventional system without CCHP (only adopting electric chillers and boilers to supply cooling and heating load; the electricity is from the utility).
- **System 2:** CCHP with 50% penetration (adopting PGUs, absorption units, electric chillers, and boilers to cooperate to supply electricity, cooling, and heating).
- **System 3:** CCHP with 100% penetration (only adopting PGUs and absorption units to supply energy).

Table 9.5 Configurations of three cases

System	Equipment	Amount	Rated output (kW)
Conventional system (without CCHP)	Electric chiller 1	6	6330
	Electric chiller 2	4	3165
	Boiler	4	7000
CCHP with 50% penetration	ICE	5	4401
	Absorption unit	5	3931
	Electric chiller 1	4	6330
	Electric chiller 2	2	3165
	Boiler	2	7000
CCHP with 100% penetration	ICE	15	4401
	Absorption unit	15	3931

Table 9.6 The facility price, energy price, and other parameters of the CCHP system [25]

Parameter			Symbol	Unit	Value
Facility price		IC engine	C_{pgu}	\$/kW	971
		Absorption unit	C_{ab}		172
		Electric chiller	C_{ec}		139
		Boiler	C_b		43
Energy prices	Natural gas		EC_f^t	\$/kWh	0.039
	Cooling/heating		EC_c^t/EC_h^t,	\$/kWh	0.04
	Electricity	22:00–6:00	EC_e^t	\$/kWh	0.044
		6:00–8:00, 11:00–18:00, 21:00–22:00			0.089
		8:00–11:00, 18:00–21:00			0.151
The CO_2 emission conversion factors		Natural gas	$\mu_{CO_2,gas}$	g/kWh	220
		Electricity from grid	$\mu_{CO_2,e}$		968
The discount rate			i	%	4.9

The configurations of three cases are presented in Table 9.5. Table 9.6 listed the basic parameters to calculate the economic and environmental criteria based on the simulation results.

2. **Comparison of Economic and Environmental Performance in Three Systems.**

Based on the simulation results, the economic and environmental performance is shown in Table 9.7. The comparison of the payback period and CO_2 emissions of the three different systems is shown in Fig. 9.16. As the penetration of the CCHP increases, carbon emission is reduced, but the economic performance becomes worse since the payback period is extended because of the large investment cost.

Table 9.7 The economic criteria of the three systems

Systems	ATP ($)	IN ($)	Payback period (year)
Conventional system (without CCHP)	2.28	8.3	4.01
CCHP with 50% penetration	5.35	29.8	6.5
CCHP with 100% penetration	7.09	74.2	14.67

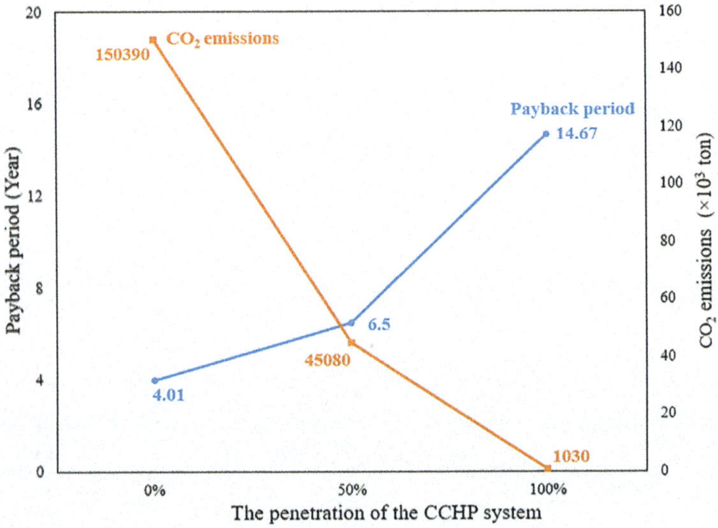

Fig. 9.16 The economic and environmental performance of the CCHP system with different penetrations

3. **Impact of Different Factors on the Economic Performance of CCHP System.**

In this section, we analyzed the economic influence factors of the CCHP system and compared the economic criteria of the three CCHP systems with different penetrations with various changes in these factors. Investment cost is one of the main factors affecting the economics of the CCHP system. In the future, as the cost of the ICE decreases, its economy will gradually improve. Energy prices determine the operation cost and profit of the CCHP system, which directly reflect the economic performance of the CCHP system. Supportive policy is one of the most effective measures that can contribute to the economics of the CCHP system. At present, an investment subsidy is available in Shanghai [26]. This investment subsidy is a direct grant provided by the government according to the installed capacity of the CCHP system during construction. With reasonable subsidies, the attraction of the investment into and installed capacity of the CCHP system can be improved. With an emphasis on energy-saving and emission reduction, the implementation of a carbon tax can increase the operation cost of an energy system, because the energy becomes more expensive when imposing taxes on fossil energy

Table 9.8 The changes in factors on the economic performance of CCHP system

Factors		Changes
Penetration of CCHP system		0%, 50%, 100%
Investment cost of ICE decrease		0% to 50%
Energy prices	Electricity price	−50% to 50%
	Natural gas price	−50% to 50%
Investment subsidy		0 to 2 times the current subsidy1
Carbon tax		0 to 50 $/ton [30]

consumption. Compared with the conventional system, the advantage of the CCHP system would be more obvious after employing a carbon tax due to the low emission character.

Therefore, how the above factors affect the economics of the CCHP system is discussed below.

The influencing factors and values were shown in Table 9.8.

After simulation and analysis, we can get some conclusion as follows:

(a) **Impact of the Investment Cost of ICE.**

The ICE is much expensive than other equipment, which extend the payback period of the CCHP system. With the decline of investment cost of ICE in the future, the payback period of CCHP system is effectively reduced. Compared with the conventional system, the investment cost of ICE needs to be decreased by more than 48% to achieve a shorter payback period.

(b) **Impact of Energy Prices.**

Energy prices determine the profits. The increase of electricity price can bring greater benefits for CCHP system. When the electricity price increases by more than 11.5%, the payback period of CCHP system can be shorter than the conventional system. But the change of natural gas price is less helpful to reduce the economic gap between the CCHP system and conventional system. The payback period of CCHP system is still longer than conventional system; even gas price reduces to 50%.

(c) **Impact of the Investment Subsidy.**

Incentive policy is an effective way that contributes to the economy of the CCHP system. Under the current investment subsidy, the payback period of the CCHP system with 50% penetration is almost the same as that of the conventional system. When the investment subsidy increases to 1.7 times, the 100% CCHP system will achieve better economic performance than the conventional system.

(d) **Impact of the Carbon Tax.**

The introduction of a carbon tax can highlight the superiority of the low-emission characteristics of the CCHP system. With the increase of the carbon tax, the economy of the conventional system (without CCHP) decreases rapidly. When the carbon tax is more than 12.5 \$/ton, it gives an economic advantage to the 50% CCHP system even without other incentive policies. The more the carbon tax increases, the environmental advantage is more prominent with the penetration of CCHP system. When the carbon tax reaches 25.5 \$/ton, the economic performance of the 100% CCHP system is superior to that of the conventional system (without CCHP).

9.4.1.3 Summary

Taking a typical CCHP system of an amusement park resort in Shanghai, China, as a research case, the economic and environmental criteria of the CCHP system were analyzed based on its dynamic payback period and carbon emissions. The results indicated that CCHP system could bring significant environmental benefits, but as the penetration of the CCHP system increases, the economic performance becomes worse. Adjusting the energy price and implementing subsidy policies could help to address the problem. Among them, increasing the carbon tax is the best way to accelerate the promotion of CCHP system.

9.4.2 Promotion and Utilization of the Distributed Energy Resource System: A Case Study of Emergency Power System

Power outages will not only cause huge economic losses but also threaten people's lives. Therefore, improving the reliability and safety of the power system is significant for developing the modern power grid. Emergency power system (EPS) can provide critical load in the power outages by installing small and localized power generators to improve the power grid reliability. The research on the reliability and economic analysis of the emergency power system is divided into two parts: first is the optimization of the emergency power system integration, and second is application of distributed generation as emergency power.

9.4.2.1 Evaluation Criteria

1. Reliability Criteria.

We use power outage loss load called expected energy not supplied (EENS) to reflect the power supply reliability of the building complex. Equation 9.26 shows the calculation of EENS. Less EENS means higher reliability.

$$\mathrm{EENS}_k^t = \lambda_k r_k L_k^t \tag{9.26}$$

Fig. 9.17 Economic criteria

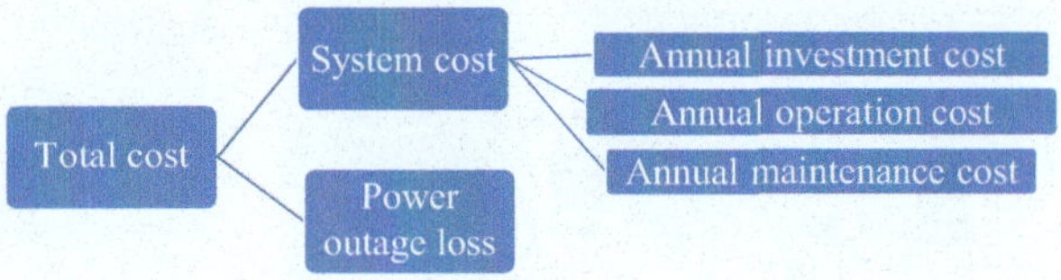

2. Economic Criteria.

The total cost is economic criterion which is combined with system cost and power outage loss, as displayed in Fig. 9.17.

System cost is divided into three parts: annualized investment cost (AIC), annualized operation cost (AOC), and annualized maintenance cost (AMC), which are calculated as Eqs. 9.6, 9.7, 9.8 and 9.9.

Power outage loss is divided into two parts: direct economic loss and indirect economic loss. The former refers to losses incurred during and after the actual outage; the latter refers to additional costs incurred by users to reduce the outage, adjust their activities, or adopt standby energy sources [27]. The direct economic loss increased sharply at the instant of power outages and increased with the duration of power outage. However, the growth rate decreases with the duration of outage [28]. The indirect economic loss increases with the duration of the outage, and the growth rate of different types of building losses is different. The composite customer damage function (CCDF) [29] is used to express the relationship between outage loss and outage duration, considering the loss characteristics of different types of users. The composite customer damage function is obtained as follows:

$$f_{\mathrm{CCDF}}\left(T_{\mathrm{outage}}\right)=\sum\nolimits_{i=1}^{n}\frac{c_i\times k\left(T_{\mathrm{outage}}\right)}{N_i} \tag{9.27}$$

Among them, i is the type i customer (the type of users include residence, industry, government, and so on), n is the number of categories of users, c_i is the proportion of power consumption of the type i customer, and $k(T_{\mathrm{outage}})$ is the unit power outage loss of the type i customer, \$/kWh, of which it can be obtained by investigation [30]. T_{outage} is the duration of power outage, h. N_i is the load rate of the type i customer.

According to Ref. [31] based on its research data, the function can be estimated as Eq. 9.28.

$$f_{\mathrm{CCDF}}\left(T_{\mathrm{outage}}\right)=18.989T_{\mathrm{outage}}^{0.4156} \tag{9.28}$$

Therefore, the formula for calculating the annualized power outage loss is calculated as follows:

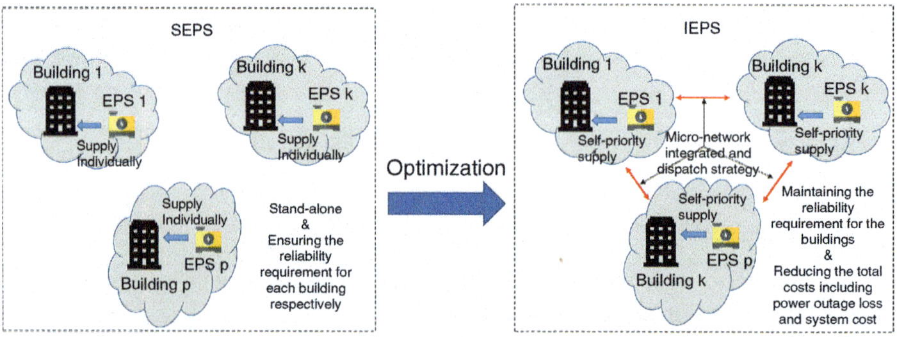

Fig. 9.18 The schematic diagram of stand-alone emergency power system (SEPS) and integrated emergency power system (IEPS)

Table 9.9 The information of the building complex

Buildings	The type of building	Average load (kW)	Critical load (kW)
Building 1	Commercial	4935	1819
Building 2	Office	4985	2165
Building 3	Central business district	7298	1473

$$C_{\mathrm{outage}_k} = \sum\nolimits_{t}^{8760} \left[f_{\mathrm{CCDF}}\left(T_{\mathrm{outage}}\right) \times \mathrm{EENS}_k^t \right] \quad (9.29)$$

9.4.2.2 Case Study

Generally speaking, EPSs usually adopt diesel generators, and they are installed independently in every building. If one of them is broken, the other cannot provide help. Since the power outage probability of each building is different, an emergency power system integration model was proposed by connecting the stand-alone emergency power subsystems with the micro-network. In this way, the integrated emergency power system can serve several buildings in a more safe and economic manner. Then, the whole system cost can be reduced, and the power supply safety of the building complex will be improved. The concept schematic diagram of the stand-alone emergency power system (SEPS) and the integrated emergency power system (IEPS) was displayed in Fig. 9.18.

The emergency power system integration model was verified in a building complex in a certain region in Shanghai, China. The building complex included three adjacent buildings, which were a commercial high-rise building (Building 1), an office high-rise building (Building 2), and a central business district building. The information of the three buildings is shown in Table 9.9. To meet the reliability

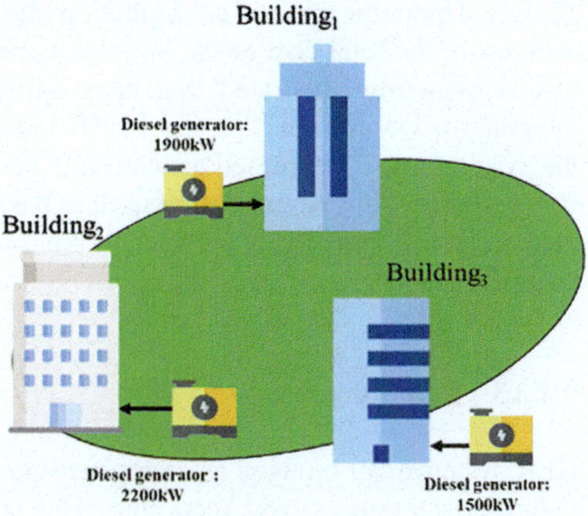

Fig. 9.19 The economic and environmental performance of the CCHP system with different penetrations

Table 9.10 The results comparison of SEPS and IEPS

Index		Stand-alone emergency power system	Integrated emergency power system with optimal configuration	Rate (%)
Cp_i(kW)	Building 1	1900	1860	−2.11
	Building 2	2200	1680	−23.64
	Building 3	1500	1470	−2.00
	Total	5600	5010	−10.54
Total EENS (MWh)		220.34	10.65	−94.23
Annual investment cost (million $)		0.134	0.120	−10.60
Annual operation cost (million $)		0.054	0.102	+88.62
Annual maintenance cost (million $)		0.003	0.005	+88.62
Outage loss (million $)		1.249	0.061	−95.15
Total cost (million $)		1.440	0.288	−80.03

requirements of the critical load of each building, EPSs are stand-alone allocated in each building. Their initially capacity is set as the critical load of its own building, shown in Fig. 9.19. The EPSs of these three buildings are adopting diesel generator, which is a common form of back-up power.

Taking the minimum expense of the building complex as the objective function, the optimal configuration of the integrated emergency power system can be obtained using the genetic algorithm. Table 9.10 displayed the results comparison of stand-alone emergency power system and integrated emergency power system. The optimal installed capacity of the IEPS is 5010 kW, which is decreased by 10.54% compared with the initial installed capacity; therefore the investment cost is saved by

10.60%. Under the optimal configuration, the total EENS could decrease by 94.23%; as a result, the total cost of the building complex can be cut down by 95.15%. The results evidently illustrated that after optimization of emergency power system integration, we can see that the EENS is sharply reduced, and at the same time, the system cost is decreased as well. It is because that when one of the subsystems fails, the other subsystems can be used as back-up to help each other. It indicates that the integrated emergency power system can save the cost while improving the reliability.

9.4.2.3 Summary

DER systems can be used as emergency power systems to improve reliability and reduce power outage loss. According to the concept of micro-network, the integrated emergency power system (IEPS) can serve multiple buildings more safely and economically in a building complex. Due to the different probabilities of distribution network failure in each building, the stand-alone emergency power system of each building can integrate and dispatch to maximize the utilization of resources. Therefore, Sect. 9.4.2 proposed an integration emergency power system (IEPS) by connecting the stand-alone emergency power subsystems with the micro-network and backing up each other to achieve the purpose of improvement in reliability and economy.

9.5 Conclusion

The high rate of world energy consumption caused by a growing population and global economic expansion coupled with rapid industrialization has necessitated a massive investment in reliable and efficient energy supply. With the rapid growth in energy demand, concerns about climate change, high prices of fossil fuels, and the depletion of fossil fuels, the countries around the world are changing the focus of the production of electricity from large, centralized power plants to local generation units scattered over the territory. As an ideal system for incorporating renewable energy sources, distributed energy resource (DER) systems allow producing energy that is consumed in proximity to the points of production, thus reducing energy costs and carbon emissions, achieving energy self-sufficiency. The outstanding performance of DER systems in reliability, economy, and environment is leading to an increased interest in its application. However, the implementation of DER systems is still hindered. Lack of a comprehensive evaluation in the design and operation of the DER systems is one of the major barriers. At present, the evaluation method of DER systems is relatively simple and one-sided, which cannot reasonably and accurately evaluate the comprehensive benefits of DER systems. Therefore, this study proposed a DER system composed of photovoltaic, energy storage, and gas engine, and its grid stabilization and carbon reduction potentials were analyzed. Focusing on these advantages, a multi-criteria evaluation method was established to optimize the system. Finally, different case study scenarios of the DER utilization were demonstrated. It is hoped to improve the core competitiveness of the DER and promote its development.

References

1. BP (2020) Statistical review of world energy. Bp:66
2. Hannah R, Roser M (2020) Emissions by sector. https://ourworldindataorg/emissions-by-sector. Zugegriffen: 1 Nov 2020
3. Höök M, Tang X (2013) Depletion of fossil fuels and anthropogenic climate change – a review. Energy Policy 52:797–809. https://doi.org/10.1016/j.enpol.2012.10.046
4. Wei D, Chen A, Sun B, Zhang C (2016) Multi-objective optimal operation and energy coupling analysis of combined cooling and heating system. Energy 98:296–307. https://doi.org/10.1016/j.energy.2016.01.027
5. Beck F, Martinot E (2004) Renewable energy policies and barriers. Encycl Energy:365–383. https://doi.org/10.1016/b0-12-176480-x/00488-5
6. Wang J, Mao T, Dou C (2016) Configuration optimization with operation strategy of solar-assisted building cooling heating and power system to minimize energy consumption. Energy Procedia 88:742–747. https://doi.org/10.1016/j.egypro.2016.06.063
7. Wang JJ, Jing YY, Bai H, Zhang JL (2012) Economic analysis and optimization design of a solar combined cooling heating and power system in different operation strategies. Proceedings of the 2012 7th IEEE conference on industrial electronics and Applications ICIEA 2012, pp 108–112. https://doi.org/10.1109/ICIEA.2012.6360706
8. Wang JJ, Jing YY, Zhang CF (2010) Optimization of capacity and operation for CCHP system by genetic algorithm. Appl Energy 87:1325–1335. https://doi.org/10.1016/j.apenergy.2009.08.005
9. CO_2 emission coefficient by electric utility (2020). https://ghg-santeikohyo.env.go.jp/files/calc/r02_coefficient.pdf. Zugegriffen: 5 Aug 2021
10. Carbon dioxide emission coefficient of urban gas (2020). http://www.saibugas.co.jp/business/others/co2_emission_factor.htm. Zugegriffen: 5 Aug 2021
11. Sun Y, Mao X, Yin X et al (2021) Optimizing carbon tax rates and revenue recycling schemes: model development, and a case study for the Bohai Bay area. China J Clean Prod 296:126519. https://doi.org/10.1016/j.jclepro.2021.126519
12. Carattini S, Carvalho M, Fankhauser S (2018) Overcoming public resistance to carbon taxes. Wiley Interdiscip Rev Clim Chang 9:1–26. https://doi.org/10.1002/wcc.531
13. Raesaar P, Tiigimägi E, Valtin J (2006) Assessment of electricity supply interruption costs under restricted time and information resources. Proceedings of the 2006 IASME/WSEAS International Conference on energy and environmental Systems Chalkida, Greece, 8–10 May 2006, pp 409–415
14. Vásquez P, Vaca Á (2012) Methodology for estimating the cost of energy not supplied-Ecuadorian case. Proceedings of the 2012 6th IEEE/PES transmission and distribution: Latin America conference and exposition T D-LA 2012. https://doi.org/10.1109/TDC-LA.2012.6319047
15. TOU electricity price in Higashida District. http://www.kyuden.co.jp/agreement_rate_gyomukijia-1_h26_5.html. Zugegriffen: 5 Aug 2021
16. Li Y, Gao W, Ruan Y (2018) Performance investigation of grid-connected residential PV-battery system focusing on enhancing self-consumption and peak shaving in Kyushu, Japan. Renew Energy 127:514–523. https://doi.org/10.1016/j.renene.2018.04.074

17. Cost of solar power generation in Japan. https://www.renewable-ei.org/pdfdownload/activities/Report_SolarCost_201907.pdf. Zugegriffen: 5 Aug 2021
18. New energy related technology trend survey. http://www.city.takatsuki.osaka.jp/ikkrwebBrowse/material/files/group/82/newe-h14.pdf. Zugegriffen: 5 Aug 2021
19. A study on expansion of stationary batteries and its application to aggregation services. https://www.meti.go.jp/meti_lib/report/H28FY/000479.pdf. Zugegriffen: 5 Aug 2021
20. Natural gas price of power generation industry in Kitakyushu. http://www.saibugas.co.jp/business/rates/index.htm. Zugegriffen: 5 Aug 2021
21. Zhang X, Zeng R, Deng Q et al (2019) Energy, exergy and economic analysis of biomass and geothermal energy based CCHP system integrated with compressed air energy storage (CAES). Energy Convers Manag 199:111953. https://doi.org/10.1016/j.enconman.2019.111953
22. Feng L, Dai X, Mo J, Shi L (2019) Performance assessment of CCHP systems with different cooling supply modes and operation strategies. Energy Convers Manag 192:188–201. https://doi.org/10.1016/j.enconman.2019.04.048
23. Lin B, Jia Z (2018) The energy, environmental and economic impacts of carbon tax rate and taxation industry: a CGE based study in China. Energy 159:558–568. https://doi.org/10.1016/j.energy.2018.06.167
24. Ren H, Lu Y, Zhang Y et al (2018) Operation simulation and optimization of distributed energy system based on TRNSYS. Energy Procedia 152:3–8. https://doi.org/10.1016/j.egypro.2018.09.050
25. Li Y, Tian R, Wei M et al (2020) An improved operation strategy for CCHP system based on high-speed railways station case study. Energy Convers Manag 216:112936. https://doi.org/10.1016/j.enconman.2020.112936
26. Dong R, Xu J (2015) Impact of differentiated local subsidy policies on the development of distributed energy system. Energ Buildings 101:45–53. https://doi.org/10.1016/j.enbuild.2015.05.001
27. Wang L, Singh C, Tan KC (2007) Reliability evaluation of power-generating systems including time-dependent sources based on binary particle swarm optimization. 2007 IEEE congress on evolutionary computation, pp 3346–3352. https://doi.org/10.1109/CEC.2007.4424904
28. Adefarati T, Bansal RC (2017) Reliability and economic assessment of a microgrid power system with the integration of renewable energy resources. Appl Energy 206:911–933. https://doi.org/10.1016/j.apenergy.2017.08.228
29. Chao W, Zheng XU, Peng GAO et al (2007) Reliability index framework for reliability evaluation of bulk power system. Proc CSU 27:42–48
30. Chowdhury A (2001) Application of customer interruption costs in transmission network reliability planning. IEEE Trans Ind Appl 37:53–60
31. Swinand, Peter G, Natraj, Ashwini (2019) The value of lost load (voll) in European electricity markets: uses, methodologies, future directions. International conference on the European energy market (EEM) https://doi.org/10.1109/EEM.2019.8916400

Chapter 10
Application Potential and Implication of Hydrogen Energy in Distributed Energy System

Fanyue Qian

10.1 Economic and Potential Analysis of Hydrogen Energy Equipment in Distributed Energy System

10.1.1 The Significance of Hydrogen Energy for Energy Environment

1. *Hydrogen energy will be an important way to absorb renewable energy*

Since hydrogen must be produced from hydrogen-containing substances such as water and fossil fuels, it is a secondary energy source. At present, high-efficiency and low-cost hydrogen production is the focus of attention of countries around the world. Using renewable energy to produce hydrogen can both reduce production costs and achieve the purpose of protecting the environment. It is the most effective way to produce hydrogen. At the same time, this method can effectively alleviate the current consumption problems caused by the continuous development of renewable energy.

Hydrogen and electricity can be said to be complementary to each other during the transformation of the energy system. The use of electrolysis devices to achieve hydrogen production from renewable energy power is conducive to the integration of highly volatile renewable energy power (VRE) into the energy system. At the same time, large-scale use of solar energy, wind energy, and other renewable energy to produce hydrogen and the development of large-scale, low-cost VRE facilities in marginal areas with rich solar or wind energy resources, dedicated to hydrogen production, can realize the reuse of wind and light and energy conversion, improve

F. Qian (✉)
School of Mechanical and Energy Engineering, Tongji University, Shanghai, China

W. Gao (ed.), *Distributed Energy Resources*, Green Energy and Technology,
https://doi.org/10.1007/978-3-031-21097-6_10

the utilization rate of renewable energy, and reduce waste of clean energy [1]. Although batteries and demand-side measures can provide short-term flexibility, hydrogen is the only large-scale technology that can be used for long-term energy storage. It can use the existing natural gas network, salt caves, and barren gas fields to store energy for a long time at a lower cost. Through the hydrogen produced from renewable energy, a large amount of renewable energy can be led from the power sector to the end-use sector. Renewable energy power can be used to produce hydrogen, which in turn can provide energy for sectors that have difficulty to achieve decarbonization through electrification and achieve sustainable energy development.

2. *Hydrogen energy helps deep decarbonization in various industries*

As a transportation medium for renewable energy, hydrogen can realize the long-distance transmission of renewable energy; promote the interconnection between electricity and construction, transportation, and industry; and help in areas where electrification is difficult (transportation, industrial, construction departments that rely on existing natural gas pipeline networks, etc. make more use of renewable energy to reduce carbon dioxide emissions Fig. 10.1).

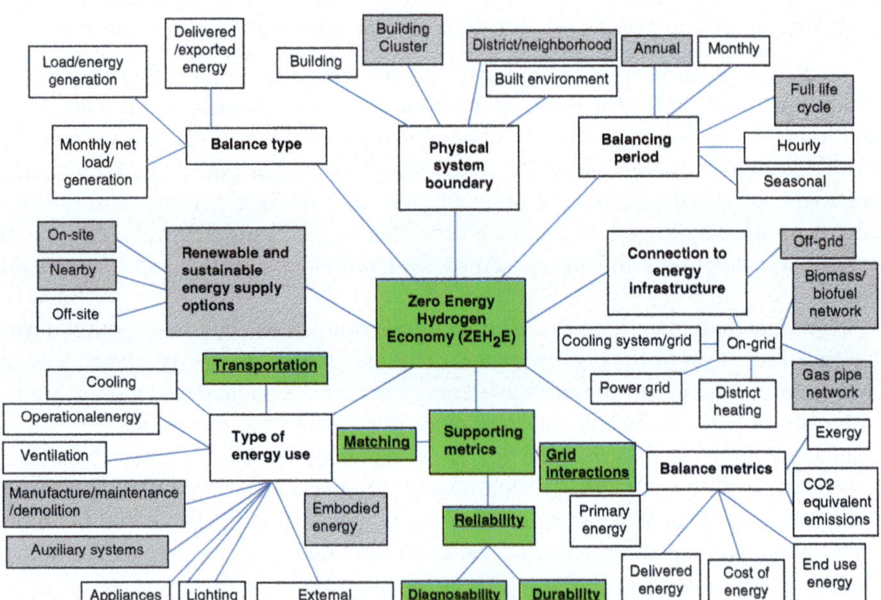

Fig. 10.1 Mind map for the definition of zero-energy hydrogen economy [2]

10.1.2 Hydrogen Energy Equipment

10.1.2.1 Fuel Cell

Fuel cell is one of the most widely used methods of hydrogen. The scope of application includes fuel cell vehicles, household fuel cell water heaters, and large fuel cell cogeneration [3]. Compared with other energy systems, fuel cells have the following advantages: (1) high energy conversion efficiency [4], (2) modularization, and (3) short construction period and flexible start and stop [5].

Fuel cells can be divided into the following five types: alkaline fuel cells (alkaline fuel cell), proton exchange membrane fuel cells (proton exchange membrane fuel cell), phosphoric acid fuel cells (phosphoric acid fuel cell), molten carbonate fuel cells (molten carbonate fuel cell), and solid oxide fuel cell. According to the gradual increase of the operating temperature from 50 °C to 1000 °C, it is low, medium, and high temperature. PEMFC and SOFC are considered to be the future development direction due to their special operating temperatures. At the same time, the proton exchange membrane fuel cell is also the most mature fuel cell currently developed and has a wide range of applications in home power, mobile power, distributed power, and vehicle power.

10.1.2.2 Hydrogen Vehicle

Hydrogen vehicles are an important part of the new energy vehicles that are being vigorously promoted worldwide. New energy vehicles include pure electric vehicles (PEV), hybrid electric vehicles (HEV), fuel cell electric vehicles (FCV), hydrogen engine vehicles (HEV), and so on. At present, pure electric vehicles and hybrid electric vehicles have improved after many years of promotion, occupying a certain share of the automobile market. According to the statistics of the International Energy Agency, in 2017, electric vehicles accounted for 2.2% of the total sales market share of automobiles, of which Norway accounted for the highest proportion, and its sales market share of electric vehicles reached 39%.

Among them, as hydrogen energy [6] and fuel cell technology have become the major strategic direction of energy and power transformation in the world, the next stage of new energy vehicles will focus on the technical innovation and application promotion of FCVs. Figure 10.2 shows a comparison between FCV and other vehicles. The number of stars varies between 1 and 5, reflecting the advantages or disadvantages of the car in this comparison. The outstanding part with red stars represents the advantages of FCV and EV, as well as the problems FCV is facing at present. The main advantage of FCV and EV is that there is no carbon dioxide emission during driving. In addition, the fuel filling of FCV is fast. The main problems FCV faces are high cost and imperfect supporting facilities.

Project	FCV	EV	PHV	HV
CO2 emission	★★★★★ (Carbon dioxide is not emitted while driving)	★★★★★ (Carbon dioxide is not emitted while driving)	★★★★	★★★
Maximum driving distance	★★★★★ (760km)	★★ (200km)	★★★★★ (Over 500km)	★★★★★ (Over 500km)
Durability	★★★★★	★★★★	★★★★★	★★★★★
Vehicle price	★★★★★ (Just beginning to popularize)	★★★	★★★	★★★★
Complementary facilities	★ (Just beginning to popularize)	★★★★ (Mass popularization)	★★★★ (Mass popularization)	★★★★★
Energy replenishment time	★★★★★ (3 minutes)	★ (Normal 8 hours Fast 20-30 minutes)	★ (Normal 4 hours)	★★★★★ (2-3 minutes)

Fig. 10.2 Carbon emissions and economic chart of various vehicles [7]

Table 10.1 Annual energy utilization and cost of fuel cells

Project	Value	Unit
Hot water demand	20.10	GJ
Gas consumption	979.5322	Nm^3
Electricity produce	2938.60	kWh
Electricity cost reduce	75,169.3	Yen
Gas cost	14584	Yen
Total energy cost	60,585.3	Yen

10.1.3 Development Potential of Energy Equipment in Demand Side

10.1.3.1 Household Fuel Cell

According to Japan's statistics on the use of domestic fuel cells in 2009, the average annual carbon dioxide emission reduction of each 700 W fuel cell is 1330 kg. According to the trial calculation and relevant gas price policies of Tokyo Gas Company for fuel cells, the annual energy utilization and cost of fuel cells are as follows (Table 10.1).

The energy cost reduction after using fuel cells is about 60,000 Yen. Using the same hot water demand, according to the heat pump trial model of Tokyo Gas Company, the energy cost reduction is about 50,000 Yen. According to the carbon dioxide emission coefficient of 0.418 kg/kWh for electricity and 2.2455 Nm^3/kWh for gas (Statistics publicity of Japan's Ministry of Economy and Industry in 2018), the carbon dioxide emission after using the heat pump increased by about 155 kg. Although the energy cost of using fuel cells is reduced more than that of heat pumps, it helps to reduce carbon dioxide emissions. However, due to the high purchase cost of fuel cells, this part of the revenue cannot be recovered within the service life of fuel cells.

Therefore, the concept of carbon tax should be introduced to re-measure the economic benefits of domestic fuel cells by converting carbon dioxide emission reduction into economic benefits. According to the "World Energy Outlook" put forward by the International Energy Agency (IEA) in 2014, it is expected that by 2020, the carbon emission prices of all countries will be around 2.17 Yen/kgCO_2 and will reach 15.22 Yen/kgCO_2 by 2040 (6.52 Yen/kgCO_2 by 2025, 10.87 Yen/kgCO_2 by 2030).

On the one hand, the introduction of carbon tax will increase the cost of electricity and hydrogen; on the other hand, because the equipment manufacturing stage will also produce carbon dioxide emissions, it will cause the price of equipment to rise. The carbon dioxide emissions during the production of 1 kW fuel cell are 275 kg [8], and the carbon emissions during the heat pump production process account for 2–3% of the carbon emissions throughout the life cycle [9]. The fuel cells and heat pumps used in the study are natural gas and electricity, respectively. The CO_2 emissions of different power generation types in the table are all conversions of carbon emissions from production, use, scrapping, and recycling throughout the life cycle. The carbon emission factors of electricity and gas are 418–769 g/kWh (Kansai-Okinawa) and 2.245 kgCO_2/Nm^3, respectively [10].

According to Japan's fuel cell development strategy, the dynamic payback period of fuel cell and heat pump shown in Tables 10.2 and 10.3 is calculated. The benchmark yield is 4.5%.

In the above table, cases with recoveries greater than 15 years under all carbon taxes are not shown in the table. The dynamic payback period of more than 15 years is displayed as "15+," indicating that the service life of the equipment is exceeded. It can be seen that the revenue of fuel cell will rise with the increase of carbon tax, while that of heat pump is the opposite. Fuel cells need to be reduced to 1,000,000 Yen to recover costs, while heat pumps cost 750,000 Yen.

Table 10.2 Payback period of domestic fuel cell

	Carbon tax (Yen/kg-CO_2)				
Price (Yen)	0	2.17	6.52	10.87	15.22
1,000,000	15+	15+	15+	15.05	13.58
800,000	13.58	12.75	11.37	10.27	9.36
600,000	8.10	7.66	6.92	6.31	5.81

Table 10.3 Payback period of heat pump

	Carbon tax (Yen/kg-CO_2)				
Price (Yen)	0	2.17	6.52	10.87	15.22
750,000	15.78	15+	15+	15+	15+
700,000	13.58	13.72	14.02	14.32	14.64
650,000	11.79	11.91	12.16	12.42	12.68
600,000	10.14	10.24	10.44	10.66	10.88

At the same time, when the cost of fuel cell is reduced to 800,000 Yen, its economic benefit is the same as that of heat pump in the period of carbon tax of 10.87 Yen/kgCO_2. When the cost of fuel cell reaches the target of 600,000 Yen, or when the cost is 800,000 Yen and carbon tax is 15.22 Yen/kgCO_2, the economic benefit of fuel cell will exceed that of heat pump.

10.1.3.2 Hydrogen Fuel Cell Vehicle (FCV)

In the aspect of hydrogen fuel cell vehicles and electric vehicles, the same method is used to calculate the dynamic payback period of them. If it is assumed that the hydrogen used by the fuel cell is produced from renewable energy sources, the CO2 emissions of 1 kW of hydrogen are 10–24 g (conversion efficiency is 0.7 [11]).

The price of hydrogen fuel cell vehicles is expected to fall to 5,300,000 Yen from the current 7,600,000 Yen. The price range of electric vehicles is set at 3700000–3300000 Yen with reference to different automobile brands. The calculation parameters are shown in Table 10.4, and the calculation results are shown in Tables 10.5 and 10.6.

Table 10.4 Parameters of hydrogen fuel cell vehicle and electric vehicle

	Hydrogen fuel cell vehicle	Electric vehicle
100 km energy consumption	0.99 kg	15.7 kWh
Unit Price	1011 Yen/kg	23 Yen/kWh
Cost (Yen)	1000	361
Vehicle selling price (10^4 Yen)	760	570
100 km CO_2 emission(kgCO_2)	0.4–0.96 (wind-hydroelectric)	9.53
CO_2 emissions during the manufacturing stage (gCO_2/km) [12]	45–55	55–100

Table 10.5 Payback period of hydrogen fuel cell vehicle

	Carbon tax (Yen/kgCO_2)				
Price (Yen)	0	2.17	6.52	10.87	15.22
5,700,000	15+	15+	14.91	13.20	11.85
5,500,000	11.54	10.80	9.56	8.58	7.79
5,300,000	6.20	5.85	5.24	4.75	4.35

Table 10.6 Payback period of electric vehicle

	Carbon tax (Yen/kgCO_2)				
Price (Yen)	0	2.17	6.52	10.87	15.22
3,600,000	15+	15.26	14.49	13.80	13.17
3,500,000	12.19	11.89	11.34	10.83	10.37
3,400,000	9.174	8.97	8.57	8.21	7.88
3,300,000	6.51	6.37	6.11	5.86	5.64

Because hydrogen fuel cell vehicles and electric vehicles can reduce carbon dioxide emissions when they are used, their income increases with the rise of carbon tax, but the rise of hydrogen fuel cell vehicles is higher. At the same time, when the cost of hydrogen fuel cell vehicles is reduced to 5,700,000 Yen, which is the medium level of current vehicles, it begins to have economic competitiveness. When the target of JPY5300000 is reached, its economic benefits will exceed that of electric vehicles.

It can be seen that the introduction of carbon tax will promote the promotion of fuel cell due to the reduction effect of carbon dioxide emission of fuel cell. At the same time, it is predicted that by 2030, when the price of fuel cell has been reduced to a certain extent and the price of carbon tax has been stabilized at a higher level, fuel cell will realize the same or even higher economic benefits as the existing conventional energy system and have market competitiveness.

10.2 Economic and Potential Analysis of Hydrogen Vehicle-to-Grid (V2G) System

10.2.1 Research Status of Hydrogen Vehicle-to-Grid (V2G) System

Many countries in the world have formulated the development goals, strategic planning, R&D investment, road map, and other related policies for hydrogen energy utilization and fuel cells, among which are Japan, Germany, the United States, and other leading countries. Taking Japan as an example, a demonstration of a hydrogen gas pipeline network was set up in Kitakyushu City from 2011 to 2014, including verified research on hydrogen fuel cell vehicles. As a part of the smart community demonstration project in Kitakyushu City, the FC vehicle-to-house (FCV2H) power supply from the FCV to the housing was verified by the external power supply function in emergencies and the peak cut effect in the power supply and demand tightening. The current concept of FCV2G is shown in Fig. 10.3. FCV forms an energy aggregator through buildings with large parking lots to participate in V2G services with the grid.

At present, the research on FCV2G is still relatively small, but there is much V2G and V2H research with pure electric vehicles and hybrid electric vehicles, which have great reference value. Nathaniel S [13] provided a comprehensive and current review of concepts, recently published studies, and demonstration projects/deployments of V2X technology from around the world. Triviño-Cabrera [14] studied dynamic pricing and V2G market-driven models for driving habits and revenue of electric vehicle fleets. P H Hashemi-Dezaki [15] proposed a comprehensive method to evaluate the system's reliability based on PHEV Monte Carlo simulations. M Alirezaei [16] aimed to investigate the role of vehicle-to-home technology in satisfying the energy requirements for a net-zero energy building and indicated

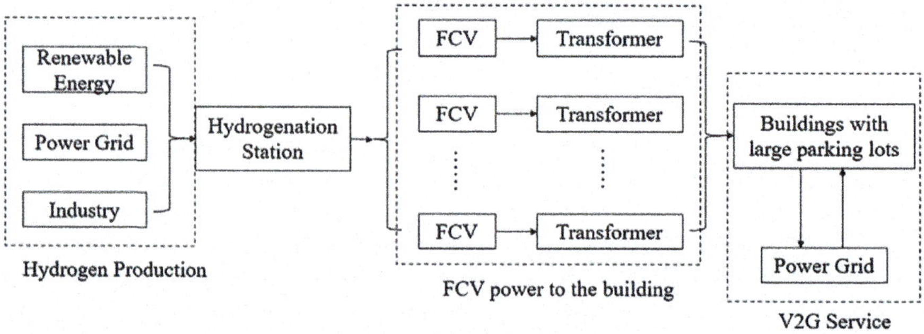

Fig. 10.3 The current concept of FCV2G

that this system can not only reduce the monetary value of the required grid electricity but also earn money to compensate for the installation costs of other technologies.

FCV will occupy an important position in the field of new energy vehicles in the future. Compared with other new energy vehicles, FCV has lower carbon emissions, larger capacity, and higher discharge power, which is consistent with the V2G service. In order to illustrate the economic and environmental effects of FCV's participation in V2G, firstly, this part of the paper chooses a large shopping mall in Japan as the research scenario, after obtaining its annual electricity consumption. Then, replacing part of the FCVs with EVs is considered, as well as the discharges of vehicles to buildings and power grids, using buildings as agents for all vehicles to provide V2G services for power grids. A genetic algorithm (GA) is used to find the best discharge price, the choice of vehicle discharging under the condition of the highest economic benefit, and to analyze the change of building income under different FCV ratios and EV charging demand.

10.2.2 Hydrogen V2G Research Combined with Case Study

10.2.2.1 Case Study and Data Processing

The research objective of this chapter is a large shopping mall in Kitakyushu, Japan. The average daily vehicle visit rate (proportion of visiting vehicles to maximum in a year) for the target area is shown in Fig. 10.4. As the changes in visiting vehicles from Monday to Friday are close, in subsequent simulations, Monday will represent the other weekdays. The target building has two parking lots, which can accommodate 2000 and 500 vehicles, respectively. Through simulation, the parking lot with 500 vehicles is more suitable for this research.

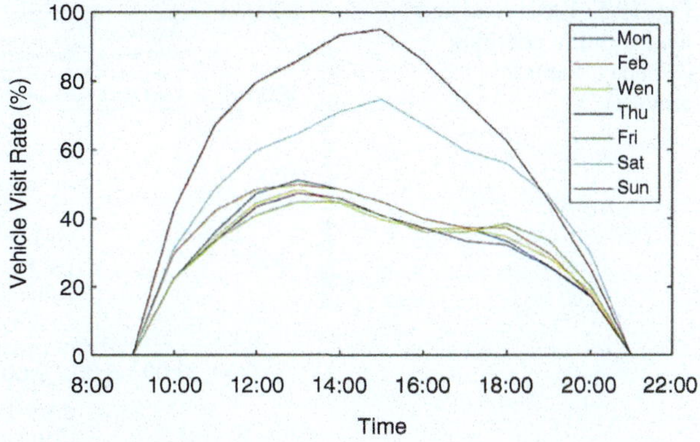

Fig. 10.4 Vehicle visit rate in 1 week

10.2.2.2 Background Parameter Setting

The target shopping mall is located in Kitakyushu City, Japan, which is within the scope of the Kyushu power grid. Therefore, the price we choose comes from the official website of Kyushu Electric Power. Similar to Sect. 10.1.3, 4320 Yen/t-CO_2 (about 40$/t-$CO_2$) is chosen as the initial carbon emission pricing, and its sensitivity analysis will be carried out in the follow-up.

Considering that EVs will still dominate the new energy vehicles for a long time, all the V2G vehicles will be divided into six situations: 100% FCV, 80% FCV + 20% EV, 60% FCV + 40% EV, 40% FCV + 60% EV, 20% FCV + 80% EV, and 100% EV. At the same time, the investment cost of FCV [17] and EV [18] is also used to calculate the user loss per discharge.

10.2.3 Case Study Results of Hydrogen V2G

If the charging demand of EV is considered, it is assumed that there are some EVs that don't participate in the V2G but need to be charged. Through simulation, the different situations on Monday, Saturday, and Sunday are shown in Fig. 10.5. As a result of the difference in the number of vehicles visited, there are also differences in the charging demand on Monday, Saturday, and Sunday, with Monday being the smallest and Sunday the largest. The results of optimization on Monday, Saturday, and Sunday are shown in Table 10.7.

As can be seen from Table 10.7, because of the addition of EV charging demand, the gap between the peak and the valley during the V2G period has been widened, thus increasing peak-cutting income. On Monday, because of the higher electricity price for buildings, the total income of the introduction of FCV has been improved, but the effect on peak reduction is smaller than that of V2G. The increase of total income mainly comes from the smaller discharge loss of FCV, the benefits of carbon emission reduction, and income from providing V2G services.

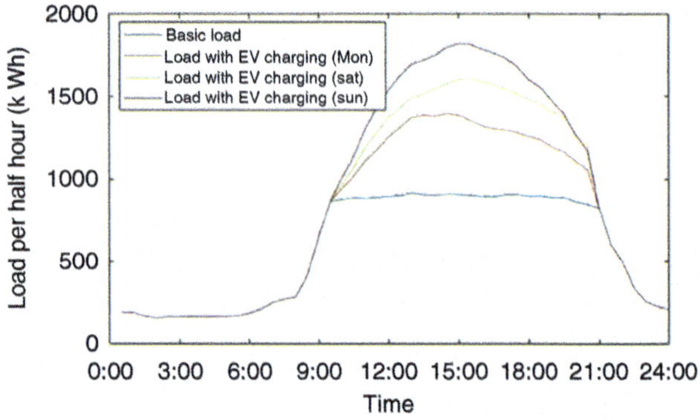

Fig. 10.5 Basic load and load with EV charging (Monday, Saturday, Sunday)

Table 10.7 The results of optimization on Monday, Saturday, and Sunday with EV charging demand

Monday ($Price_{dis}^{EV}$=25.85Yen/kWh, $Price_{dis}^{FCV}$=32.57Yen/kW)						
Proportion of FCV	100%	80%	60%	40%	20%	0%
Proportion of EV	0%	20%	40%	60%	80%	100%
Peak-cutting income (Yen)	2206	6123	10,320	13,427	13,133	12,678
V2G income (Yen)	78,601	70,940	63,098	66,345	58,036	49,136
Total spending (Yen)	−47,894	−42,606	−37,152	−43,895	−37,958	−31,389
Total income (Yen)	32,913	34,458	36,266	35,877	33,210	30,425
Saturday ($Price_{dis}^{EV}$=22.90Yen/kWh, $Price_{dis}^{FCV}$=28.92Yen/kW)						
Proportion of FCV	100%	80%	60%	40%	20%	0%
Proportion of EV	0%	20.0%	40.0%	60.0%	80.0%	100.0%
Peak-cutting income (Yen)	2211	7769	9972	9101	9428	1123
V2G income (Yen)	17,408	49,328	53,941	46,303	40,184	630
Total spending (Yen)	−15,799	−47,096	−53,989	−47,336	−43,255	−1187
Total income (Yen)	3820	10,001	9925	8068	6357	567
Sunday ($Price_{dis}^{EV}$=22.95Yen/kWh, $Price_{dis}^{FCV}$=28.89Yen/kW)						
Proportion of FCV	100%	80%	60%	40%	20%	0%
Proportion of EV	0%	20.0%	40.0%	60.0%	80.0%	100.0%
Peak-cutting income (Yen)	2220	9710	13,329	12,454	11,164	1416
V2G income (Yen)	18,170	61,700	70,762	61,336	47,869	884
Total spending (Yen)	−16,116	−58,626	−70,764	−62,882	−51,057	−1600
Total income (Yen)	4274	12,785	13,328	10,908	7977	699

In the case of 80%FCV, the overall benefit is about 3.8% higher than that of 20% FCV, of which 53.4% is lower than that of peak cutting and 22.2% is higher than that of V2G. V2G service income occupies the dominant position, so the optimization goal tends to make more vehicles participate in V2G services. At the same time, due to the reduction of EV charging demand, the peak-cutting income is reduced, resulting in the situation of maximum revenue at 60%FCV + 40%EV. So, EV and FCV are not completely in conflict, and EV charging demand can bring the benefit of FCV.

10.3 Economic and Potential Analysis of Regional Distributed Hydrogen Energy System (RDHES)

10.3.1 Research Status of RDHES

The compound system of hydrogen energy and other energy sources, especially the application of hydrogen energy in regional distributed energy systems (RDES), is still in the theoretical research stage. The RDES refers to the establishment of an energy system on the user's side, and at the same time generating electricity, it will also provide users with multiple composite energy sources such as cold and heat [19]. At present, the core areas of conventional RDES are mainly gas-fired combined heat and power [20], photovoltaic and energy storage integration [21], etc. The equipment involved mainly includes internal combustion engines, absorption lithium bromide, heat pump, photovoltaic, and battery. At present, the research and application promotion of these devices and core fields are relatively mature, and many more mature theoretical systems have been derived, including cross-research in different directions or cross-domains with electric vehicles V2G [22], electricity price incentive mechanism [23], and urban spatial structure [24]. Due to large-scale production, the cost of related energy equipment has also been greatly reduced [25]. At present, the RDES has become a high-quality investment project receiving high attention in the capital market [26].

There are few studies on the combination of hydrogen energy systems and regional distributed energy sources, and more attention is focused on fuel cells [27, 28], hydrogen fuel cell electric vehicles [29], and large-scale hydrogen storage [30, 31] and the interaction between these areas [32, 33]. At present, there are three main ways to combine hydrogen energy systems with regional distributed energy sources. One is to replace equipment with similar functions, that is, use fuel cells to replace internal combustion engines and gas turbines as the core equipment for combined heat and power. Kang [34] studied the operating characteristics and optimization models of the SOFC-engine composite system applied to distributed energy sources and confirmed the reliability of the proposed model through comparison with experimental data. Yang [35] proposed a distributed energy system based on the new utilization of LNG cold energy combined with SOFC. Through modeling, it was verified that the system can achieve higher thermoelectric efficiency, and the sensitivity of parameters such as fuel cell utilization rate and fuel supply capacity was conducted.

The second is based on the ability of hydrogen energy to be stored across seasons and uses the hydrogen storage system as a backup storage system for RDES [36]. Figure 10.6 shows the practical application concept of hydrogen energy as a storage medium for photovoltaic power. Mehrjerdi [38] proposed a photovoltaic-hydrogen storage P2P model for distributed energy systems in homes and buildings and proved that this model can effectively improve the system's revenue. Hemmati [39] proposed the improvement of the hybrid energy storage of hydrogen storage and storage battery. The optimization analysis after combining with photovoltaic proved that it can increase the overall income by 28%.

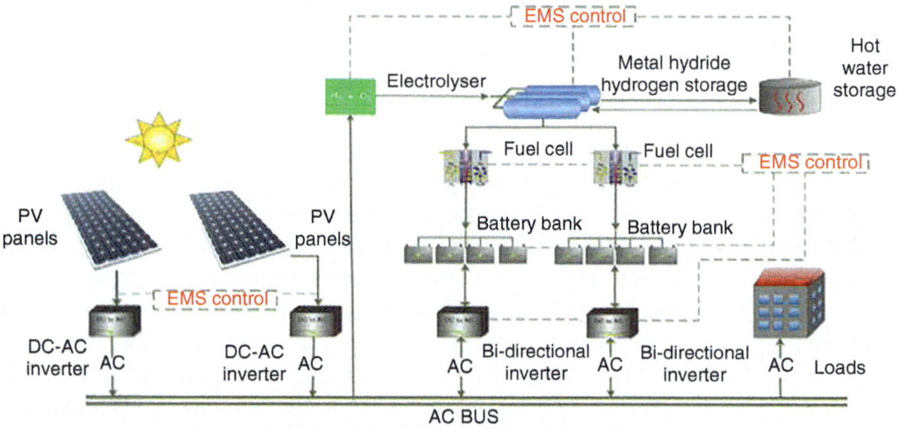

Fig. 10.6 Example of a real hydrogen fuel cell system design. The main electrical components are duplicated in the interest of redundancy and reliability [37]

The third is the design of RDES [40] and operation optimization analysis [41] using hydrogen energy equipment as the main core. Virji [42] analyzed the economic benefits of combining the hydrogen energy system with wind power and photovoltaic when the island is off-grid based on the actual data of an island in Hawaii. Martin [43] introduced user evaluation into the design of the hydrogen energy system and studied the improvement space of the hydrogen energy system design.

This part first selects ten buildings of different building types in the Higashida area of Kitakyushu City, Japan, as the research goal, the hydrogen distributed energy system is the research object, and the conventional distributed energy system is the comparative reference. After that, a RDES optimization model is established, and the genetic system is used to design and optimize the conventional system and the hydrogen energy system, respectively, and the comparison of the two systems under different carbon taxes is obtained. Through the analysis of the results of different types of buildings, the adaptability of the hydrogen RDES is studied.

10.3.2 RDHES Research Combined with Case Study

10.3.2.1 Case Study and Data Processing

In this paper, two RDES, hydrogen and conventional, are used. Among them, conventional RDES include internal combustion engine (ICE), absorption chiller-heater (AC), heat pump (HP), cold and heat storage (CHS), photovoltaic (PV), and battery. Its energy input is electricity and natural gas, and its energy output is electricity, cold, and heat, as shown in Fig. 10.7.

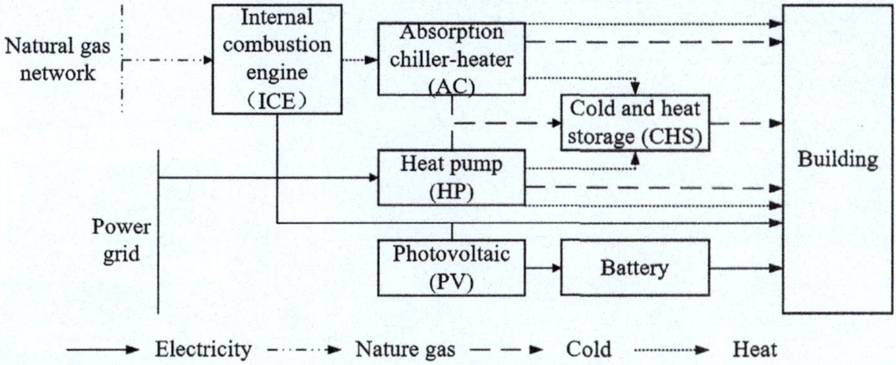

Fig. 10.7 Conventional RDES

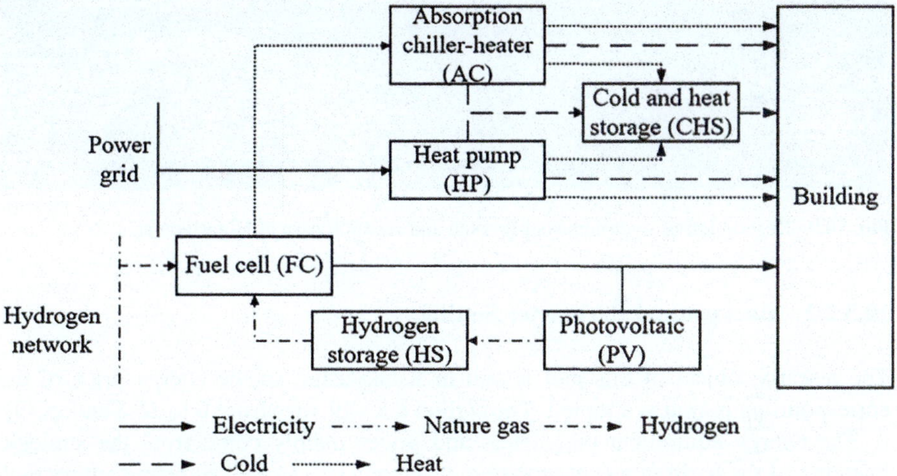

Fig. 10.8 Hydrogen RDES

Hydrogen RDES include fuel cell (FC), absorption chiller-heater (AC), heat pump (HP), cold and heat storage (CHS), photovoltaic (PV), and hydrogen storage (HS). The energy input is electricity and hydrogen, and the energy output is electricity, cold, and heat, as shown in Fig. 10.8.

The target area selected in this article is the Higashida area in Kitakyushu City, Japan, which serves as a key renovation and application test area for an environmentally friendly city. This paper selects different types of buildings on the hydrogen energy supply route as the research goal (represented by red dots in Fig. 10.9) to analyze the application of hydrogen energy in the RDES.

This article selects five building types, museum, hospital, commercial, residential, and office, in the area and selects two buildings as research objects for each type. There are also some differences between the two buildings in the same type, which is helpful for verifying the adaptability of the hydrogen energy system to different building types and different load situations.

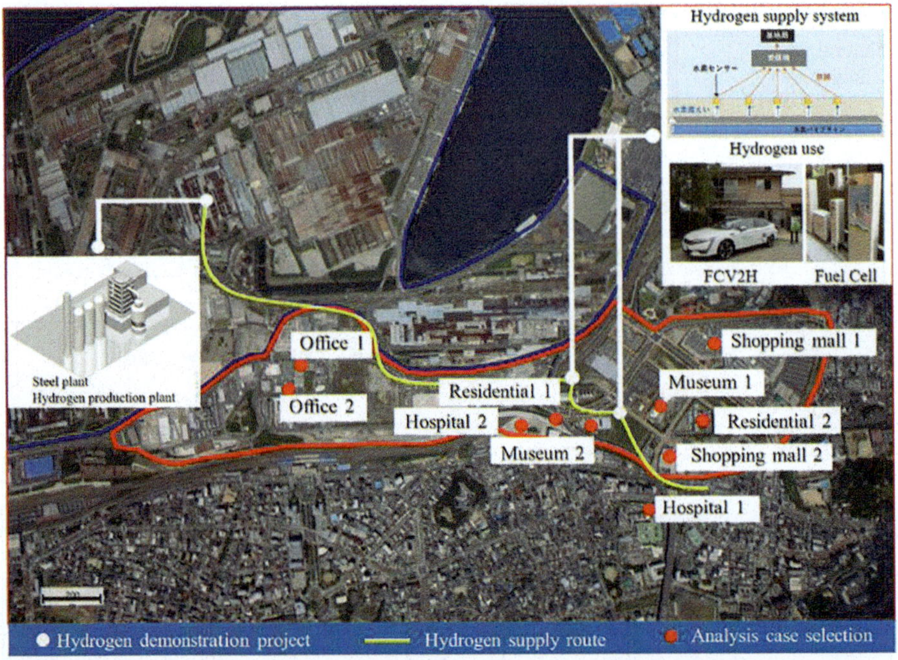

Fig. 10.9 Description of hydrogen supply route and research case in Higashida area

10.3.2.2 Background Parameter Setting

The research object of this part is still in Kitakyushu, so the energy price of the corresponding region is adopted. The carbon tax will vary from 0 to 15 Yen/kgCO_2.

The energy equipment parameters and prices mainly come from the strategic planning of the hydrogen energy system in Japan [16], as well as manufacturer data (Table 10.8). According to the above price and performance parameters, the genetic algorithm will be used to perform the optimization calculation. Finally, the total cost is obtained through the calculation of equipment investment, energy costs, and carbon tax, and the system design and optimization are aimed at the lowest total cost.

10.3.3 Case Study Results of RDHES

10.3.3.1 Optimization Results

First, take Museum 1 as an example to explain the results. In order to highlight the comparison between ICE and FC, battery, and HS, exclude the impact of PV and CHS on the overall revenue; the RDES using only HP, PV, and CHS will be used as a reference value. The optimization results show that the variables fall on the boundary, the PV permeability is 40%, the installed capacity of the heat pump is 2726 kW, and the total energy storage of CHS is 5454 kWh.

Table 10.8 Energy equipment parameters

ICE	Unit price	3101 Yen/kW	CHS	Unit price	120 Yen/kWh
	Power efficiency	0.45		Type	Water storage
	Thermal efficiency	0.4		Storage loss	2% per 24 hours
FC	Unit price	13298 Yen/kW	PV	Unit price	5040 Yen/kW
	Power efficiency	0.4		Type	Polysilicon
	Thermal efficiency	0.45		Power efficiency	18%
AC	Unit price	3101 Yen/kW	Battery	Unit price	3896 Yen/kWh
	Cold COP	1		Type	Sodium-sulfur
	Heat COP	0.9		Storage loss	0.95
HP	Unit price	775 Yen/kW	HS	Unit price	3571 Yen/kWh
	Cold COP	3.5		Storage loss	0.7
	Heat COP	3.5		Energy source	Photovoltaic
	Storage loss	0.95			

Then calculate the total cost of hydrogen energy and conventional RDES under different carbon taxes separately, compare with the baseline value, and calculate the total cost reduction based on the baseline value. The results are shown in Fig. 10.10. In the case of a small carbon tax, the hydrogen energy system cannot generate revenue. The optimization results show that the values of PV, HP, and CHS fall on the boundary, and the values of FC, AC, and HS are zero. Until the carbon tax was 12 Yen/kgCO_2, the total cost reduction began to occur, but the final reduction was still less than the conventional RDES. That is, under the carbon tax level of 2040, the benefit of the hydrogen RDES is still less than the conventional RDES. The carbon tax will continue to increase until 21 Yen/kgCO_2, and the total cost reduction will be similar to that of the conventional RDES. At this time, the installed capacity of the two systems is shown in Table 10.9. When the carbon tax is 21 Yen/kgCO_2, FC has just begun to have economic benefits. In contrast to conventional RDES, ICE has occupied a large proportion of power generation.

Considering future large-scale production, the price of FC will fall further, and the investment of FC will be adjusted to the same level as ICE. The optimization results of hydrogen energy and conventional regional energy system with different carbon taxes are obtained, as shown in Fig. 10.11. The change of FC installed capacity with carbon tax is shown in Fig. 10.12. The change of hydrogen energy system with carbon tax is greater than that of conventional system. When the carbon tax changes in the 12–15 stages, the hydrogen energy system has surpassed the benefits of the conventional system.

Hydrogen energy systems have a stronger carbon tax dependency than conventional systems, and the effectiveness of hydrogen energy systems depends largely on the increase in carbon taxes. At the same time, based on the cost of the current hydrogen energy system, the establishment of a carbon tax can continue to provide system benefits based on the existing PV and CHS, but the benefits provided are less than other energy systems of the same type. The cost of hydrogen energy systems still needs to be further reduced.

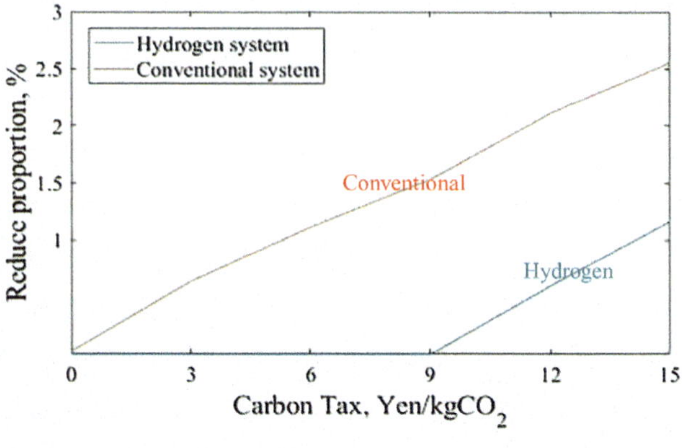

Fig. 10.10 The proportion of total cost reduction under different carbon taxes

Table 10.9 Optimal results with a carbon tax of 21 Yen/kgCO_2

Conventional RDES					
ICE (kW)	AC (kW)	HP (kW)	CHS (kWh)	Battery (kWh)	Total cost (10^8 Yen)
397	447	2279	26	91	1.05954
Hydrogen RDES					
FC (kW)	AC (kW)	HP (kW)	CHS (kW)	HS (kW)	Total cost (10^8 Yen)
60	67	2659	657	365	1.05952

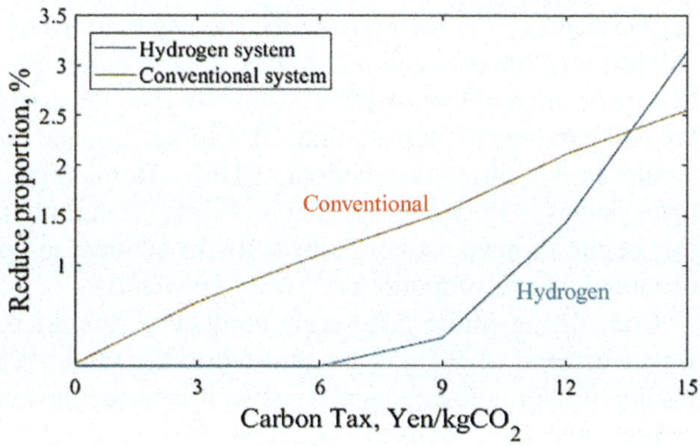

Fig. 10.11 The proportion of total cost reduction under different carbon taxes

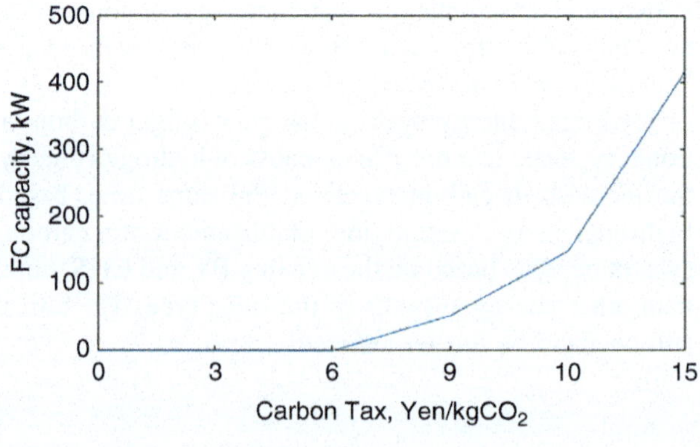

Fig. 10.12 FC installed capacity under different carbon taxes

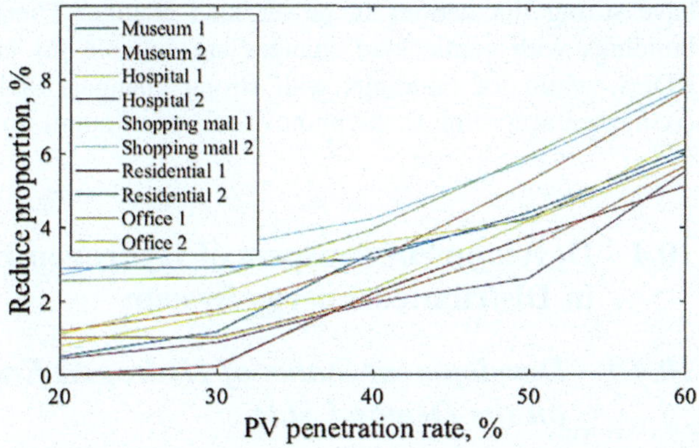

Fig. 10.13 Reduce proportion of HRDES in ten buildings under different PV penetration

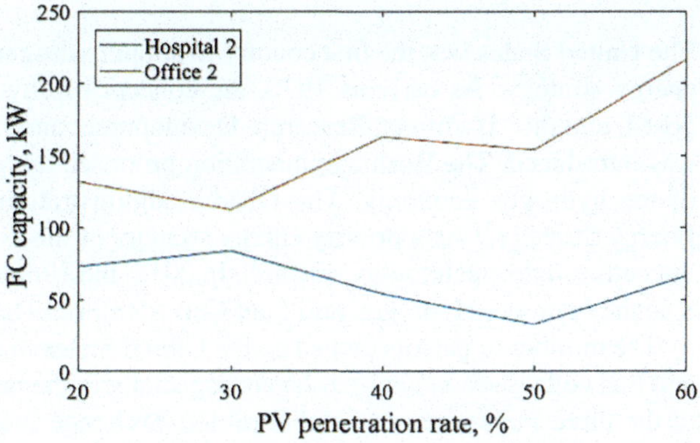

Fig. 10.14 FC capacity in case Hospital 2 and Office 2 under different PV penetration

10.3.3.2 Building Adaptability Analysis

To adapt to the adaptability of different building types to hydrogen energy RDES, the optimization calculations in ten buildings under different PV penetration are carried out (Fig. 10.13).

Overall, the increase in PV permeability will bring about an increase in the economic benefits of hydrogen RDES. However, the lifting effect is different for different types of buildings. At the same time, for example, case Hospital 2 and Office 2, the improvement effect suddenly drops under a certain PV penetration rate. The reason is that the increase of PV installed capacity will affect FC's power generation revenue, so that in some types of buildings, the installed capacity of FC cannot be increased with the increase of PV, as shown in Fig. 10.14.

The increase in installed photovoltaic capacity will have an inhibitory effect on the increase in installed FC capacity. The increase in the amount of waste light will increase the installed capacity of HS, reduce the energy cost of hydrogen, and thus increase the installed capacity of FC. As a result, the installed capacity of FC shows periodic fluctuations with the increase of PV penetration rate. According to the power load change characteristics, buildings that are prone to this phenomenon have strong fluctuations in power load change. Therefore, it is considered that buildings with gentle load changes are suitable for large-scale use of hydrogen RDES, while for buildings with strong changes, their hydrogen RDES can be configured according to the photovoltaic penetration limit (20–40%).

10.4 Development Prospect of Hydrogen Energy in Distributed Energy System

10.4.1 Development Status of Hydrogen Energy on the Demand Side

10.4.1.1 United States

The United States was the first country to adopt hydrogen energy and fuel cells as an energy strategy. As early as 1970, the concept of "hydrogen economy" was proposed, and the "Hydrogen Research, Development and Demonstration Act of 1990" was introduced. The Bush administration proposed a blueprint for the development of the hydrogen economy. The Obama administration issued a "Comprehensive Energy Strategy." As a priority energy strategy of the United States, fuel cells carry out cutting-edge technology research. In 2018, the United States announced October 8 as the National Hydrogen and Fuel Cell Memorial Day.

The number of patents owned by the United States in the field of hydrogen energy and fuel cells is second only to Japan, especially in the number of technology patents in the three major areas of global proton exchange membrane fuel cells, fuel cell systems, and on-board hydrogen storage 50%.

On November 6, 2019, the Fuel Cell and Hydrogen Energy Association (FCHEA) released an executive summary report on the US hydrogen economic roadmap. The report said that by 2050, hydrogen will account for 14% of US energy demand. The strong hydrogen industry will strengthen the US economy. The United States plans to realize the application of hydrogen energy in the fields of small passenger cars, forklift trucks, distributed power sources, household cogeneration, and carbon capture from 2020 to 2022.

10.4.1.2 Japan

Japan also attaches great importance to the development of fuel cell and hydrogen energy industry and proposes to "become the first country in the world to realize 'hydrogen society.'" The government has successively issued policies such as "Japan's rejuvenation strategy," "Energy strategic plan," "Basic strategy of hydrogen energy," and "Strategic Road map of hydrogen energy and fuel cell" and planned the technical route to realize hydrogen energy society. In 2018, Japan held the first Global Ministerial Conference on hydrogen energy, which was attended by energy ministerial government officials from more than 20 countries and the European Union, and took the Tokyo Olympic Games as an opportunity to promote fuel cell vehicles and build a hydrogen energy town. In the integrated energy system solution proposed by Tokyo Gas Company, it is explained that on the basis of the traditional integrated energy supply (electricity, gas, and heat) system, a hydrogen energy supply network covering the whole society will be built. At the end of the energy network, different energy use equipment, energy source conversion, and storage units together constitute the terminal integrated energy system.

At present, the commercial applications of fuel cells in Japan mainly include household fuel cells, cogeneration fixed power stations, business/industrial fuel cells, and fuel cell vehicles. Especially in the application of demand side fuel cell cogeneration, Japan is at the leading level in the world. Japan has built more than 9000 regional distributed integrated energy systems, of which a considerable part includes fuel cells. As early as 2001, Kitakyushu City University in Japan established the world's first normally operating fuel cell regional integrated energy system. In 2011, Kitakyushu, Japan, established the world's first hydrogen energy community project to demonstrate the whole process of hydrogen energy production, transmission, and application. The research object in this chapter also mainly comes from the buildings in this hydrogen energy community.

Household cogeneration system ene-farm is an energy system in which fuel cells are used efficiently in the home. It produces hydrogen through natural gas reforming and then injects the hydrogen into the fuel cell to generate electricity. At the same time, the heat generated during power generation is used to supply heating and hot water, and the overall energy efficiency can reach 90%. Japanese manufacturers of household fuel cell cogeneration system (ene-farm) mainly include Aisin Seiji, Panasonic, and Toshiba. The power generation efficiency of the products of the three companies can reach 40%, the total efficiency is more than 90%, the durability time is more than 80,000 hours, and the start-up time is only 1–2 minutes, which can be connected to the grid as required. In terms of sales volume, in May 2017, the global sales volume of ene-farm exceeded 200,000 sets. In this regard, the Japanese government proposes to achieve the sales target of 5.3 million units equivalent to 10% of Japanese households by 2030.

10.4.1.3 Germany

Germany is the most representative country in Europe to develop hydrogen energy and integrated energy system. In 2008, E-Energy project selected six pilot areas for a 4-year e-energy technological innovation promotion plan, with a total investment of about 140 million euros, including smart power generation, smart grid, smart consumption, and smart energy storage. The integrated development of hydrogen energy, renewable energy, and demand side energy system is an important part of Germany's sustainable energy system and low-carbon economy. In 2006, the national innovation plan (NIP) for hydrogen energy and fuel cell technology was launched, with a total investment of 1.4 billion euros from 2007 to 2016, and funded more than 240 enterprises, 50 scientific research and education institutions, and the public sector; the second phase of work will be carried out in 2017–2019, with a planned investment of 250 million euros.

With the support of scientific research projects, Germany has established a leading position in the field of hydrogen energy and fuel cells. The scale of hydrogen production from renewable energy is the first in the world, and the supply and manufacturing scale of fuel cells is the third in the world. The rapid development of hydrogen technology has also promoted the proportion of renewable energy in Germany. In the first quarter of 2020, the proportion of renewable energy power generation in Germany exceeded half of the national power generation for the first time, reaching 52%.

German operator is the world's second largest hydrogenation network, with 60 hydrogenation stations in operation, second only to Japan. The world's first hydrogen fuel cell train has been put into commercial operation in Germany, with a mileage of nearly 1000 km. It is planned to add 14 hydrogen fuel cell trains by the end of 2021. In addition, Germany has installed more than 300 communication base stations using fuel cells as backup power, and the cumulative use of domestic fuel cells has reached 80,000 hours.

10.4.1.4 People's Republic of China

As the largest country in hydrogen energy production in the world, China's planning and development strategy for hydrogen energy and fuel cells basically focuses on the layout of hydrogen refueling stations and the R&D and industrial promotion of hydrogen energy vehicles. At present, the relevant policies of hydrogen energy have been popularized and applied in the field of fuel cells in vehicles, but it is still blank in the field of civil construction.

Although China has initially formed a comprehensive energy supply system for the all-round development of coal, electricity, oil, natural gas, and new energy, and the consumption structure has gradually developed to clean and low carbon, the structural problems are still prominent. Therefore, the state pays great attention to the development of hydrogen energy and fuel cell industry. Due to the long industrial chain of hydrogen energy, it covers many links such as hydrogen preparation, storage and transportation, hydrogenation infrastructure, fuel cell, and its application. Compared with developed countries, China is still relatively backward in hydrogen energy independent technology research and development, equipment manufacturing, and infrastructure construction. However, due to the support of the huge industrial system, China's hydrogen energy output ranks first in the world. In April 2020, China's National Energy Administration issued "The Energy Law of the People's Republic of China (Exposure Draft)," and hydrogen energy was listed as secondary energy. In September, Shanghai issued the Shanghai fuel cell vehicle development plan. So far, hydrogen energy and fuel cells have entered a high-speed development stage in China.

"The White Paper on China's Hydrogen Energy and Fuel Cell Industry" shows that hydrogen energy will become an important part of China's energy system. It is estimated that by 2050, the proportion of hydrogen energy in China's energy system will reach 10%, the demand for hydrogen will be close to 6000 tons, and the annual economic output value will exceed 10 trillion yuan. There are more than 10,000 hydrogenation stations in China; hydrogen energy will be widely used in transportation, industry, and other fields; the output of fuel cell vehicles reaches 5.2 million units/year, the fixed power generation unit reaches 20,000 units/year, and the production capacity of fuel cell system reaches 5.5 million units/year.

10.4.2 Prospect of Hydrogen Energy in Distributed Energy System

In order to achieve the goals of the Paris Agreement, all countries in the world are in the stage of gradual transformation of the energy system. In addition to increasing the supply proportion of clean energy and renewable energy on the supply side, the establishment of a regional comprehensive energy system on the demand side can form the integration of energy production, supply, and marketing through the optimization and coordination of energy production, transmission, distribution, conversion, storage, and consumption and promote the deep integration of source network load storage. Through comprehensive integrated energy distribution, cascade utilization of capacity, and network optimization of energy transmission, we can realize comprehensive, efficient, and safe utilization of energy. According to the "Energy in urban areas: fully stimulating the potential of energy efficiency and renewable energy" released by the United Nations Environment Program, regional integrated energy system is the most effective way to realize sustainable heating and cooling in many cities around the world.

Hydrogen energy and fuel cell technology can realize deep decarbonization in the fields of industry, transportation, and electric power because they do not produce carbon dioxide in the process of use. Among them, the fuel cell cogeneration technology is also considered to have great application potential in the demand side regional integrated energy system, which is one of the important supports for the low-carbon transformation of the energy system. Therefore, many countries at home and abroad have formulated policies related to the utilization of hydrogen energy and fuel cells in recent years. However, due to the low technical maturity, high equipment cost, and insufficient supporting facilities in the whole process of hydrogen energy production, storage, transportation, and application, most of the development strategies focus on the layout of hydrogen refueling stations and the promotion of hydrogen energy vehicles and less on the application of fuel cells on the demand side.

The deep electrification of industry and transportation, as well as the technological progress and cost reduction of renewable energy, can effectively promote the further decarbonization of the energy system, which is considered to be the key path to realize the global low-carbon transformation of energy. However, the increase of electrification rate has led to the increasing demand for electricity. The increasing popularity of renewable energy has also brought great pressure and challenges to the planning and operation of the whole energy system because of its intermittent characteristics. To deal with the above two problems, the distributed energy system on the demand side can greatly alleviate the scheduling pressure of the energy network. At the same time, the development of hydrogen storage and hydrogen energy utilization technologies will help renewable energy break through the existing bottleneck and realize the application of "long-distance and cross season." Therefore, the application and promotion of hydrogen energy in distributed energy is to combine the low-carbon advantages of the two, which fully meets the core requirements of global green development and is bound to become a key development field in the future.

However, the application of hydrogen energy and fuel cell technology in distributed energy is in its infancy in the world. At present, there are only a few demonstration projects, most of which are still in the stage of theoretical research. At the same time, the distributed energy system involves many disciplines. The system has high complexity and strong coupling, and the overall design and operation and maintenance are difficult. Although fuel cells are experiencing an unprecedented stage of rapid political and industrial development, there is still a gap between the current equipment cost and the mainstream energy system, and the data and experience accumulation in the application process are not sufficient. Therefore, although the application of hydrogen energy in the field of distributed energy has strong environmental advantages, it cannot be popularized in a large area until the equipment cost is greatly reduced. According to the research in this chapter and the promotion policies of various countries, we believe that the development of hydrogen energy in the field of distributed energy needs to rely on the high proportion of renewable energy, the promotion of carbon tax, and the liberalization of energy market. It is expected that by 2030, the above development conditions will be basically mature, and the cost of hydrogen energy equipment will drop to the level close to the mainstream equipment. At this time, the development of hydrogen energy in the field of distributed energy will increase explosively.

References

1. Glenk G, Reichelstein S (2019) Economics of converting renewable power to hydrogen. Nat Energy 4(3):216–222
2. D'Agostino D (2015) Assessment of the progress towards the establishment of definitions of nearly zero energy buildings (nZEBs) in European member states. J Build Eng 1:20–32
3. Baolian Y (2003) Fuel cells-principles, technology, applications (in Chinese)

4. Rodi W et al (1997) Status of large eddy simulation: results of a workshop. Trans Am Soc Mech Eng J Fluids Eng 119:248–262
5. Liu S, Meneveau C, Katz J (1994) On the properties of similarity subgrid-scale models as deduced from measurements in a turbulent jet. J Fluid Mech 275:83–119
6. Acar C, Dincer I (2014) Energy and exergy analyses of a zero emission power plant for coproduction of electricity and methanol. In: Progress in exergy, energy, and the environment. Springer, Cham, pp 145–156
7. Fuel Cell Promotion Office of Japan Agency for Natural Resources and Energy. (2014) About fuel cell automatic vehicles. 13–14. https://www.meti.go.jp/committee/kenkyukai/energy/suiso_nenryodenchi/suiso_nenryodenchi_wg/pdf/003_02_00.pdf
8. Dhanushkodi SR et al (2008) Life cycle analysis of fuel cell technology. J Environ Inf 11(1): 36–44
9. Johnson EP (2011) Air-source heat pump carbon footprints: HFC impacts and comparison to other heat sources. Energy Policy 39(3):1369–1381
10. Compare electric power companies by CO2 emissions. https://power-hikaku.info/choice/eco/co2.php
11. Japan Ministry of Economy, Trade and Industry. Hydrogen/fuel cell strategy roadmap. https://www.meti.go.jp/press/2018/03/20190312001/20190312001.html
12. Up, Hydrogen Scaling. (2017) A sustainable pathway for the global energy transition. Hydrogen Council, November 2017. https://hydrogencouncil.com/wp-content/uploads/2017/11/Hydrogen-scaling-up-Hydrogen-Council.pdf
13. Pearre NS, Ribberink H (2019) Review of research on V2X technologies, strategies, and operations. Renew Sust Energ Rev 105:61–70
14. Triviño-Cabrera A, Aguado JA, de la Torre S (2019) Joint routing and scheduling for electric vehicles in smart grids with V2G. Energy 175:113–122
15. Hashemi-Dezaki H et al (2015) Risk management of smart grids based on managed charging of PHEVs and vehicle-to-grid strategy using Monte Carlo simulation. Energy Convers Manag 100:262–276
16. Alirezaei M, Noori M, Tatari O (2016) Getting to net zero energy building: investigating the role of vehicle to home technology. Energ Build 130:465–476
17. Hydrogen and Fuel Cell Strategy Council (2019) Hydrogen and fuel cell strategy roadmap, 21–22. https://www.meti.go.jp/shingikai/energy_environment/suiso_nenryo/roadmap_hyoka_wg/pdf/001_s02_00.pdf
18. Xin T (2017) Electric price guidance and interaction method for electric vehicle charging [D]. North China Electric Power University (Beijing) (in Chinese)
19. Mahmud K et al (2020) An internet of energy framework with distributed energy resources, prosumers and small-scale virtual power plants: an overview. Renew Sust Energy Rev 127: 109840
20. Wu Q et al (2016) Multi-objective optimization of a distributed energy network integrated with heating interchange. Energy 109:353–364
21. Kumar A et al (2019) Strategic integration of battery energy storage systems with the provision of distributed ancillary services in active distribution systems. Appl Energy 253:113503
22. Li C et al (2020) Economic dispatching strategy of distributed energy storage for deferring substation expansion in the distribution network with distributed generation and electric vehicle. J Clean Prod 253:119862
23. Varghese S, Sioshansi R (2020) The price is right? How pricing and incentive mechanisms in California incentivize building distributed hybrid solar and energy-storage systems. Energy Policy 138:111242
24. Wu Q et al (2018) Coupling optimization of urban spatial structure and neighborhood-scale distributed energy systems. Energy 144:472–481
25. Bustos C, Watts D, Olivares D (2019) The evolution over time of distributed energy resource's penetration: a robust framework to assess the future impact of prosumage under different tariff designs. Appl Energy 256:113903

26. Ren H et al (2019) Cost–benefit analysis of distributed energy systems considering multi-benefits and multi-stakeholders. Energy 189:116382
27. Vijay A, Hawkes A (2018) Impact of dynamic aspects on economics of fuel cell based micro co-generation in low carbon futures. Energy 155:874–886
28. Qiu D et al (2020) Numerical analysis of air-cooled proton exchange membrane fuel cells with various cathode flow channels. Energy 198:117334
29. Qian F et al (2020) Economic optimization and potential analysis of fuel cell vehicle-to-grid (FCV2G) system with large-scale buildings. Energy Convers Manag 205:112463
30. Seo S-K, Yun D-Y, Lee C-J (2020) Design and optimization of a hydrogen supply chain using a centralized storage model. Appl Energy 262:114452
31. Samsatli S, Samsatli NJ (2019) The role of renewable hydrogen and inter-seasonal storage in decarbonising heat–comprehensive optimisation of future renewable energy value chains. Appl Energy 233:854–893
32. Corgnale C et al (2019) Hydrogen storage in a two-liter adsorbent prototype tank for fuel cell driven vehicles. Appl Energy 250:333–343
33. Zhou Y, Ravey A, Péra M-C (2020) Multi-mode predictive energy management for fuel cell hybrid electric vehicles using Markov driving pattern recognizer. Appl Energy 258:114057
34. Kang S, Ahn K-Y (2017) Dynamic modeling of solid oxide fuel cell and engine hybrid system for distributed power generation. Appl Energy 195:1086–1099
35. Yang X, Zhao H (2019) Thermodynamic performance study of the SOFC-STIG distributed energy system fueled by LNG with CO2 recovery. Energy 186:115860
36. Reuß M et al (2017) Seasonal storage and alternative carriers: a flexible hydrogen supply chain model. Appl Energy 200:290–302
37. Eriksson ELV, MacA Gray E (2017) Optimization and integration of hybrid renewable energy hydrogen fuel cell energy systems–a critical review. Appl Energy 202:348–364
38. Mehrjerdi H (2020) Peer-to-peer home energy management incorporating hydrogen storage system and solar generating units. Renew Energy 156:183
39. Hemmati R, Mehrjerdi H, Bornapour M (2020) Hybrid hydrogen-battery storage to smooth solar energy volatility and energy arbitrage considering uncertain electrical-thermal loads. Renew Energy 154:1180
40. Fonseca JD et al (2019) Trends in design of distributed energy systems using hydrogen as energy vector: a systematic literature review. Int J Hydrog Energy 44(19):9486–9504
41. Maghami MR et al (2020) Hybrid energy management with respect to a hydrogen energy system and demand response. Int J Hydrog Energy 45(3):1499–1509
42. Virji M et al (2020) Analyses of hydrogen energy system as a grid management tool for the Hawaiian isles. Int J Hydrog Energy 45(15):8052–8066
43. Martin A, Agnoletti M-F, Brangier E (2020) Users in the design of hydrogen energy systems: a systematic review. Int J Hydrog Energy 45:11889

Chapter 11
Future Prospect of Distributed Energy System

Yafei Wang, Yao Zhang, and Weijun Gao

11.1 Introduction of Virtual Power Plant

The word "virtual" has been heard a lot recently. Virtual reality, which allows you to experience immersive virtual spaces, is already well known, and "virtual currencies" such as Bitcoin are a hot topic. In fact, the word "virtual" is also starting to be used in the energy sector. What is becoming virtual in the energy sector is the power plant itself. In most people's minds, tall chimneys, huge plants, and endless stretches of "white smoke," coupled with dense high-voltage lines, this is the image of power plants should be, as shown in Fig. 11.1, a traditional power system with energy flow from the generation side to the consumption side, that is, the power generation side needs to ensure the stability of the grid while generating electricity, and the demand side only needs to turn on the switch when electricity is needed. However, in the

Y. Wang (✉)
iSMART, Qingdao University of Technology, Qingdao, China

Faculty of Environmental Engineering, The University of Kitakyushu, Kitakyushu, Japan

Y. Zhang (✉)
School of Mechanical and Electrical Engineering, Shaanxi University of Science and Technology, Shaanxi, China
e-mail: zhangyao@sust.edu.cn

W. Gao
The Engineering Academy of Japan, Chiyoda City, Tokyo, Japan

The University of Kitakyushu, Kitakyushu, Japan

Qingdao University of Technology, Qingdao, China

International Research Center for Green Buildings and Low-Carbon Cities, Zhejiang University, Hangzhou, China

Xi'an Jiaotong University, Xi'an, China

W. Gao (ed.), *Distributed Energy Resources*, Green Energy and Technology,
https://doi.org/10.1007/978-3-031-21097-6_11

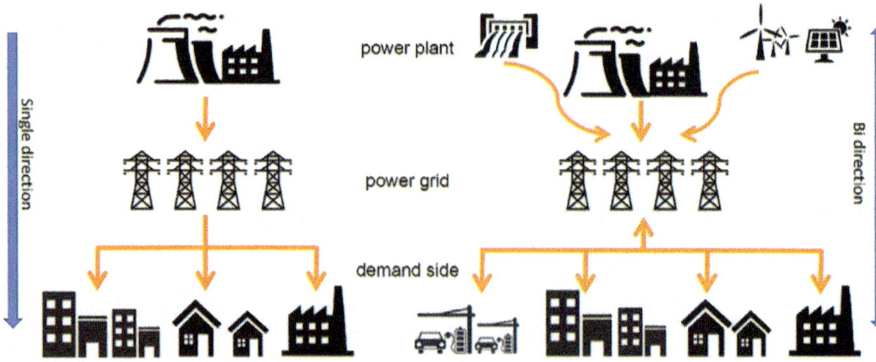

Fig. 11.1 Difference between conventional power plant and virtual power plant

twenty-first century, both the generation side and the consumption side have changed dramatically. The power generation side has introduced uncontrollable power generation technologies such as scenic power, but because of its uncontrollability and dependence on external conditions, it also poses a danger to the power grid. And the electricity consumption side is not only consuming electricity, but new generation and consumption technologies have been introduced, resulting in a bidirectional energy flow, which also brings great difficulties to the balance and stability of the grid.

Conventional energy supply systems that rely on large power plants are being challenged by the increasing popularity of distributed energy sources, including solar and wind energy. Since the amount of renewable energy generated depends heavily on the weather, energy supply will become unstable as usage expands. The new energy service concept is a solution to maintain a stable power supply. This means, for example, that decentralized energy sources such as distributed power supplies and batteries can be controlled remotely via IoT devices and function as a power plant. As a result, virtual power plant (VPP) with hierarchical zoning and elastic balancing is born! A VPP is a network of decentralized, medium-scale power generating units such as wind farms, solar parks, and combined heat and power units, as well as flexible power consumers and storage systems.

In recent years, small-scale power generation devices such as solar power and fuel cells have been widely installed in homes and offices (small, distributed power supplies). This means that we (consumers) who used to only consume electricity can now generate it just like the power company. Despite the huge potential of renewable energy, due to the instability of its power generation, the energy supply will become unstable as its use spreads. How to maximize the use of renewable energy and ensure the stability of the grid is a problem to be solved. With the widespread use of batteries, electric vehicles, heat pumps, etc., it has become possible to disperse and store energy. VPP works as aggregator that manages scattered energy sources, such as distributed power sources and storage batteries,

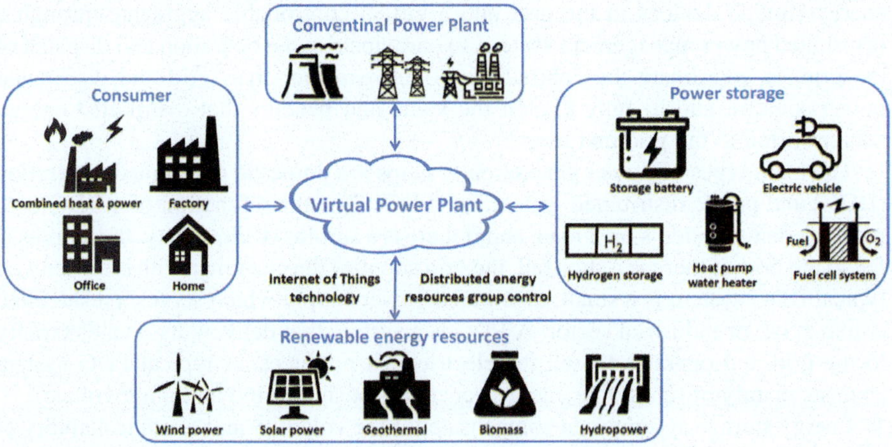

Fig. 11.2 Composition of virtual power plant

and can be remotely controlled and worked as one power plant. In this way, the electricity generated by the generating plant can be used, or when there is a surplus of electricity generated, it can be stored in a battery and used when needed, and each consumer saves electricity. VPP is expected to promote the introduction and expansion of renewable energy and contribute to a decarbonized society. The new concept of VPP comes as a solution to maintain the stability of the power supply. Figure 11.2 shows the composition of VPP; generally, VPP is related to the following three departments: power generation system, energy storage system, and communication systems. Specifically, the VPP uses advanced information and communication technology and software systems to aggregate and optimize the DER of distributed power sources, energy storage systems, controllable compounding, and electric vehicles, to participate in the power market and grid as a special power plant, which can be equated to a controllable power management system. This system can be used externally as a "positive power plant" to supply power to the system or as a "negative power plant" to consume power from the system, playing a flexible role in peak and valley reduction. In a sense, the VPP can be seen as an advanced regional centralized power management model, providing management and ancillary services to the distribution and transmission grids.

11.2 Operating Principle of the Virtual Power Plant

Why was there no VPP before? This is because of the former single direction of the power flow, and there are very few distributed energy sources. As mentioned earlier, the VPP is proposed to integrate various distributed energy sources, including distributed power sources, controllable loads, and energy storage devices. The basic concept is to aggregate distributed power sources, controllable loads, and energy storage devices in the grid into a virtual controllable aggregate through a distributed power management system, to participate in the operation and dispatch of the grid, to coordinate the contradictions between the smart grid and distributed power sources, and to fully exploit the value and benefits that distributed energy sources bring to the grid and users.

The main types of power generation systems are domestic distributed generation (DDG) and public distributed generation (PDG). The main function of a DDG is to meet the demand side's own load, and if there is a surplus of electricity, the surplus is fed to the grid; if there is a shortfall, the grid supplies the customer with electricity. A typical DDG system is a small, distributed power supply which serves a residential, commercial, or industrial sector. A PDG is a system that delivers its own electricity to the grid and operates to sell the electricity it produces. A typical PDG system consists mainly of renewable energy generators such as wind and photovoltaic.

Energy storage systems can compensate for the volatility and uncontrollability of renewable energy generation output, adapt to changes in electricity demand, improve the weakness of the grid caused by fluctuations in renewable energy, enhance the system's ability to accept renewable energy generation, and improve energy efficiency.

The communication system is an important part of the VPP for energy management, data collection and monitoring, and communication with the power system dispatch center. By interacting with information from the grid or with other VPPs, the management of the VPP is more visualized, and it is easier for the grid to monitor and manage the VPP. The algorithm is the core of the VPP, through the input of multi-dimensional time series and information, including the price forecast of the power trading center, the node data of the transmission operator, the weather forecast data, the real-time power generation asset status data, historical data, the output optimal operation strategy to complete automatic control of power generation assets, etc.

Traditional power plant adopts the form of generation following demand; on the other hand, the VPP adopts the form of demand chasing generation. However, individual consumer often has low electricity consumption, irregular consumption patterns, and a lack of response mechanisms. Therefore, an aggregator is needed to aggregate consumers' loads to form VPP, which can then be centrally regulated. For example, the VPP operate like import/export business. It can produce and sell by itself. It is also can provide services to organizations outside the VPP (such as the grid). Internal storage or production reduction is also possible in the case of overcapacity.

The profit modes of VPP can be considered from the following aspects. VPP can be profitable in the energy market by saving the cost of services to balance electricity demand, control power plants that can control power generation within the VPP, increase power generation in anticipation of an imminent rise in electricity prices, and share this profit with participating power plants.

11.3 Development Status and Achievements of VPP

With the development of clean energy and emerging technologies, VPPs will become an important form of energy aggregation in the construction of smart grids and the global energy Internet, with a wide scope for development. However, the complementary nature of distributed power sources reduces the uncertainty of power output. Due to the large randomness, volatility, and intermittency of renewable energy output, the dynamic mix of distributed power sources needs to be addressed.

Multiple distributed units can be flexibly and dynamically combined to form a VPP. The main difference between a VPP and a microgrid is that the multiple distributed generation units that make up a VPP are not necessarily located in the same geographical area, and their scope of aggregation and interaction with the market depends on communication capabilities and reliability. Multiple distributed generation units are aggregated according to certain rules or objectives and participate as a whole in the electricity market or ancillary services market, with the benefits being allocated to each distributed generation unit. The VPP acts as an intermediary for the flexible dynamic combination of multiple distributed generation units according to rules such as dynamic combination algorithms or dynamic game theory. The real-time nature and flexibility of the dynamic combination avoid the cost problems associated with real-time imbalances and the combination deviations caused by plant shutdowns and errors in load and renewable energy output forecasting.

Depending on the control structure of the information flow transmission of the VPP, the control of the VPP can be divided into centralized control methods, decentralized control methods, and fully decentralized control methods. A VPP in this structure requires that the plant has complete information about each unit involved in the distributed operation and that its operational settings meet the different needs of the local power system. This type of VPP has great potential for achieving optimal operating modes. However, scalability and compatibility are often limited due to the constraints of specific operational realities. A VPP under centralized control has full access to all information about the distributed units within its jurisdiction and has full control over all generation or consumption units.

The VPP in the decentralized control approach is divided into several levels. The control coordination center of the VPP at the lower level controls the generation or consumption units within its jurisdiction, and the control coordination center of the VPP at that level then feeds information back to the control coordination center of the VPP at the higher level, thus forming an overall hierarchy. This one refers to the locally controlled distributed mode of operation, which constitutes an overall hierarchy in a locally controlled system. In response to the weaknesses in the previous centralized control model, the DCVPP model effectively improves on the deficiencies through a modular local operation model and information gathering model. However, the central control system still needs to be at the top of the entire decentralized virtual power generation system during operation in order to ensure safety during system operation and overall operational economy.

In the fully decentralized control approach, the VPP control and coordination center is replaced by a data exchange and processing center, which only provides information on market prices, weather forecasts, etc. In turn, the VPP is divided into autonomous, intelligent subunits that are independent of each other. These subunits are not controlled by the data exchange and processing center but only receive information from the data exchange and processing center and optimize their own operation according to the information received. This mode of operation can be considered as an extension of the decentralized controlled VPP. The central control system in the decentralized control mode is replaced by data exchange agents which provide valuable information such as market prices, weather forecasts, and data records. For small units in the decentralized control model, running in this model will be very scalable and open compared to the previous two models due to the plug and play capability in the fully decentralized controlled VPP model.

Using big data for renewable energy forecasting and increasing the speed of data processing in VPPs has also been a hot spot recently. Big data is a collection of data that cannot be sensed, acquired, managed, processed, and analyzed in an affordable time frame using traditional IT technologies, hardware and software tools, and mathematical analysis methods. Big data technologies allow for load forecasting and renewable energy output forecasting, including wind and solar. Wind energy forecasting is essential because data shows that the actual capacity of wind farms varies considerably during peak electricity consumption periods. Accurate forecasting of solar and wind energy requires analysis of large amounts of data, including meteorological data such as wind speed and cloud cover. At the same time, the use of big data technology to process various information within the VPP can effectively increase the processing speed of the data exchange and processing center, providing the data exchange and processing center of the VPP with a real-time, accurate flow of data and information from the various subsystems.

VPP participate in multiple markets for optimal dispatch and bidding. By aggregating several distributed units into one, VPPs can operate in the electricity market, taking advantage of the stable output and volume sales of traditional power plants, but also being complementary by aggregating multiple generation units. The market in which the VPP participates includes the day-ahead market, the real-time market, and the ancillary services market, which allows for a variety of market models such as day-ahead markets, bilateral contracts, balancing markets, and hybrid markets. Considering the uncertainties of renewable energy output, load, and real-time tariffs in the VPP, scheduling and bidding models can be developed in different market environments to make the VPP more widely applicable.

The VPP plays a delicate role in the relationship between the grid operator, the various electricity markets, the customers of distributed generation assets, and the energy consumers. By interacting with the data between the four players, the VPP performs an optimization function in the electricity system. There are a large amount of researches establishing a scientific cooperation mechanism based on game theory to ensure the stability of the VPP. Game theory focuses on the theory of how multiple decision-makers with conflicting interests can each make decisions that benefit themselves or the group of decision-makers, depending on their own capabilities and the information they know. Based on game theory, all generation and consumption units within the VPP and all operators outside the VPP are cooperative games. A scientific cooperation mechanism based on cooperative game theory, including cooperation between multiple generation or consumption units aggregated within the VPP and between the VPP and the integrated operator, the distribution or transmission grid, and the electricity market operator, ensures a reasonable return for all participants, keeps participants motivated to participate in the long term, and ensures the stability of the VPP.

11.3.1 VPP in Europe

At present, VPPs are still in the development stage of theoretical research and preliminary pilots. From a global perspective, the research and implementation of VPP are mainly concentrated in Europe and North America. According to data released by Pike Research, as of the end of 2009, the global VPP total capacity was 19.4 GW, of which 51% in Europe and 44% in the United States; as of the end of 2011, the global VPP capacity increased to 55.6 GW. However, the application forms of VPP in Europe and the United States are significantly different, and the VPPs in European countries also have their own characteristics. The concept of VPPs has existed for more than 20 years, emerging in the early twenty-first century in European countries such as Germany, the United Kingdom, France, and the Netherlands, and has a number of mature demonstration projects that focus on the reliable grid connection of distributed energy while constructing a stable business model in the electricity market. In Europe, VPPs can also be called grid aggregators (aggregators). Among market participants, grid aggregators are responsible for pooling the control of distributed small-scale generation plants (new energy sources) into a pool, reaching a minimum threshold for participation in the electricity market, equivalent to the role of an agent.

For Europe, take the example of Next-Kraftwerk, the largest VPP in Germany. The project of the German company Next-Kraftwerk, which realizes the management of more than 4000 distributed generation facilities, also including a portion of adjustable loads, has already realized the management of more than 7000 distributed energy and adjustable loads across five countries at the end of 2019, with a total scale of nearly three million kW. Founded in 2009, Next-Kraftwerk employs with a total of no more than 200 employees and has sales revenue of around 400 million euros and 140 GWh of traded power; the performance is outstanding. VPPs exist to help Germany achieve its goal of increasing the proportion of new energy sources connected to the grid as part of the energy transition:

- Aggregate distributed new energy generation assets.
- Automatically control generation assets through flexible algorithms.

- Reconcile deviations between electricity production and consumption and balance forecast fluctuations.
- Reduce the impact of new energy sources injected into the grid.

It obtains benefits through several aspects: the first is the lower marginal cost of photovoltaic, wind power, and other power sources, participating in market-based trading profit, the second is biomass, combustion engines, and other power sources with better regulation performance, participating in frequency regulation profit; third is the regulation of distributed power sources and controllable load, using the peak valley difference in the power market profit. For resource coordination and aggregation capabilities, the requirements are very high. The characteristics of the source network, load and storage, how to interface with the business model of the power market, how to respond in a timely manner, and efficient control are great topics.

11.3.2 VPP in China

With the advancement of the global energy Internet, three Chinese ministries and commissions (Ministry of Finance, Chinese Ministry of Propaganda, Ministry of Education) have jointly released the "the Belt and Road Initiatives" development strategies for renewable energy, with countries along the "the Belt and Road Initiatives" having abundant wind and solar resources, promoting the transmission of electricity from large renewable energy bases and the exchange of electricity between continents. The energy Internet strategy promotes the construction of cross-border power and transmission channels, actively carries out cooperation in upgrading and transforming regional power grids, gives full play to the complementary time lag and seasonal characteristics of distributed power sources in different regions, and improves the utilization rate of renewable energy and the benefits of VPPs. For China, take the example of VPP project in Huangpu District, in 2020. The largest trial runs with over 50 participating buildings, releasing power load of about 10,000 kW. How was it achieved? During peak hours of electricity consumption, the system automatically adjusts several characteristic parameters such as temperature, air volume, and speed of the central air conditioners of the buildings involved, within the VPP system, with little impact on the user experience. How do users participate in VPP projects and realize profits? Take the Huangpu project as an example, the load aggregator bids on the system platform. The subsidy price is differentiated based on response time. The user is within 30 minutes of peak shaving, the subsidy to you is 3 times the price (there is a benchmark value), between 30 minutes and 2 hours is 2 times, and longer time is lower. At present, the subsidies mainly come from the surplus of the inter-provincial renewable energy power spot trading purchase price difference, so there are still some constraints; many provinces have not started power spot trading.

11.3.3 VPP in Japan

In Japan, which wants to further increase the introduction of renewable energy, the "Japan Revitalization Strategy" decided by the government in June 2015 states that VPP will be used. The Ministry of Economy, Trade, and Industry's Energy Innovation Strategy, announced in April 2016, states that it will promote demonstration trials and commercialization of VPP technology and provide subsidies in a 5-year plan ending in fiscal 2020. The government established it to support companies working on VPP.

11.4 Barriers and Challenges of Virtual Power Plant

Distributed energy resources and controllable loads can be aggregated in VPPs to participate in bidding in day-ahead power markets, intra-day demand response markets, regulation markets, real-time electricity markets, and carbon trading markets. However, the uncertain output of distributed energy generation brings higher transaction risks to VPPs. The bidding models and strategies for VPPs have been extensively researched both nationally and internationally. The market bidding problem for VPPs consists of single day-ahead market bidding and joint market bidding.

And another problem is coordinate control of VPPs; it can be divided into two types: (1) internal dispatch, where the VPP optimizes the capacity allocation or output of multiple power sources within itself, and (2) external dispatch, where the VPP is optimally dispatched by the grid. The rapid development of communication and computer technology has enabled VPPs to communicate in real time. Most of the current grid dispatch and monitoring uses supervisory control and data acquisition system and energy management system; these systems have disadvantages such as slow processing speed and limited information that can be stored in the system. The system is weak in online analysis and most of the time requires human brain experience to solve problems. These weaknesses make these systems unsuitable for the increasingly complex and changing power systems of today's world. VPPs require advanced computer algorithms to collect, organize, classify, and process massive amounts of information in the smart grid in real time, providing operators with assistance in decision-making based on data and analysis. In addition, the VPP needs to have fast guidelines and simulation capabilities for real-time monitoring and analysis of the current state of the system to help the VPP respond and forecast quickly.

The key technology of VPP is still immature and has yet to be effectively solved. For China's national, national, and electric conditions, the following problems need to be solved:

1. Reasonable positioning of resource functions, including identification of regulable distributed energy resources on the distribution network side; analysis of the technical and economic characteristics of various types of distributed energy resources; and reasonable positioning of potential distributed energy resources in combination with specific distribution network operation constraints, clarifying the type of services they provide, response speed, response frequency, etc.
2. Need to develop supporting software and hardware technologies.
3. Need to stimulate active participation of all parties, including establishing a cooperation mechanism for VPP operation-related participants and establishing an incentive mechanism conducive to mobilizing grid enterprises to participate actively.

If the optimal strategy of the VPP is not managed uniformly by the distribution network operator, i.e., the operator of the VPP takes the liberty of minimizing all its short-term costs, including production costs and transaction costs with the grid, thus forcing the provision of their heat and power, this will cause the voltage and current of the distribution network to exceed the allowable limits. As a result, VPPs are still not widely used around the world.

VPPs have the technical characteristics of diversity, synergy, and flexibility to meet the future needs of new power systems such as green, flexible, multi-interactive, and highly market-oriented operations and are an important technical support, as well as providing a full participation mechanism for the development of the energy storage industry. With the certainty of the vision of the double carbon target, VPPs are bound to usher in a good development in the world.

11.5 Introduction of Smart Grids

The growing problems of conventional energy shortages and environmental pollution have become the most significant challenges facing the sustainable development of human society. According to a report by the World Resources Institute, from 1990 to 2014, worldwide greenhouse gas emissions (GHG) by sector continued to increase beyond the problems that have already been noted over the last two decades. Moreover, the GHG released during energy production accounts for the majority of total GHG emissions. In the International Energy Outlook 2016 (IEO2016) Reference case, world energy-related CO_2 emissions will increase from 32.3 billion metric tons in 2012 to 35.6 billion metric tons in 2020 and to 43.2 billion metric tons in 2040. Awareness of the environmental impact and the carbon footprint of all energy sectors continues to increase [1].

In conjunction with this phenomenon, decarbonization has been proposed by many researchers and unions. Two elements are highlighted as important for decarbonization: improved energy efficiency and increased shares of renewable energy. Efficient energy usage at all stages of the energy chain from production to final consumption is meaningful for the reduction of GHG and therefore the mitigation of climate change.

In addition, conventional energy shortages and serious environmental pollution problems have compelled many countries to develop environmentally friendly renewable energy so that they can reduce their dependence on conventional energy resources, realize reductions in environmental pollution caused by the increasing energy demand, and ensure sustainable social and economic development. However, compared to conventional energy sources, many renewable energy sources exhibit randomness and intermittency. A large amount of renewable energy generation in a power system, whether in large-scale centralized systems or small-scale distributed systems, can adversely impact the safety and reliability of traditional power systems. Therefore, sophisticated control systems are needed to facilitate the connection of sources to the highly controllable grid [2]. In such cases, SGs play a pivotal role in renewable-based low-carbon energy systems while providing an essential platform to enable renewable energy generation in the central grid [3, 4].

SGs are therefore essential to realizing decarbonization in the energy sector. In addition, the call for nuclear-free production in the energy sector has appeared many times in recent years owing to the vulnerabilities and insecurities of nuclear power, although nuclear resources are environmentally friendly. In other words, the emissions of GHG from the energy sector can be eliminated with technologies that are now available or foreseeable [5]. This can be realized while creating a much more effective energy system than before. The conventional electricity grid has no potential to provide enough services to address energy needs and the integration of RE at the scale required to meet the clean energy demand for the future [6]. Therefore, the introduction of SGs is essential to reduce GHGs.

The basic characteristics of SGs are summarized in Table 11.1, in which compatibility, flexibility, and high efficiency are the basic characteristics of SGs. Serviceability and safety are auxiliary characteristics of SGs, and interoperability represents the backbone. Figure 11.3 shows the interoperability layer of SGs.

Table 11.1 Key characteristics of SGs

Characteristics	Contents
Compatibility	Accommodates all generation options; deployment includes the integration of various types of distributed resources (renewable energy, small-scale combined, power and energy storage, etc.)
Flexibility	Flexible power resources; enables informed participation by customers (demand response); new controllable loads (electric vehicles)
High efficiency	Digital information and advanced technologies; dynamic optimization of power resource allocation; enhanced system operating efficiency
Serviceability	Enables new products, services, and markets; provides the necessary power quality for a range of needs; integration of smart appliances and consumer devices
Safety	Self-healing ability; resiliency to disturbances, attacks, and natural disasters
Interoperability	Deployment of smart metering; control and visualization technologies; information communication technologies

Source: [16, 17]

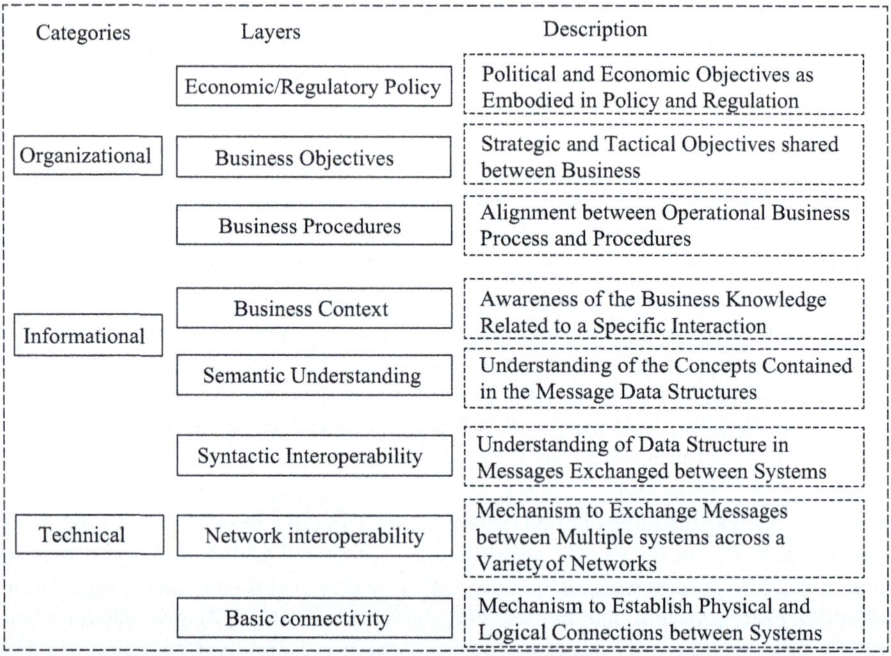

Fig. 11.3 The interoperability layer of SGs

SGs play an important role in current and future grid construction [7]. Moreover, the critical issues concerning smart grids can mainly be defined as technologies (including equipment, skills, systems, services, infrastructure, software, and components) that are currently available or are expected to become available in the near future [8, 9]. According to previous studies and many academic surveys concerning SGs [10–15], the relevant technologies can be mainly divided into technologies for the transmission system, technologies for the distribution system, and technologies for the demand side. In addition, these technologies can be classified by function as follows: (1) monitoring and control technologies for power transmission and distribution systems, (2) energy management technologies for the demand side, (3) possible advanced technologies to enable the effective operation of systems, and (4) advanced interface technologies. Additional details are depicted in Fig. 11.4.

11.6 Definitions of SGs in Different Countries

A uniform definition of SGs has yet to be formed at the international scale. In the United States, SGs emphasize the reliability, safety, and operational efficiency of power systems through the strong support of digital and other advanced technologies. In addition, the United States is also devoted to the reduction of the power supply costs created by an aging power infrastructure [18]. Europe's innovative SG

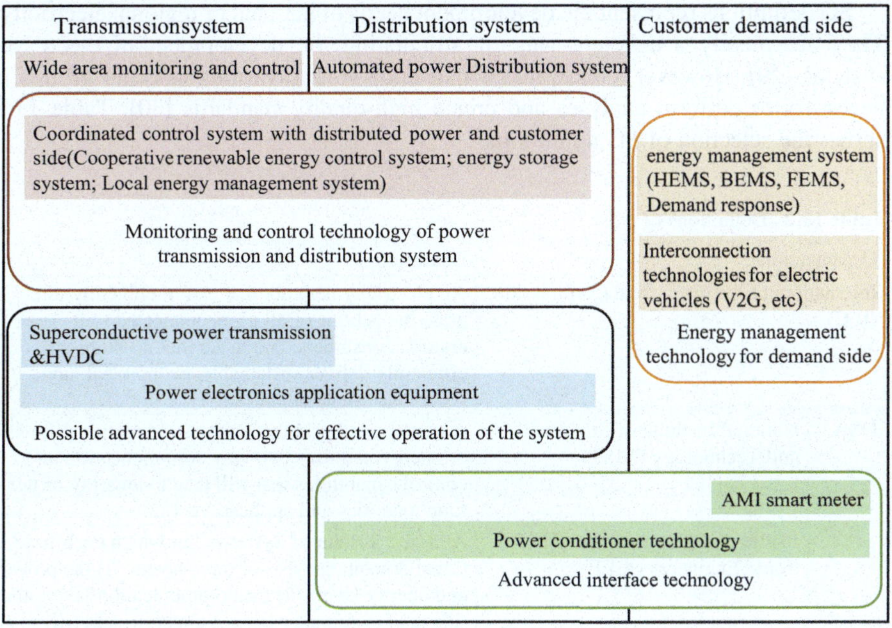

Fig. 11.4 SG technologies

scheme attempts to reconcile two approaches of renewable energy development, namely, large-scale centralized approaches and small-scale, local and decentralized approaches, to realize a transition toward a fully low-carbon electricity system [19] while attempting to realize energy trading between European countries. To promote such objectives, the European Commission also monitored SG projects, proposed guidelines for the cost-benefit analysis of SG projects and smart meter deployment, investigated the complexity features of smart energy grids, and evaluated the social dimensions of SG projects [3, 4, 20–23]. In Japan, because its energy self-sufficiency is a mere 4% [24], the focus of SG plans is to build renewable-friendly power grids. Moreover, the Great East Japan Earthquake that struck on March 11, 2011, and the subsequent nuclear power plant accident have prompted the Japanese government to adopt reforms targeted at the power system; here, SGs provide stable power supplies and optimize overall grid operations from power generation to the end user [25, 26]. Moreover, Japan has developed SGs to achieve the CO_2 emission reductions stipulated in the Kyoto Protocol. In China, high electricity consumption and multiple electricity load structures have appeared with the rapid development of the economy and increasingly large populations, which result in a high demand for a strong SG [27]. China prefers to renovate traditional power systems with modern information technology while establishing a highly automated and widely distributed network for energy exchange to solve its energy balance problem [28].

SG definitions represent the development needs of national or regional electricity. They are merely a different way of articulating the development of electricity systems [29]. However, conceptual consistency among various working groups is necessary to perform analyses and create high-quality standards [30]. Table 11.2 shows the selection of SG definitions.

Table 11.2 Definitions of SGs

Organizations		Definition
International Electrotechnical Commission (IEC)		An electricity network that can intelligently integrate the actions of all users connected to it—generators, consumers, and those that do both—to efficiently deliver sustainable, economical, and secure electricity supplies [31]
USA	National Institute of Standards and Technology (NIST)	A modernized grid that enables bidirectional flows of energy and uses two-way communication and control capabilities that will lead to an array of new functionalities and applications [32]
	Institute of Electrical and Electronic Engineers (IEEE)	A large "System of Systems," in which each functional domain consists of three layers: (i) the power and energy layer, (ii) the communication layer, and (iii) the IT/computer layer. Layers (ii) and (iii) above are the enabling infrastructure that makes the existing power and energy infrastructure "smarter" [32]
EU	European Commission (EC)	Energy networks that can automatically monitor energy flows and adjust to changes in the energy supply and demand accordingly, reach consumers and suppliers by providing information on real-time consumption, and better integrate renewable energy [33]
	European Regulators' Group for Electricity and Gas (ERGEG)	An electricity network that can efficiently integrate the behavior and actions of all users connected to it—including generators, consumers, and those that do both—to ensure an economically efficient, sustainable power system with low losses and high levels of quality and security of supply and safety [34]
Japan	Japan Smart Community Alliance (JSCA)	A system that can promote the greater use of renewable and unused energy and local generation of heat energy for local consumption and contribute to the improvement of energy self-sufficiency rates and reduction of $CO2$ emissions. Provides a stable power supply and optimize overall grid operations from power generation to the end user [35]
China	State Grid Corporation of China (SGCC)	An integration of renewable energy, new materials, advanced equipment, information technology, control technology, and energy storage technology, which can realize digital management, intelligent decision-making, and interactive transactions of electricity generation, transmission, deployment, usage, and storage [27]

Although six different definitions of SGs exist, they are reasonably consistent. In brief, the principal parameters of SGs can be summarized as follows [29, 36]: (1) digitization, two-way communication, and automatic monitoring; (2) accommodating all generation and storage options, integrating renewable and energy storage in the electricity network; (3) self-healing from power disturbance events with necessary maintenance of self-adaptive networks; (4) enabling the creation of new products, services, and markets; (5) demand response and energy management system for lowering peak demand and overall load; (6) enabling active participation of end users and provide more user options; and (7) optimizing assets and operating efficiently and resiliently.

11.7 Development Status and Achievements of SGs

11.7.1 Infrastructure of SGs

Because the internal environment and the basic conditions of existing grids are different in each selected country, their development priorities and routes for SGs must be different. Five essential aspects were analyzed to compare the infrastructures of SGs for various countries. Concerning the organizational and incentive perspective, America and Europe follow the most similar modes. Both have government-dominated systems, making SG development a national development strategy through legislation. Moreover, the government provides special funds to support research and demonstration projects. Japan uses a special operation mode wherein government, industry, and academic institutions jointly promote SGs. The government (METI) provides policy supervision and creates a favorable business environment; industry (NEDO) and academic institutions lead research and demonstration projects [37]. In China, SGs were included in the national development strategy and in the 12th Five-Year Plan. The government supports the research work in the form of national science and technology projects. The demonstration and construction of SGs has been led by the grid company [38].

Concerning supervision and tariffs, America and Europe have strong regulations and market mechanisms: they have applied a terminal price; have experience implementing real-time pricing (RTP), critical peak pricing (CPP), peak time rebate (PTR), and time-of-use (TOU) prices [39]; and have facilitated the implementation of automated demand response and distributed energy. Japan has only introduced various types of TOU price systems. Some regional pilot programs have implemented CPP prices. China continues to employ time-invariant electricity pricing, although some regional pilot programs have introduced TOU pricing. Clearly, there is a large gap between China and the other two countries and Europe in terms of electricity tariff mechanisms. In terms of customer features, China exhibits significant differences from other countries. In America, Europe and Japan, residents consume a higher proportion of electricity, generally approximately

30%. However, in China, industry consumes a higher proportion of electricity. Residential electricity consumption accounts for a relatively small proportion, generally approximately 15.1% in 2014.

Concerning energy structure, America and Europe have a uniform distribution and rational structure. In Japan, most of the energy resources rely on imports from abroad, which creates a large demand for the utilization of renewable energy. In addition, renewable energy in these two countries and Europe is provided to the grid mainly in a distributed manner. In addition, all these countries utilize fuel- and gas-fired generators. Unlike in America, Europe, and Japan, the energy and load distribution of China is unbalanced; electricity must be transmitted over long distances, ranging from 1000 and 3000 km, and must be provided with a wide-range configuration due to power resources (mostly located in northern and western China) being far from power demand centers (usually located in central and eastern China). Renewable energy, such as wind power, is provided to the grid mainly in concentrated forms. The power grid is dominated by thermal power plants instead of gas, oil, and other plants, which can achieve higher efficiencies.

Concerning management, the management of SGs in America and Europe is decentralized. Moreover, the formulation of widely recognized standards is a time-consuming process because there are many stakeholders and suppliers involved in the electricity system. In Japan, there are only ten electricity companies divided up by area, and grid ownership and management is relatively decentralized. The government directs energy consumption, and power companies must implement corresponding policies. Compared to power grids in America, Europe, and Japan, China's power grid has a relatively concentrated ownership and control structure, which favors unified planning and scheduling.

11.7.2 Policy Support

In Europe and the above three countries, there are numerous SG stakeholders due to their wide variety of functions and applications [40]. Both power company providers and common stakeholders are integral aspects of SGs. Therefore, how to form a balance under such complex situations is essential to the continued development of SGs. Here, policy plays an important guiding role in this process. Clear policy can be used to establish a good external development environment for SGs while facilitating the coordination and cooperation of all parties involved in SGs. SGs are extremely complex systems with many components, including generation, transmission, substation, distribution, consumption, power dispatch, and information platforms. Any policy related to these areas can be further related to SGs [41]. Table 11.3 summarizes the policy support for SGs among the three selected countries and Europe.

It is evident that America, Japan, Europe, and China published several regulations to support the development of SGs in the early stages. SG policies in America and Europe are mainly established by states. In America and Europe, the construction of

Table 11.3 Policy support for SGs

Countries	Time issued	Policy support	Main contents
America	2007	EISA [42]	Commitment to allocate state-owned special funds; support NIST to compile standards system for SGs; coordination of nationwide SG standardization
	2009	ARRA [43]	Commission Department of Energy grants totaling $4.5 billion of government funds; motivate domestic private investment into SGs; support the research and demonstration of SGs
Europe	2008	EEPP [44]	Indicate that green technology plays a key role in the economic recovery plan; stipulate considerable portion of funds should be used for electricity interconnection and offshore wind projects
	2009	Third Legislation for Further Liberalization of the Electricity and Gas Markets [45]	States to further liberalize the electricity markets to facilitate greater supplier competition and consumer choice
	2010	European Council summits [46]	Encourage the investigation of energy infrastructure, research, and innovation projects; guarantee the security of EU energy supply system; address climate change
	2012	Energy Efficiency Directive 2012/27/EU, the European Commission [33]	Clear focus on achieving the overall energy efficiency target of reducing primary energy consumption by 20% by 2020
	2014	Regulation [EU] No 333/2014 and No 517/2014 on Carbon dioxide emissions [34]	Establishes updated policies and strengthens the existing climate policies
Japan	2009	Policy Package to Address the Economic Crisis [47]	Advocate the development of solar power generation, energy-saving appliances, and low-fuel-consumption cars; attempt to realize an installed solar power generation capacity of 28 million kW by 2020
	2010	Report of the next-generation power transmission and distribution network [48]	Emphasis on the stability of power systems; set encouragement strategies for improving distribution systems and developing battery technologies; guarantee a power supply during system accidents
	2010	New Strategic Energy Plan [49]	Establish 3E energy viewpoint (energy security; environment; economic efficiency); promote the development of SGs

(continued)

Table 11.3 (continued)

Countries	Time issued	Policy support	Main contents
	2014	Fourth Basic Energy Plan [50]	Concentrates on the policy objectives of energy reliability, security, affordability, efficiency, and reduced emissions
China	2006	Renewable energy pricing and cost-sharing (pilot management scheme) [51]	Stipulate two types of pricing mechanisms: government-directed pricing and government-guided pricing
	2006	Management measures on auxiliary service in power system 2006 [52]	States that power plants will provide two kinds of ancillary services: basic ancillary services and paid ancillary services
	2010	Renewable energy law [53]	Stipulate full protective purchase of renewable electricity
	2012	Twelfth Five-Year Plan related to major science and technology industrialization projects of Smart Grid [54]	Present the development ideas and principles of SG; establish general development objectives; perform nine key tasks
	2013	Financial subsidy for distributed PV project [55]	Give subsidies for distributed PV based on resource conditions related to distributed energy in SGs

Source: [42–55]

SGs is organized and guided by legislative actions, and the government releases policy framework reports. Among them, the United States has focused on formulating policy related to the upfront investment in an attempt to motivate private investment and stimulate the long-term involvement of all stakeholders to promote the development of SGs. Europe, as an active promoter in the reform of world energy generation, has concentrated their SG policies on low-carbon programs. Energy supply security is their objective. Japan has focused on the deployment of renewable energy in an attempt to improve their energy self-sufficiency while affirming the importance of SGs in its new energy structures. The government of China, as representative of emerging countries, has issued a series of support policies for renewable energy development and energy savings. Since 2010, the implementation of SGs has been promoted as a national strategy.

11.7.3 Investment

Investment is the economic foundation for the development of SGs. Generally, the landscape of SGs is highly dynamic and rapidly changing, and emerging economies are major players in SG investment [56]. Table 11.4 shows the investment situation of SGs in various countries.

Table 11.4 Investments in SG

Countries	Forecast SG investments	Funding for SG development	Smart meters deployments and plans (number)
America	$338 to 476 billion by 2030 [59]	$7 billion in 2009 [61]	8 million in 2011; 60 million by 2020 [62]
Europe	$79 billion by 2020 [58]	$261 billion	45 million already installed; 240 million by 2020 [58]
Japan	$7.4 billion by 2016 [57]	$ 849 million in 2009 [61]	82 million by 2023 [63]
China	$101 billion [60]	$ 7.3 billion in 2009 [61]	360 million by 2030 [64]

Source: [57–64]

Clearly, America, Europe, and China have invested substantial funds in the development of SGs. In contrast to those countries and Europe, Japan invested only $849 million to develop its SG in 2009. Prior to the Fukushima nuclear crisis, the energy supply structure of Japan was stable and secure. At that time, some people even believed that a SG did not need to be developed. Therefore, in terms of past funding for SGs, the investment by Japan has not been as strong as that of other countries. However, the promotion of efficiency and reliability of energy over the last several years has ultimately forced all utilities to make plans for the development of SGs. The investment in SGs in Japan will increase from approximately $1 billion in 2011 to $7.4 billion in 2016 [57].

Europe has invested the most money ($261 billion) in the development of SGs. Europe also forecasts that it will increase its investment in SG development to $79 billion by 2020 [58]. Table 11.4 shows that America will increase investment to $338 to $476 billion in its SG implementation. The costs allocated to transmission and substations are between 19% and 24% of the total cost, the costs allocated to distribution are between 69% and 71%, and the costs allocated to consumer systems are between 7% and 10% [59]. In China, SGCC is solely responsible for the development of nationwide SGs. A total of $101 billion will be provided to support future SG development [60]. Compared to Europe, America and China will provide greater funding for the future study of SGs. Concerning smart meter deployment, America and Europe have already deployed a large number of smart meters; in contrast, China remains in the planning stage. In Japan, ten major utilities have planned to begin widespread smart meter rollouts between 2016 and 2024, by which time an expected 82 million units will be in place for residential and low-use customers [63].

Concerning funding allocation, we only analyze America and Europe because Japan and China have relatively simple stakeholder systems for SGs. Funding in America is mainly provided by SGIG (Smart Grid Investment Grant).

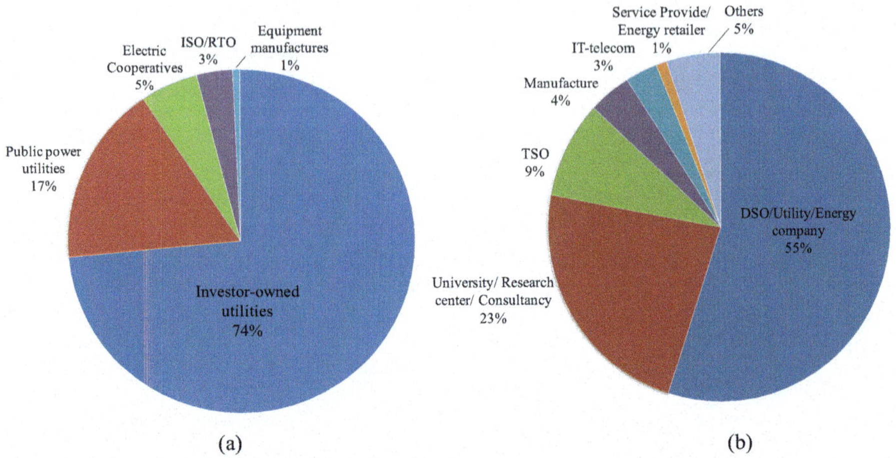

Fig. 11.5 Investment allocation in America and Europe. (**a**)SGIG funds by type of recipient. (**b**) Distribution of EC funding across leading organizations

Figure 11.5a illustrates SGIG funds by type of recipient. Clearly, investor-owned utilities account for the largest proportion, followed by public power utilities.

Figure 11.5b shows the distribution of European funding across leading organizations. In contrast to America, DSO/utility/energy companies spent the greatest proportion of the funds, reaching 55%, followed by university/research center/ consultancy.

11.7.4 Projects and Achievements

Over the past decade, various countries have strengthened their SG research with increasing demonstrations and policies. The current situation of SG projects plays an important role in terms of formulating a clear future direction for SGs.

Table 11.5 shows the categories for the classification of SG projects in the above three countries and Europe.

The table shows that the pilot projects in America, Europe, and Japan concentrate on the deployment of advanced and digital electricity systems (introducing smart meters, etc.) while focusing on the application of renewable energy and distributed generation. In addition, America, Europe, and Japan also attempt to provide a greater number of new services to customers such as smart houses and storage batteries. However, China continues to focus on developing its electricity infrastructure and ensuring a "unified, strong and smart grid network."

Figure 11.6 presents the proportion of project numbers by technology application in America and Europe. The figure shows that both America and Europe have attached great importance to the study of smart network management, aggregation of DR and VPP, and smart consumers and smart house technologies. In addition, Europe prefers to research the integration of DER and electric vehicle technologies, and America prefers to study smart meters.

Table 11.5 Categories for the classification of SG projects in various countries and the European Union

America	European Union	Japan	China
Advanced metering infrastructure	Smart network management	Renewable energy generation	Generation
Electric transmission systems	Integration of DER		Transmission
Electric distribution systems	Integration of large-scale RES	Renewable energy utilization	Transformation
Integrated and crosscutting systems	Aggregation (demand response, VPP)		Distribution
Customer systems	Smart customer and smart home	Electricity network	Utilization
Storage demonstration	Electric vehicles applications	Customer systems	Dispatch
Equipment manufacturing	Other (please specify)	Transportation	

Source: [19, 65, 66]

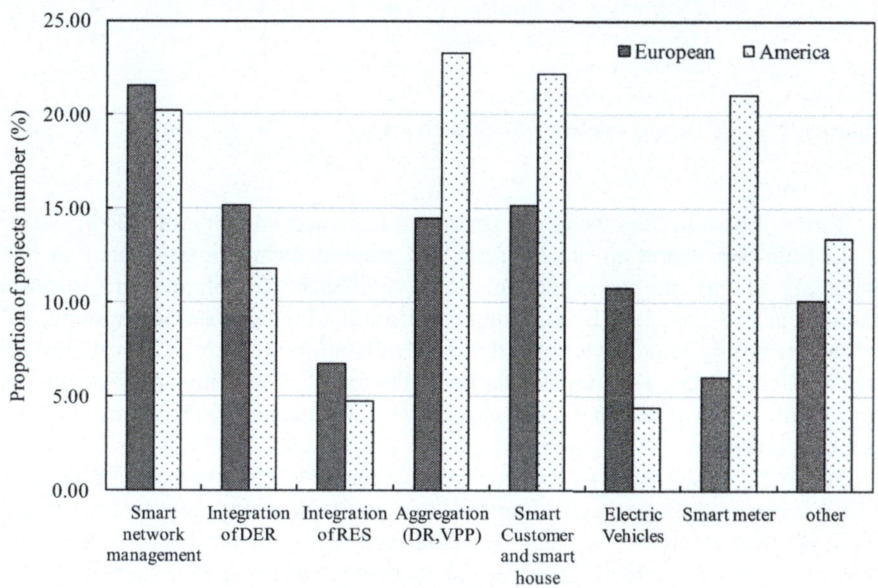

Fig. 11.6 Proportion of the projects by technological application

Table 11.6 Demonstration projects for SGs in Japan source [66]

		Yokohama	Toyota	Kyoto	Kitakyushu
Renewable energy generation	Solar power generation	•			•
	Wind power generation				•
	Cogeneration				•
Renewable energy utilization	Solar heat	•			
	Biomass			•	•
	Waste energy (rubbish, water reclamation, sewage sludge)	•		•	•
	By-product energy	•			•
	Regional air conditioning	•			
Electricity network	Superconducting and transmission network	•			•
	Storage batteries	•	•	•	•
Customer systems	CEMS・HEMS・BEMS・FEMS	•	•	•	•
	Demand response(DR)	•	•	•	•
	Two-way communication	•	•	•	•
	Smart house/building/industry	•	•	•	•
	Storage battery	•	•	•	•
Transportation	EV・PHV	•	•	•	
	EV charging infrastructure	•	•	•	•
	Advancement of the transportation system	•	•	•	•
Consult	Comprehensive solution	•			
University	Industry-university cooperation	•	•	•	•

Table 11.6 shows the detailed demonstration project situation of SGs in Japan.

In four demonstration areas, Japan has studied technologies related to SGs, including energy use visualization, home appliance control, demand response, family electric vehicles (EVs), and optimization of power storage systems [67] while attempting to achieve optimal energy utilization through an energy management system (EMS). The implementation of the four demonstration projects not only represents technical tests but also helps create new business models toward providing new services.

The SGCC also strongly promotes the development of SG projects in China. The SGCC organized 26 provincial branches to conduct SG pilot projects, arranging in total 303 projects across 32 categories, and completed 269 projects across 29 categories. Moreover, the SGCC constructed distribution automation systems in 64 central urban spaces, which enhanced the intelligence of the distribution network. In addition, it also built 360 electric vehicle charging stations and put them into operation in 26 provinces, thereby stimulating the rapid development of industries related to electric vehicles.

11.7.5 Standardization

Regarding the existing SG standards, the recommended core standard is IEC TC 57. According to the number of recommendations, we select four standards from TC 57 that cover the matters of greatest importance according to most experts. One IEEE standard is also considered; however, it does not have much impact on the world-wide scale [40]. The standardization of SGs in various countries is shown in Table 11.7. Clearly, America and Europe play leading roles in international standardization efforts due to their rich experience and mature SG development system, whereas Japan and China play a supporting role because of their SG development statuses. NIST and SGIP are responsible for the standardization work for SGs in America. Under the guidance of EISA 2007, NIST developed the SG standards system in 2009 and published the "Framework and roadmap for smart grid interoperability standards," version 1.0 and version 2.0 [74, 75]. In addition, to facilitate America SG standardization and attract stakeholders to participate in the standardization efforts, NIST subsequently established SGIP. With the impetus and guidance of SGIP, America has completed a series of SG standard revision works and have formulated a standard library (Category of Standards, CoS) [76]. Although these standards have not yet been widely recognized, the SGIP's work has been highly valued by the Federal Energy Regulatory committee (FERC). These standards have also laid a firm foundation for the internationalization of the America SG standards. The Joint Working Group (CEN, CENELE, ETSI) is responsible for the development of the SG standardization roadmap of Europe [84]. That group transformed into a permanent organization, namely, the Smart Grid Coordination Group (SGCG), which continues to be responsible for coordinating and guiding the European SG standardization efforts. The key points of European SG standardization can be categorized into two aspects: setting European standards for electric vehicles and electric meters and making recommendations to international standardization organizations [80]. The SG standardization effort of Japan was led by the Japanese Industrial Standards Committee (JISC), which mainly focused on the standardization of the related technology in favor of cooperation among various companies [81]. SGCC was responsible for China's SG standardization. In 2010, SGCC published the "Framework and roadmap for strong and smart grid standards" [83] and released a series of detailed SG standards that represent an important component of SG international standardization.

11.8 Barriers and Challenges of SGs

In America, there are four main problems facing SG development. First, although the government has fully affirmed the progress of SG development, industry and the public remain skeptical. Industry and the public believe that the development status of the current SG remains slow and that SGs have not achieved what the government

Table 11.7 Standardization of SGs in various countries

Countries	Leading organization	Time issued	Main SG standards	Main contents
IEC TC 57 (International Electrotechnical Commission Technical Committee 57)		2003	IEC 60870 [41]	Defines systems used for controlling electric power transmission grids and other geographically widespread control systems.
		2006	IEC 62351 [70]	Focuses on power system management and associated information exchange—data and communication security
		2007	IEC 61970 [71]	Addresses the application program interfaces for energy management systems (EMS)
		2009	IEC 62357 [72]	Focuses on power system control and associated communications—object model, service facilities, and protocol architecture with references
IEEE (Institute of Electrical and Electronic Engineers)		2009	Draft Guide for Smart Grid Interoperability of Energy Technology and Information Technology Operation with the Electric Power System [73]	Focuses on smart grid interoperability consisting of consistent terminology, characteristics, functional descriptions, evaluation criteria, and appropriate development activities
America	NIST (National Institute of Standards and Technology)	January, 2010	Framework and roadmap for smart grid interoperability standards Release 1.0 [74]	Identifies 19 standards that require priority revision; establishes SGIP to attract various stakeholders to participate in the standardization efforts
		February, 2012	Framework and roadmap for smart grid interoperability standards Release 2.0 [75]	
	SGIP (Smart Grid Interoperability Panel)	–	CoS (Catalog of Standards) [76]	Continued research on conceptual architectural framework and SG interoperability panel (SGIP); focuses on the standards identified for

(continued)

Table 11.7 (continued)

Countries	Leading organization	Time issued	Main SG standards	Main contents
				implementation and cybersecurity strategy; discusses framework for SG interoperability testing and certification
Europe	Joint Working Group (CEN/CENELE/ ETSI)	2009	Mandate CEN/CENELEC M/441 [77]	Focal topics of standardization including terminology, systems aspects, data communication reference architecture, and data communication interfaces with the focus on sectional standards including DMS, SCADA, data models, and ERP interfaces
		October, 2010	Report on standards for smart grids V1.0 [78]	Provides a conceptual model and reference architectural principles; establishes the framework of the SG architecture model (SGAM); sets the Europe standards for electric vehicles and electric meters; makes recommendations to the IEC and other international standardization organizations
		June, 2011	Final report of the CEN/CENELE/ETSI Joint Working Group on Standards for Smart Grids [79]	
	SGCG (Smart Grid Coordination Group)	October, 2012	First, Set of Standards Version1.1 [80]	
Japan	JISC (Japanese Industrial Standards Committee)	December, 2012	Twenty important items for international standardization of SGs [81]	A total of 8 review groups select 20 important items for the international standardization of SGs, including energy management and demand response systems, distribution automation systems, smart meter systems, and technologies related to electric vehicles. METI encourages the application of HEMS and ECHONET, realizing collaboration among different companies
	METI (Ministry of Economy, Trade and Industry)	February, 2012	Recommend HEMS and ECHONET lite [81]	
		2010	Japans Roadmap to International Standardization for Smart Grid and Collaborations with other Countries [82]	

(continued)

Table 11.7 (continued)

Countries	Leading organization	Time issued	Main SG standards	Main contents
China	SGCC (Smart Gird Corporation of China)	August, 2010	Framework and roadmap for strong and smart grid standards [83]	Establishes an SG standard system based on integrated planning, generation, transmission, substations, distributions, communication and 2 other professional branches involving 26 technical areas and consisting of 92 standard series; compiles 220 enterprise SG standards and 841 national and industry standards

Source: [44, 45, 48–51, 68, 69]

promised the public. These promises include realizing two-way communication between users and grids, allowing users to manage their own energy production and consumption and providing more employment opportunities for the community. Second, there is a substantial problem in how power and utility companies communicate with users who have already installed smart meters. Because America simultaneously promoted new technologies and introduced new electricity tariffs, consumers have a poor understanding of the results of pilot projects. Moreover, high quality and how to interact with customers and persuade them to recognize the value and real benefits of SGs have become important subjects in the continued development of the SG. Third, with encouragement in the form of financial support provided by the government, SG research and demonstration efforts have seen smooth progress; however, the investment enthusiasm by power companies and other private businesses is not high, which affects the future development of SGs to a certain extent. Fourth, America must strongly invest in SGs to meet the policy requirements for the full deployment of the SG by 2030. However, in the present circumstances, it is difficult for America to continue to invest in SGs. Actually, the power asset ownership and management of many American power companies are relatively decentralized, and investments in electrical equipment are very high and have long life cycles. This results in each power company having to perform a cost-effectiveness analysis before making investment decisions. Therefore, subsequent follow-up funding is difficult to obtain.

In Europe, the situation is similar to that in America: the most significant bottleneck in the development of SGs is follow-up funding problems; this issue is even more serious than it is in America. According to estimates by Pike research, the investments in SGs of EU countries will reach 79 billion Euros by 2020. Moreover, due to the impact of the European credit crisis, the speed of and capital investment in

Europe SG development continue to face uncertainty. Moreover, the interoperability of Europe SGs is not high. As new applications of SGs are realized, the roles and responsibilities of various stakeholders remain uncertain. A clear cost-benefit-sharing mechanism also remains unclear.

Japan's SG pilot project realized many achievements and essentially met its original targets. However, scaling up small-scale pilot projects into large-scale practical applications remains a substantial problem. In addition, although Japan's SG-related technologies, especially battery technology, are world class, America continues to have the power to establish basic SG international standards. Actively participating in and promoting the development of international standards remain a significant challenge to Japan.

In China, the most significant obstacle encountered in SG development is the lack of a clear national policy and roadmap. Although a series of documents have been issued to facilitate the implementation of SGs, planning and related standards that guide specific actions have not been introduced, making various SG stakeholders feel anxious and confused. In addition, China continues to need to break through technical barriers. Chinese power companies view developing renewable energy resources as an opportunity but face many challenges in certain technical aspects. The traditional Chinese grid is powered by thermal power plants and hydropower plants, and the design of the grid network is expected to remain stable for a long time. However, with the introduction of wind, solar, and other clean energy generation types, grid technologies have begun to exhibit various issues. Addressing the transition between the old and new grid represents a substantial challenge to China's SG development.

The United States and Europe still need to establish and improve relevant laws and regulations for the long-term development of SGs in terms of managing the risks and benefits from the perspective of policy. They must do this while developing technical standards that can be accepted by industry and achieving the integration of different equipment manufactured by different companies. This would encourage the active participation of various private enterprises and power companies while attracting subsequent investment. In addition, strengthening communication with users is also very important and can help SG developers understand the lack of SG process. The development of SGs in Japan should continue to strengthen the links and interaction between the three institutions (government, industry, and academic institution) and attempt to create a new method that can help promote SGs while providing new services in line with the needs of users. This would enhance their understanding of SGs, encourage and stimulate the active participation of local residents and local businesses, and finally push the widespread development of SGs. What is more, Japan should continue to concentrate on international cooperation related to SGs, seeking to use its technological advantage to obtain the right to influence international standard-setting. The Chinese government should focus on implementing a reform of the power grid, therein actively improving the infrastructure of the power industry so that it can be coordinated with the development of SGs. In addition, the Chinese government should give strong support to SGs through policy and continue to improve the relevant standards and provide favorable conditions for the development of SGs.

SG is still a relatively new concept despite its great accomplishments. To realize the scale-up and industrialization of smart grids, significant further effort is still required. According to the above analysis, SG development can be improved in three respects. The first is the end user side. Enhancing the feedback of end users is essential to the process of developing SGs. This factor has the most effective and direct influence on the successful implementation of SGs. The related technology must be constantly upgraded to meet the changes in users' motivation. Only in this way can end users adapt the technology according to their own needs and expectations. Human habits must also be seriously investigated to improve the efficiency and flexibility of energy management systems. In addition, users' knowledge of SGs plays an important role during the promotion of SGs. With better understanding of SGs, end users can take an active part in SG development.

For example, improved understanding can help them correctly install and properly configure the smart devices while improving the awareness of energy management. The second is the technology side. Simple technology should be explored to coordinate with the consumers; more products and services should be provided to meet the requirements of different stakeholders. The challenge of developing good technology is realizing not only the data communication but also the ability to enable various roles to have the opportunity for continual involvement in the adaptation and customization process of SG. The last is the policy side. SGs involve a series of new technologies that have great potential in future power grid development. To build a secure, economic, clean, transparent, and compatible power grid in the future, we should take measures to propel the development of smart grids [36]. In a word, SGs represent not only energy system innovation but also institutional innovation. The healthy development of SGs is closely related to clear policy support and unified technical regulations under the background of power marketization. A generally approved standard system is also a key factor for orderly development of SGs owing to the complexity of SGs, which involve many industries and technical areas.

References

1. El-hawary ME (2014) The smart grid—state-of-the-art and future trend. Electr Power Compon Syst 42(3–4):239–250
2. Berger LT, Iniewski K (eds) (2012) Smart grid – applications, communications and security. Wiley, Hoboken
3. Colak I, Fulli G, Sagiroglu S, Yesilbudak M, Catalin-Felix C (2015) Smart grid projects in Europe: current status, maturity and future scenarios. Appl Energy 152:58–70
4. Vincenzo G, Meletiou A, Covrig CF, et al (2013) Smart grid projects in Europe: lessons learned and current developments [R]. Publications Office of the European Union, Luxembourg
5. Makhijani A (2007) Carbon-free and nuclear-free. Science for Democratic Action
6. Zahedi A (2011) Developing a System Model for Future Smart Grid," Proceedings in 2011 IEEE PES Innovative Smart Grid Technologies Conference, ISGT Asia 2011, Perth, 13–16 Nov 2011, pp. 1–5
7. Sun Q, Ge X, Liu L, Xu X, Zhang Y, Niu R et al (2011) Review of smart grid comprehensive assessment systems. Energy Procedia 12:219–229

8. Evaluation of critical and emerging security technologies for the elaboration of a strategic research agenda. Available at: http://www.etcetera-project.eu/wp-content/uploads/2012/01/D1.2-Validated-Critical-Technology-List-revised.pdf. Accessed 01-09-2015
9. Colak L, Sagiroglu S, Fulli G, Yesilbudak M, Covrig C (2016) A survey on the critical issues in smart grid technologies. Renew Sust Energ Rev 54:396–405
10. Luthra S, Kumar S, Kharb R, Ansari F, Shimmi SL (2014) Adoption of smart grid technologies: an analysis of interactions among barriers. Renew Sust Energ Rev 33:554–565
11. Arif SM, Bouzguend M (2013) Effects of smart grid technologies on capacity and energy savings – a case study of Oman. Energy 54:365–371
12. Zahurul S, Mariun N, Grozescu IV, Tsuyoshi H, Mitani Y, Othman ML et al (2016) Future strategic plan analysis for integrating distributed renewable generation to smart grid through wireless sensor network: Malaysia prospect. Renew Sust Energ Rev 53:978–992
13. Di Santo KG, Kanashiro E, Di Santo SG, Saidel MA (2015) A review on smart grids and experiences in Brazil. Renew Sust Energ Rev 52:1072–1082
14. Haidar AMA, Muttaqi K, Sutanto D (2015) Smart grid and its future perspectives in Australia. Renew Sust Energ Rev 51:1375–1389
15. Thakur J, Chakraborty B (2015) Intelli-grid: moving towards automation of electric grid in India. Renew Sust Energ Rev 42:16–25
16. International Energy Agency (IEA) (2011) Smart grid roadmap. Available at: www.iea.org/publications/.../smartgrids_roadmap.pdf
17. Locke G, Gallagher PD (2010) NIST framework and roadmap for smart grid interoperability standards release 1.0. National Institute of Standards and Technology (NIST)
18. International Energy Agency (IEA). Smart grid roadmap. Available at: https://www.iea.org/publications/freepublications/publication/smartgrids_roadmap.pdf
19. Yuan J, Shen J, Pan L, Zhao C, Kang J (2014) Smart grids in China. Renew Sust Energ Rev 37: 896–906
20. European Renewable Energy Council (2008) Renewable energy technology roadmap: 20% by 2020. Belgium, Brussels
21. (2012) Guidelines for conducting a cost-benefit analysis of smart grid projects. Joint Research Centre of European Commission. Available at: http://ses.jrc.ec.europa.eu/publications-list/reports
22. (2012) Smart energy grids and complexity science. Joint Research Centre of European Commission. Available at: http://ses.jrc.ec.europa.eu//publications-list/reports
23. (2012) The social dimension of smart grids. Joint Research Centre of European Commission. Available at: http://ses.jrc.ec.europa.eu//publications-list/reports
24. Hamasaki H, Kanudia A (2014) System analysis of Japanese renewable energy. Fujitsu Research Institute
25. Ina H, Umino K, Rudd J, Tokumoto N (2015) Japan approves final set of power market reforms. Jones Day
26. Japan Smart Community Alliance. Available at: www.smart-japan.org/english/index.html
27. State Grid Corporation of China (SGCC) (2010) Power gird twelfth five years plan report of State Grid Corporation of China [R]
28. Yu YX, Luang WJ (2011) Basic philosophy of smart grid. J Tianjin Univ 44:377–384
29. Electricity system development: a focus on smart grids overview of activities and players in smart grids. United Nations Economic Commission for Europe (UNECE). Available online: https://www.unece.org/fileadmin/DAM/energy/se/pdfs/eneff/eneff_h.news/Smart.Grids.Overview.pdf. Accessed 23 Jan 2017
30. Arndt S, Sheveleva T, Goeker C (2015) Smart grid terminology development—crossing the boundaries of terminology standardization. Energy Sustain Soc 5:20
31. International Electrotechnical Commission. www.iec.ch/smartgrid/background/explained.htm
32. Chakrabarti DS. Smart grid: concepts and deployment [J]. Smart Grid 18
33. European Commission. http://ec.europa.eu/energy/en/topics/markets-and-consumers/smart-grids-and-meters. Accessed 1 Feb 2017

34. European Regulators' Group for Electricity and Gas. Annex 3: Evaluation of Responses—ERGEG Conclusions Paper on Smart grids; Technical Report for Council of European Energy Regulators ASBL: Brussels, Belgium, 2009. Available online: http://www.energy-regulators.eu/portal/page/portal/EER_HOME/EER_CONSULT/CLOSED%20PUBLIC%20CONSULTATIONS/ELECTRICITY/Smart%20Grids/CD/E10-EQS-38-05a_SmartGrids_EoR_10-Jun-2010.pdf. Accessed 1 Feb 2017
35. Japan smart community alliance. www.smart-japan.org/english/index.html. Accessed 19 Jan 2017
36. YanShan Y, Yang J, Chen B (2012) The smart grids in China – a review. Energies 5:1321–1338
37. New Energy and Industrial Technology Development Organization (NEDO) (2012) Current status and future prospects of the smart community. Available at: http://www.nedo.go.jp
38. State Council of China (2011) The 12th five-year-plan for national economy and social development. Available at: http://news.xinhuanet.com/politics/2011-03/16/c_121193916.htm
39. Newsham GR, Bowker BG (2010) The effect of utility time-varying pricing and load control strategies on residential summer peak electricity use: a review. Energ Policy 38:3289–3296
40. Uslar M, Rohjans S, Bleiker R, González J, Specht M, Suding T, Weidelt T (2010) Survey of smart grid standardization studies and recommendations [C]. Innovative Smart Grid Technologies conference Europe (ISGT Europe), 2010 IEEE PES
41. International Electrotechnical Commission (2001) International standard IEC 60870-5-101: telecontrol equipment and systems. Part 5–101: transmission protocols, companion standard for basic telecontrol tasks: amendment 2 [M]. IEC
42. Energy Independence and Security Act (EISA) (2007). Available at: http://www.gpo.gov/fdsys/pkg/BILLS-110hr6enr/pdf/BILLS-110hr6enr.pdf
43. U.S Department of Energy. American Recovery and Reinvestment Act of 2009: Smart Grid Investment Grant Program (SGIG) progress report. Available at: http://www.smartgridnews.com/artman/publish/Projects_R_D/Strategic-R-D-Opportunities-for-the-Smart-Grid-5671.html
44. European Economic Recovery Plan 2008. Available at: http://eur-lex.europa.eu/legal-content/EN/TXT/?uri=CELEX:52008DC0800
45. (2009) Third legislative for further liberalization of the electricity and gas markets. Available at: http://www.europarl.europa.eu/ftu/pdf/en/FTU_5.7.2.pdf
46. European Council summits 2010. Available at: http://www.consilium.europa.eu/uedocs/cms_data/docs/pressdata/en/ec/113591.pdf
47. Policy package to address the economic crisis. Available at: http://www.kantei.go.jp/jp/asophoto/2009/04/090410kikitaisaku.pdf
48. Report of the next-generation power transmission and distribution network. Available at: http://www.enecho.meti.go.jp/committee/council/electric_power_industry_subcommittee/001_038/pdf/038_009.pdf
49. Basic energy plan. Available at: http://www.enecho.meti.go.jp/category/others/basic_plan/pdf/100618honbun.pdf
50. Fourth basic energy plan. Available at: http://www.enecho.meti.go.jp/en/category/others/basic_plan/pdf/4th_strategic_energy_plan.pdf
51. NDRC (2006) Renewable energy pricing and cost-sharing (pilot management scheme). Available at: http://www.gov.cn/ztzl/2006-01/20/content_165910.htm
52. State Electricity Regulatory Commission (SERC) (2006) Management measures on auxiliary service in power system. Available at: http://www.law-lib.com/law/law_view.asp?id=178504
53. National People's congress of China (NPC) (2009) Renewable energy law (amendment). Available at: http://www.npc.gov.cn/huiyi/cwh/1112/2009-12/26/content_1533216.htm
54. MOST. Twelfth five plan related to major science and technology industrialization projects of Smart Grid. Available at: http://pg.jrj.com.cn/acc/Res/CN_RES/INDUS/2012/5/7/4713ab4a-6013-4233-b0a0-1a2ba738805d.pdf
55. MOF (2012) Ministry of Finance of the People's Republic of China notice on the implementation of the annual solar PV building projects demonstration in 2012
56. Gangale F, Mengolini A, Onyeji I (2013) Consumer engagement: an insight from smart grid projects in Europe [J]. Energy Policy 60(Sept):621–628

57. (2015) Top markets report smart grid. U.S. Department of Commerce, International Trade Administration, Industry & Analysis (I & A). Available at: http://trade.gov/topmarkets/pdf/Smart_Grid_Top_Markets_Report.pdf
58. Pike Research (2011) Smart grids in Europe, pike Research clean tech market intelligence. Available at: http://www.pikeresearch.com/research/smart-grids-in-europe. Accessed: Feb 2011
59. Electric Power Research Institute (EPRI) (2011) Estimating the costs and benefits of the smart grid – a preliminary estimate of the investment requirements and the resultant benefits of a fully functioning smart grid. Available at: http://my.epri.com/portal/server.pt?space=CommunityPage&cached=true&parentname=ObjMgr&parentid=2&control=SetCommunity&CommunityID=404&RaiseDocID=000000000001022519&RaiseDocType=Abstract_id. Accessed: Apr 2011
60. Zypryme Research & Consulting, China. Rise of the smart grid, special report by Zypryme's Smart Grid Insights. Available at: http://www.zpryme.com/SmartGridInsights/China_Rise_of_the_Smart_Grid_January_2011_Zpryme_Research.pdf. Accessed: Feb 2011
61. SmartGridnews.com. Smart grid snapshot: China tops stimulus funding. SmartGridNews.com; 2010. Available at: http://www.smartgridnews.com/artman/publish/Stimulus_News_Digest_Products/Smart-Grid-Snapshot-China-Tops-Stimulus-Funding-1827.html. Accessed: Mar 2011
62. Smartmeters.com (2011) Smart meters deployment looks strong for 2011. Available at: http://www.smartmeters.com/the-news/1476-smart-meters-deploymentlooks-strong-for-2011.html. Accessed: Jan 2011
63. http://www.metering.com/smart-meters-japan-utilities-to-accelerate-residential-rollouts-in-2015/
64. Reportlinker (2012) Smart Grid Technology investment: forecasts for 2012–2030, Electrical Equipment Industry
65. Giordano V, Bossart S. Assessing Smart Grid benefits and impacts: EU and US Initiatives. Joint report of the EC JRC and US DOE. Available at: https://ec.europa.eu/jrc/sites/default/files/eu-us_smart_grid_assessment_final_report_-online_version.pdf
66. Yun KL (2012) A study of cooperation networks and new business in the smart grid. J Cent Reg Aff Fukushima Univ 9. Available at: http://ir.lib.fukushima-u.ac.jp/dspace/bitstream/10270/3654/1/18-177.pdf
67. Morozumi S. On-going smart grid implementations and pilots in Japan. New Energy and Industrial Technology Development Organization Available at: http://sgc2014.ieee-smartgridcomm.org/sites/sgc2014.ieee-smartgridcomm.org/files/u29/IEEE_SGComm14_Presentation_WS_1_S3_Morozumi.pdf
68. European Council summits 2010
69. State Electricity Regulatory Commission (SERC) (2006) Management measures on auxiliary service in power system; Electricity supervision market
70. 62351-1 TS, Ed.1: Data and communication security – Part 1: Introduction and overview, IEC 57/802/DTS-2006
71. 61970-301 Ed. 1: Energy management system application program interface (EMS-API) – Part 301: Common information model (CIM) base, IEC Std, 2007
72. 62357 Second Edition: TC 57 Architecture – Part 1: Reference Architecture for TC 57 – Draft, IEC Std., 2009
73. P2030/D670 IEEE draft guide for smart grid interoperability of energy technology and information technology operation with the Electric Power System (EPS), and end-use applications and loads, 2011
74. (2010) NIST special publication 1108. NIST framework and roadmap for smart grid interoperability standards, Release 1.0. Available at: http://www.nist.gov/public_affairs/releases/upload/smartgrid_interoperability_final.pdf
75. (2012) NIST special publication 1108R2. NIST framework and roadmap for smart grid interoperability standards, Release 2.0. Available at: http://www.nist.gov/smartgrid/upload/NIST_Framework_Release_2-0_corr.pdf
76. Catalog of Standards. Available at: http://sgip.org/Catalog-of-Standards

77. CENELEC (2009) Smart Meters Coordination Group: Report of the Second Meeting held on 2009-09-28 and Approval of SM-CG Work Program for EC Submission
78. CEN/CENELE/ETSI JWG (2010) Report on standards for smart grid V1.0. Available at: www.nbn.be/nl/JWG-smartgrid-report_V1-0.pdf
79. CEN/CENELEC/ETSI Joint Presidents Group (2011) Final report of the CEN/CENELEC/ETSI Joint Working Group on Standards for Smart Grids. Available at: http://www.etsi.org/WebSite/document/Report_CENCLCETSI_Standards_Smart%20Grids.pdf
80. CEN-CENELEC-ETSI Smart Grid Coordination Group (2012) CEN-CENELEC-ETSI: Smart Grid Coordination Group-Smart Grid Reference Architecture Report 2.0. https://www.researchgate.net/publication/263264218_CEN_-CENELEC_-_ETSI_Smart_Grid_Coordination_Group_-_Smart_Grid_Reference_Architecture_Report_20
81. Takeshi S, Yasuro S, Yoshihiro OHBA (2013) Special reports: trends in international standardization of smart grids and Toshiba's approach. Available at: http://www.toshiba.co.jp/tech/review/2013/08/68_08pdf/a06.pdf
82. Yamamoto T (2010) Japans roadmap to international standardization for smart grid and collaborations with other countries, In: The 9th Northeast Asia Standardization Cooperation Forum
83. State Grid Corporation of China (2012) Framework and roadmap for strong & smart grid. Available at: http://esci-ksp.org/wp/wp-content/uploads/2012/03/China_State_Grid_Framework_and_Roadmap_for_SG_Standards.pdf
84. CEN/CENELEC/ETSI Joint Working Group on standards for Smart Grids. New ETSI-CEN-CENELEC approach for rapid SG deployments; 2014. Available at: http://africasmartgridforum2014.org/fr/expert/sessionb1/laurent-schmitt-new-etsi-cen-cenelec-en.pdf

Index

W. Gao (ed.), *Distributed Energy Resources*, Green Energy and Technology,
https://doi.org/10.1007/978-3-031-21097-6

The manufacturer's authorised representative in the EU is Springer Nature Customer Service Centre GmbH, Europaplatz 3, 69115 Heidelberg, Germany. If you have any concerns regarding our products, please contact ProductSafety@springernature.com

Printed and bound by CPI Group (UK) Ltd, Croydon, CR0 4YY
28/09/2026
02224085-0005